Complex Population Dynamics

MONOGRAPHS IN POPULATION BIOLOGY

EDITED BY SIMON A. LEVIN AND HENRY S. HORN

(List continued after index)

Complex Population Dynamics: A Theoretical/ Empirical Synthesis

PETER TURCHIN

PRINCETON UNIVERSITY PRESS
Princeton and Oxford

Copyright © 2003 by Princeton University Press

Published by Princeton University Press, 41 William Street, Princeton,
New Jersey 08540

In the United Kingdom: Princeton University Press, 3 Market Place,
Woodstock, Oxfordshire OX20 1SY

Library of Congress Cataloging-in-Publication Data

Turchin, Peter, 1957–

Complex population dynamics : a theoretical/empirical synthesis / Peter Turchin.
p. cm.—(Monographs in population biology)

Includes bibliographical references (p.).

ISBN 0-691-09020-3 (cloth : alk. paper) —
ISBN 0-691-09021-1 (pbk. : alk. paper)

1. Population biology. I. Title. II. Series.
QH352 T83 2003
577.8′8—dc21   2002022721

British Library Cataloging-in-Publication Data is available

This book has been composed in Times Roman

Printed on acid-free paper. ∞

www.pupress.princeton.edu

Printed in the United States of America

10 9 8 7 6 5 4 3 2 1

Contents

PART II
DATA

PART III
CASE STUDIES

Preface

Complex population dynamics such as limit cycles and chaos are intrinsically fascinating. Why do organisms become extremely abundant one year, and then apparently disappear a few years later? Why do population outbreaks in certain species in certain locations happen more or less regularly, while in other locations such eruptions are irregular, or never occur at all? One would not find a general and definitive answer to these questions in any of the standard ecology textbooks. Partly this is because population ecologists do not, as yet, have a definitive, as well as empirically tested, general theory of complex population dynamics. On the other hand, much progress toward such a theory has been made. One goal of *Complex Population Dynamics* is to review these developments, and proffer tentative answers to the question of why populations oscillate in at least some of the best-studied examples.

But there is a deeper reason for studying population oscillations. Could these complex dynamics have simple causes, at least in some cases? If we are successful in answering this question, then we shall surely gain a better understanding of population dynamics in general. Physicists were able to formulate general laws of physical motion by studying periodic orbits of planets. Perhaps ecologists will be able to formulate general laws of population dynamics by studying periodic oscillations in population density. I think this is actually the case; in fact, on closer examination, it turns out that there is a lot in common between the two disciplines—classical (Newtonian) mechanics and population dynamics. If so, then population ecology may be on the brink of maturity, rapidly becoming a quantitative and predictive science. Constructing and defending this argument is the second broad theme of the book.

The third and final theme is methodological. Population dynamics have been studied in a variety of ways: mathematical models, statistical analyses of time-series data, and field experiments. The paradigmatic approach of how one "does ecology" is the manipulative experiment. Yet, the actual experience of solving the puzzle of population cycles suggests that experiments in isolation do not result in serious progress. Significant progress in understanding population mechanisms occurs only when we combine the three major approaches in a synthetic whole.

Although this book is about a synthesis of mathematical models and empirical ecology, I am assuming very little mathematical background on the part of readers. Essentially, readers need to know what a derivative is, and have some experience with differential and difference equations. There is little explicit algebraic manipulation in the text, and what there is focuses almost completely on model development. This does not mean that this is an elementary text, however. There is too much ground to cover, and therefore I assume that readers are conversant with basic ecological concepts such as functional responses, Allee effects, and general types of species interactions. (I do, however, provide definitions of these concepts in the glossary.)

Complex Population Dynamics can serve two functions. First, it is intended for the audience of working scientists: graduate students and researchers in ecology. It is to their judgment that I submit my view of how the synthesis in population dynamics is shaping up. I hope that they will find the developments I review in the book as I exciting as I find them, and will join in the endeavor of making the "grand synthesis" a reality.

Second, it can be used as a textbook for a course in population dynamics or population ecology. True, the main focus is on population oscillations, but on the way to complex dynamical behaviors the book covers the basics of simple dynamics. And, I argue, the study of complex dynamics advances us substantially along the road of figuring out general principles of population ecology.

Finally, I need to say what this book is not about. First, and most important, I barely touch on the spatial aspects of population ecology. I wish I could review the recent exciting developments in complex spatiotemporal dynamics, but there is simply not enough space to do that in this volume. Second, although all general approaches discussed in the book are based on an explicit incorporation of stochasticity, my main focus is on *endogenous* mechanisms (those that involve population density feedbacks). Thus, dynamics in which populations largely respond to exogenous perturbations are relatively neglected. Again, the main reason is not because I think such dynamics are uninteresting or unimportant, but because I lack space to do full justice to this topic. Third, this is not a text on ecological theory. Although I extensively review certain classes of population models, I included only those parts of theory that are useful to me in studying population oscillations. And, as mentioned above, my main focus in the theoretical part is on model development. Thus, I devote little space to the discussion of how to solve the developed models. And, fourth, I limit the empirical case studies to the terrestrial field ones, because excellent summaries of fisheries (Quinn and Deriso 1999) and laboratory (Mueller and Joshi 2000) dynamics already exist.

ACKNOWLEDGMENTS

First, and foremost, I want to express deep gratitude to my coworkers: Steve Ellner, Ilkka Hanski, Alan Berryman, John Reeve, Andy Taylor, George Batzli, and Lauri Oksanen. I would also like to thank the Complex Population Dynamics working group at the National Center of Ecological Analysis and Synthesis: Cherie Briggs, Steve Ellner, Bruce Kendall, Ed McCauley, Bill Murdoch, Roger Nisbet, and Simon Wood. Chapter 9, on the larch budmoth, is the result of this collaborative effort. Thanks are also due to Werner Baltensweiler and Andreas Fischlin, who generously shared their data and insights.

Many people provided extensive comments on previous book drafts, or draft chapters: Alan Berryman, Lev Ginzburg, Roger Nisbet, Peter Hudson, Alexei Sharov, Peter Kareiva, Mauricio Lima, Peter Abrams, Jason Matthiopoulos, Larry Mueller, Robert Moss, Bertram Murray, Lennart Hansson, Ilkka Hanski, Bogumila Jêdrzejewska, Xavier Lambin, Rolf Peterson, Eric Post, Aaron King, Dennis

Murray, Stan Boutin, François Messier, Mark Boyce, Christophe Luczak, Michel Crête, Damien Joly, and Patrick Foley.

Finally, many thanks to Dennis Chitty and Bertram Murray, with whom I had a very enjoyable exchange of ideas about philosophy of science, even though disagreeing on practically every issue.

Mathematical Symbols

The general scheme is to use capital Roman-alphabet letters for variables, and lowercase Roman and Greek letters for parameters. There are a few exceptions to these rules, primarily for variables or parameters that have traditional symbols associated with them (e.g., t). Variables used in discrete-time models are subscripted with t, as in the symbol for population at time t: N_t. In continuous-time models, I usually omit explicit time dependence, but when I want to emphasize it, I use t in parentheses: $N(t)$. A list of symbols commonly used in the book follows.

STATE VARIABLES IN CONTINUOUS-TIME MODELS

t	Time
N	Population density in general, or of herbivore in trophic models
P	Population density of predators, parasitoids, or parasites
Π	Parasitism rate
V	Biomass density of vegetation
H	Host density
X	Generic state variable ("some factor X")
Q	Index of plant quality
Y	Index of individual quality (e.g., average weight)

STATE VARIABLES IN DISCRETE-TIME MODELS

N_t	Population density at time t
X_t	Generic state variable, or some intrinsic variable at time t

P_t Parasitoid density at time t
Π_t Parasitism rate at time t
Q_t Index of plant quality at time t

TIME-SERIES ANALYSIS VARIABLES AND PARAMETERS

$r(t)$ Realized per capita rate of population change
                 (continuous case)
r_t Realized per capita rate of population change
                 (discrete case)
Y_t Log-transformed N_t
Z_t Vector of state variables
n The number of data points in a data series
ϵ_t Exogenous influence
σ^2 Variance of process noise
σ^2_{obs} Variance of observation noise
Λ_∞ Global Lyapunov exponent
$\{\Lambda_t\}$ Distribution of local Lyapunov exponents
R^2 Coefficient of determination in regression
                 (proportion of variance explained)
R^2_{pred} Coefficient of prediction
τ Base lag, or time delay

GENERIC CONSTANTS AND PARAMETERS

$a, b, c \ldots$ Generic constants
θ Exponents

PARAMETERS OF EXPONENTIAL AND LOGISTIC TERMS

β Per capita birth rate
δ Per capita death rate
r_0 Intrinsic rate of population increase, general or
                 herbivore
v_0 Intrinsic rate of population increase, vegetation
u_0 Initial regrowth rate of vegetation
s_0 Intrinsic rate of population increase, predators

k	Carrying capacity in the logistic term, general or herbivore
m	Carrying capacity in the logistic term, vegetation
κ	Carrying capacity in the logistic term, predator

FUNCTIONAL RESPONSE (FR) PARAMETERS

a	Attack rate (mechanistic form)
h	Handling time (mechanistic form)
w	Wasted time (predator interference)
c	Saturation killing rate (phenom. form)
d	Half-saturation constant of the hyperbolic FR (phenom. form)
a	Saturation rate, herbivore (phenom. form)
b	Half-saturation constant, herbivore (phenom. form)
g	Saturation rate in the sigmoid FR
h	Half-saturation constant in the sigmoid FR

NUMERICAL RESPONSE PARAMETERS

μ	ZPG predator intake
χ	Prey-predator conversion rate
η	ZPG herbivore intake
ξ	Vegetation-herbivore conversion rate
δ_0	Consumer death rate in the absence of resource

PART I
THEORY

CHAPTER 1

Introduction

1.1 AT THE SOURCES

Population dynamics is the study of how and why population numbers change in time and space. Thus, population dynamicists document the empirical patterns of population change and attempt to determine the mechanisms explaining the observed patterns. Temporal population dynamics is not the only subject that population ecologists study. Among other things, they are also interested in statics (what sets the level around which populations fluctuate) and population structure (e.g., age distribution). More recently, there has been a lot of progress in spatiotemporal dynamics of populations. Nevertheless, population dynamics in time has been at the core of population ecology ever since the origins of the discipline during the 1920s (Kingsland 1995), largely as a result of efforts of Charles Elton, Alfred Lotka, Vito Volterra, and A. J. Nicholson.

1.1.1 The Puzzle of Population Cycles

Abrupt and seemingly inexplicable changes in population numbers have fascinated and puzzled humanity from prehistoric times. The Bible records the effects of locust swarms and mice "plagues" on humans. Hunters and trappers surely knew about periodic changes in populations of furbearing mammals and game birds. Norwegians have long been aware of mysterious invasions by lemmings (Stenseth and Ims 1993a). Nordic folklore has provided the basis of the modern myth of lemmings marching off to the sea to commit mass suicide, as popularized by Walt Disney's *White Wilderness.*

The scientific study of population oscillations begins with the work of Charles Elton (Stenseth and Ims 1993a; Lindström et al. 2000). In 1923 the young Elton passed through the Norwegian town of Tromsø on his way back from a zoological expedition to the Spitsbergen. In a Tromsø bookstore, he noticed *Norges Pattedyr* (Norwegian mammals) by Robert Collett. Although Elton could not read Norwegian, he noticed a very curious—apparently periodic—pattern in the abundance of Norwegian lemmings. With some of the last of his money, Elton bought the book, brought it with him back to Oxford, and had it translated into English. In 1924, Elton published the pioneering article "Periodic Fluctuations in the Number of Animals: Their Cause and Effects" (Elton 1924), based largely on Collett's data (Stenseth and Ims 1993a; Crowcroft 1991).

About the same time, Elton read *The Conservation of the Wild Life of Canada* by Gordon Hewitt, which contained graphs of the annual fur returns of the Hudson's Bay Company showing remarkably regular oscillations in the numbers of lynx and snowshoe hare pelts (Crowcroft 1991:4). Elton was appointed biological consultant to the Hudson's Bay Company in 1925, and examined the company's records to trace the dynamics of Canada lynx populations back to 1736. The results of this research were eventually published in 1942 (Elton and Nicholson 1942). A second line of attack consisted of empirically studying fluctuations in the numbers of British voles, using Oxford as a base (Crowcroft 1991:6). While Elton and his group were engaged in these empirical studies, momentous changes were occurring in the field of theoretical ecology.

1.1.2 Modeling Nature

By a curious coincidence, the mathematical study of population oscillations started practically at the same time as Elton was puzzling over lemming cycles (Lotka 1925; Volterra 1926). The two traditions, the empirical and the mathematical, although having started almost simultaneously, developed largely separately. Only three-quarters of a century later we are starting to see a true synthesis.

Theory is important because there is a tendency for common phenomena to be overlooked or misinterpreted in the absence of a well-known body of theory (Abrams 1998:211). One ecological illustration of this tendency is the meager experimental evidence for apparent competition that Holt (1977) could marshal in the article where he proposed the concept, compared with the large body of evidence reviewed by Holt and Lawton (1993) seventeen years later (Abrams 1998). So it was at the beginning of the study of population cycles. In his first paper on population cycles, Elton wrote: "It will be shown in the body of this paper that the periodic fluctuations in the numbers of certain animals there dealt with, must be due to climatic variations" (Elton 1924:119). When Volterra's 1926 article appeared in *Nature*, Julian Huxley, Elton's former tutor at Oxford, brought it to him, and Elton immediately realized its importance. The generation of population cycles through endogenous causes was new and unexpected (Kingsland 1995:127).

1.1.3   The Balance of Nature

Whereas the study of population oscillations originated with the empirical work of Elton and the theoretical work of Lotka and Volterra, time-series analysis of population fluctuations can be traced to the famous debate about population regulation, which crystallized at the 1957 meeting in Cold Spring Harbor. One of the protagonists in the debate was A. J. Nicholson, who developed the theory of population regulation by density-dependent mechanisms (Nicholson 1933, 1954). Nicholson's views were supported by Elton, who wrote, "it is becoming increasingly understood by population ecologists that the control of populations, i.e., ultimate upper and lower limits set to increase, is brought about by density-dependent factors" (Elton 1949:19). Andrewartha and Birch (1954:649) disagreed: density-dependent factors "are not a general theory because ... they do not describe any substantial body of empirical facts." The debate reached a peak at the Cold Spring Harbor Symposium (Andrewartha 1957; Nicholson 1957). It has continued ever since, reaching another peak of intensity during the 1980s (a review in Turchin 1995b), although currently some consensus is apparently beginning to emerge (section 5.4).

An interesting thing happened while the regulation debate was raging. First, empirical ecologists began collecting long-term data on population fluctuations of a wide variety of organisms. It is curious that a lot of long-term data sets were started during the 1940s and 1950s (i.e., just when the debate was at one of its peaks!). Next, quantitative ecologists started analyzing these time-series data (Moran 1953; Bulmer 1974; Berryman 1978; Royama 1981; Potts et al. 1984; Turchin 1990) using, in the beginning, such linear approaches as the Box-Jenkins time-series analysis. Then, ecologists (most notably, Robert May) participated in the nonlinear dynamics revolution (Gleick 1988). When physicists invented the new technique of attractor reconstruction in time-delayed coordinates (Takens 1981; Packard et al. 1980), some ecologists began applying it to ecological time series (Schaffer 1985). Classical time-series analyses and non-linear dynamics approaches were eventually merged in a synthetic approach to the analysis of ecological data (these approaches will be discussed in part II), and applied to issues ranging beyond mere density dependence. Presently, we are seeing how these nonlinear time-series methods are being merged with the theoretical tradition (see chapter 8), and there are also promising beginnings of the synthesis between the population-regulation analyses and experimental approaches (Cappuccino and Harrison 1996).

1.2   GENERAL PHILOSOPHY OF THE APPROACH

Most ecologists do their science without giving much thought to the broad philosophical issues underlying what they do. Among those ecologists who do worry about philosophical foundations, the most vocal, and not afraid of making strong recommendations, are the Popperians (e.g., Chitty 1996; Murray 2000; Lambin et al. 2002). Other ecologists take the view that there are many ways of doing ecology, and one should not be too dogmatic about it (e.g., Fagerstrom 1987; Pickett et al. 1994). I believe that such philosophical discussions are important, because they affect how we do ecology. Furthermore, one of the broad themes of this book is methodological (see the preface): what are the best approaches to solving the puzzle of population cycles? Thus, I need to describe the philosophical basis of the general approach that I advocate.

While Popper's idea that all theories have to be testable in order even to be called scientific seems quite reasonable to me, I find the rest of his philosophy of science, at least as expounded by his eco- logical disciples, not to be a very useful way of doing science. I am particularly bothered by the emphasis of Popperians on falsification- ism as *the* way of doing science. First, the view that data are "hard facts" is untenable for methodological and psychological reasons (see Fagerstrom 1987 for a very clear discussion of this point). Thus, it is not true that in any contest between theory and data, it is theory that should necessarily lose. Second, I don't think that ecologists are in the business of rejecting theories. "Ecologists, like many others, do not reject theories for the futile reason that they are wrong; theo- ries are retained until better ones emerge" (Fagerstrom 1987). A very good idea of how futile a rejectionist program can be is conveyed by the book of Dennis Chitty (1996), *Do Lemmings Commit Suicide?* There Chitty relates how a consistent application of the rejectionist approach led him to reject all hypotheses that could be tested, leaving him with the explanation that nobody could figure out how to test.

I think that we (ecologists) are, instead, in the business of deciding which of the available alternative theories is the best, or "least wrong" (I shall make this idea more precise later in this section). One thing that any scientist has to come to terms with is that all our theories are, in the final account, wrong (the alternative of not being wrong is to become untestable, that is, nonscientific). The more explicitly we formulate our theories (which, at least in the context of population dynamics, means translating them into mathematical statements) the more wrong they become, simply because our simple theories can never capture all the complexity and detail of nature. So falsifying theories is trivial: just collect detailed data about any aspect of the theory, and you are certain to show that the theory is wrong. If you have not, it simply means that you either collected too few data points or did not measure them carefully enough.

If all our theories are a priori wrong, what can we do? Well, sci- ence is still the search for truth, but any scientific truth that we find is both approximate and tentative. Approximate, because of the reasons discussed in the paragraph above; tentative, because we have no guarantee that somebody smarter or possessing better data and analytical tools will not come up with a better "truth" sometime in the future. Therefore, we should not be in the business of rejecting theories, as ecological Popperians would have us do, but in the business of contrasting two or more theories with each other, using the data as an arbiter. The corollary of this approach is that our best theory may not explain or predict data very well, but we should still use it until we have something better. Even the theory that explains only 10% variation in the data is useful, because it sets a standard to be bettered.

In the rest of the section, I make this idea precise for the specific context of population dynamics. The basic notions are three: (1) define very carefully what you are trying to explain; (2) translate your verbal theories into explicit mathematical models (note the plural here); and (3) use formal statistical methods to quantify the relative ability of the rival models to predict data. Data may already be available, or they may be specifically collected to distinguish between predictions of the rival hypotheses (the latter constitutes *an experiment*).

1.2.1 Defining the Phenomenon to Be Explained

The broad question that I address in this book is, why do population numbers change with time? Or, to put it more succinctly, "why do populations behave as they do?" (Royama 1992:1). In any particular case study, this broad question can be broken into more specific issues. First, are dynamics of the studied population characterized by a stationary distribution of densities? (This is the issue of population regulation.) If yes, there is some characteristic mean level around which the population fluctuates, and fluctuations are characterized by a certain (finite) variance. What ecological mechanisms are responsible for setting this mean level? (This is the focus of population statics.) What mechanisms set the amplitude of fluctuations? Finally, are there detectable statistical periodicities, and what is the order and trajectory stability characterizing dynamics?

At the most general level, the phenomenon to be explained is quantified by a temporal record of population fluctuations, or time-series data. Time-series data are often available even before the beginning of the formal inquiry into dynamics of a particular population (although we often have to do with an index of population rather than an absolute measure of population density; examples include fur returns, bag records, and pheromone trap catches). If time-series data are not available, a systematic program for their collection should be initiated immediately. (One should not worry too much about limited usefulness of short time series; after all, it may take many decades to approach the solution, by which time time-series data will be long enough to be useful!)

I will call the density measurements of the "focal species" (the one whose dynamics we are trying to understand), $\{N_t\}$, the **primary data**.[1] We may have time-series data on other aspects of system dynamics available (e.g., temporal changes in mean body mass, fluctuations in the availability of food, and densities of predators or parasitoids). Such **ancillary data** may be extremely useful, but are secondary in the sense that we do not require that our explanation of the focal species dynamics would account for all of them. For example, if we are studying a forest defoliator, then a model based on plant quality does not need to explain why parasitism rates vary (perhaps parasitoids are simply responding to the oscillations of their food supply, without a detectable feedback effect on defoliator densities). Vice versa, a parasitism-based explanation does not need to account for changes in plant quality. Of course, the model based on a particular factor has to be consistent with time-series data for this factor.

A focus on the primary data permits us to use the same metric when comparing hypotheses based on very different factors. One particular metric that I will use extensively is the coefficient of prediction, R^2_{pred} (the proportion of variance in log-transformed density explained by the hypothesis). However, this is not the only metric that can be employed to quantitatively compare the performance of different hypotheses. Another approach is to first quantify the observed dynamical pattern with **probes** such as the period and amplitude of oscillations (and others, see section 6.2.2), and then to determine how well rival models predict the numerical values of probes.

---

[1] Concepts emphasized in boldface type are defined in the glossary.

Defining the problem the way I do here is not the only way to study population cycles. One alternative way, as practiced by the Canadian school (Chitty, Krebs, Boonstra, and others), is to define a "cyclic syndrome," which includes such features of dynamics as rapid changes of population density, systematic variation in body weight with the phase of the cycle, and perhaps certain changes in behavior, such as aggressiveness versus "docility." The problem with this approach is twofold. First, the "cyclic syndrome" is often defined without reference to whether population dynamics are characterized by a periodicity or not. In the cases where population dynamics are not periodic (or when there are no long-term data to determine whether there is periodicity or not), we find ourselves in a situation of studying a population "cycle" that is not a cycle by any formal definition. The second problem is that by including in the definition changes in individual quality and behavior, the Canadians tilt the field in favor of their favorite hypotheses. Suppose, for example, that periodic dynamics in a particular rodent population are driven by an interaction between rodents and their food supply. By measuring such processes as food requirements of rodents and growth dynamics of vegetation after being consumed, we may be able to construct an empirically based model that would predict the cyclic changes in rodent numbers (the primary data) very well. However, since we have not explicitly dealt with the physiological or behavioral responses of individuals to food scarcity or abundance, the model will say nothing about systematic changes of body weights with the cycle phase. Thus, the model will fail to explain the "cyclic syndrome." We could, of course, include such individual responses in the model. But this would be done at the expense of complicating the model structure, with the only yield an explanation of what really are side effects of population cycles—that individuals would be of low weight and fight more when food is scarce and population density is collapsing. This argument suggests to me that we should give logical preeminence to the primary data. I repeat that this does not mean that we should ignore various kinds of ancillary data, but neither should we necessarily aim at a theory that explains every bit of data collected about the focal population.

Before leaving the subject of problem definition, I want to reiterate that in this book I focus exclusively on nonspatial aspects of population dynamics. I recognize that movement and spatial dynamics are very important. However, one has to start somewhere, and the magnitude of the task—disentangling the mechanistic causes of temporal oscillations in any particular system—is already enormous.

1.2.2 Formalizing Hypotheses as Mathematical Models

Having defined the explanandum (what is to be explained), I now turn to the explanans (the means of explanation). The question is, what ecological mechanisms underlie temporal change in natural populations? This issue is at the core of the book.

But what do I mean by ecological mechanisms? I believe that the most useful approach to understanding population dynamics is the reductionist one. Thus, the mechanistic basis for population ecology should be provided by the properties of entities one hierarchical level lower than populations, that is, by the behavior and physiology of individual organisms (Metz and Diekmann 1986; Caswell et al. 1997): individual consumption, growth, and reproduction rates; the probabilities of being killed by a predator or succumbing to a pathogen; characteristics of individual movement; and so on. I believe that such **methodological individualism** is a valid principle, but in practice it is not always possible, nor desirable, to follow this reductionist program to the logical extreme.

For example, when studying a predator-prey interaction, we need not follow each individual predator while keeping track of its size, sex, hunger level, spatial position, and so on. We might instead summarize this wealth of information with just a few numbers, for example, the number of predators in each size class at any given time, or the density of predators in each patch, or even, most simply, the density of all predators. The mapping here is "many to one," because many potential descriptions in individual terms will map to a single number or set of numbers at the population level. Thus, an understanding of predator-prey dynamics may be approached in two steps. In the first step, the investigator performs a careful study of the individual predation process and attempts to summarize it with simple relationships, such as the functional response curve. In the second step, functions summarizing behavior and physiology of individuals serve as building blocks in a population dynamics model.

Ecological **mechanisms**, as used in this book, can refer both to detailed descriptions of what individuals do and to functions summarizing salient features of individual behaviors. I agree that it is more satisfying to build fully mechanistic explanations of population dynamics that are firmly based on what individuals do. However, it is not necessary to do it in one step. The history of population ecology shows that such concepts as "population density," "functional response," and "density-dependent population growth rate" turn out to be very useful conceptualizations for connecting population dynamics to individual-based explanations. Thus, we should continue employing these concepts, while keeping in mind their limitations.

The next step is to decide how to connect specific ecological mechanisms to testable predictions. I will require that the answer take the form of a **fully specified model**. The main reason for this requirement is that translating each rival hypothesis into an explicit model will allow us to perform quantitative cross-comparisons between different hypotheses. In other words, we shall be able to say which model explains the data better.

Constructing a fully specified model is done in three steps. First, we choose the mathematical framework and, most important, the **state variables**. Mathematical framework is often suggested by the biology of the system. For example, if we are dealing with a forest defoliator who has one generation a year, then we should probably use the discrete (difference) equations. If, on the other hand, we deal with a large ungulate population, in which the time step at which reproduction occurs (one year) is a small fraction of an average life span, then a continuous differential equations framework provides a good approximation (we might also consider adding seasonality explicitly to the model).

State variables are typically determined by the verbal hypothesis on which the model is based. For example, if we are modeling the interplay between the individual quality and dynamics, then a minimal model would have two state variables: population density and average individual quality. If we think that taking an average of quality is too restrictive, then we might explicitly model discrete quality classes (e.g., the numbers of "poor"-quality and "high"-quality individuals).

Alternatively, we might employ a partial differential equations framework, and model variation in individual quality smoothly. The choices of mathematical framework and state variables are not independent of each other.

The second step is to choose **functional forms**. These are specific functions that relate state variables and their rates of change to each other. One example is the functional form of the self-limitation term—we could choose to model it using the logistic model, or theta-logistic, and so forth. Another example is the functional response: depending on what we know about the modeled system, we may choose Type I, II, III, or ratio dependence.

The third step is determine the values of **parameters**. Examples are the intrinsic rate of population increase, the carrying capacity, the searching rate, the handling time, and so forth. This task can be accomplished in three basic ways. One is to use the information about the natural history of organisms to deduce the parameter values or, more likely, to deduce the interval where plausible values should be found. The second approach of obtaining parameter values is by fitting models to time-series data (see chapter 8). The disadvantage of this method is that if we wish to use the time-series data to test model predictions, such a test would not be as rigorous, since a degree of circularity is involved. The third way is to design a short-term experiment and directly measure the parameter. This is the best way, but the most laborious one. A short-term experiment may also be designed to measure a whole function, thus providing the empirical foundation for the functional form choice.

These three steps take the model builder progressively from general to specific issues. With each successive step the freedom of choice (or the degree of arbitrariness) increases. The choice of state variables is largely determined by the nature of the hypothesis and the mathematical framework. For functional forms, we usually have a greater latitude, but we usually are limited to discrete choices (e.g., should we use Type II or Type III functional response?). Of course, one could use a qualitative approach; that is, instead of choosing a specific function, one could just say that a function should be monotonically increasing. Such approaches, however, are more useful in building general theory (examples: Rosenzweig 1969; Oksanen et al. 1981) than in the analyses of specific case studies (but see Ellner et al. 1998 for *semimechanistic* approaches). Finally, parameter values typically vary continuously, and are least constrained by a priori considerations.

For all the above reasons, any single hypothesis can in principle be translated into an infinite number of fully specified models, depending on the choices we make at each stage. This means that rejecting a specific model in favor of another model based on a rival hypothesis may indeed indicate that the rival hypothesis is closer to the truth, but it may also indicate that we did not use the correct functional form in the first model, or perhaps misestimated a key parameter. This is not a lethal problem, since all scientific knowledge is approximate and tentative, but we should keep this caveat in mind.

1.2.3   Contrasting Models with Data

Once we have translated a set of competing hypotheses into models, we are ready to start the process of reducing this set to fewer (ideally, one) "winners." For example, suppose we are trying to understand why the population system we are studying exhibits a periodic second-order oscillation (dynamical classes are explained in chapter 5). The first basic test that each model has to pass is the ability to generate the qualitative type of dynamics characterizing the system, a periodic second-order oscillation in our example. Some models simply cannot generate second-order cycles (e.g., one-dimensional differential equation models cannot exhibit cyclic behaviors no matter what functional forms and parameters we use). We immediately eliminate such models, and by implication the hypotheses on which they are based, from the set of plausible explanations of the system's dynamics. The elimination of the verbal hypothesis is somewhat tentative, because it still may be possible to translate the hypothesis into a model (perhaps using a different mathematical framework) that would be able to generate the required qualitative type of dynamics. In any case, no rejection is final (just as no confirmation is final). However, if we do our best and still cannot translate the hypothesis into a model that generates cyclic dynamics, then we shall succeed in throwing a very grave shadow of doubt on the hypothesis. It will now be up to the advocates of the hypothesis (if there are such) to show that the hypothesis can be translated into a model that generates cycles, and that this translation can be done in a biologically reasonable way.

The next hurdle that each model in the set has to clear is the ability to predict the correct type of dynamics for biologically reasonable parameter values. Lacking good information, we may have to use a wide range for plausible values of some parameters. In such cases, the fact that the model passes this test does not count very heavily in its favor. However, we may find (by numerical explorations) that a certain parameter or parameters are critical for the ability of the model to generate the right dynamics. This means that we have a model prediction that may be tested with an experiment.

Once we are finished with these qualitative tests, we may find ourselves in a situation that none of the models managed to pass them. This means that we have to go back to the drawing board and exercise our creativity again. No cut-and-dried guidelines for generating new hypotheses exist (except, perhaps, Edison's famous dictum about 10% inspiration and 90% perspiration), which is what makes science interesting! Alternatively, there may be only one model still standing. This is a rather happy outcome, since it means that we are essentially done. Not everybody is likely to be satisfied with the conclusion, but it is no longer sufficient simply to advance a verbal hypothesis as an alternative explanation. Having a fully specified model that predicts the correct qualitative dynamics with biologically plausible parameter values substantially raises the stakes for any potential challenger—any alternative hypothesis will have to do at least as well.

A more likely outcome is that two or more models will be able to pass the qualitative tests. This means that we need to subject the remaining hypotheses to quantitative and, ultimately, experimental tests. The most rigorous and objective approach to quantitative testing is to construct the fully specified models using only ancillary data (ideally, by performing focused short-term experiments to quantify functional forms and parameter values), and then use each model to predict the primary data. Models can be compared by (1) how well they predict actual population densities and (2) how well they predict quantitative measures of population dynamics (the probes). Additionally, (3) models must describe the dynamics of other variables on which they are based, and (4) their parameters and functional forms need to be consistent with what we know about the system biology. Finally, (5) since a simpler explanation is always preferable to a more complex one, models with fewer parameters are given more weight than complicated models. Comparisons (1) and (2) can be translated into a single metric, allowing us to establish a ranking order for the models. Issue (5), although seemingly dependent on investigator judgment, can actually also be incorporated into the overall measure, using approaches based on information criteria, such as AIC (Burnham and Anderson 1998). Issues (3) and (4), by contrast, are difficult to translate into a common quantitative metric. For example, a parasitoid-host model for a forest defoliator may predict parasitism rates better than a food quality model for the same system predicts the changes in food quality. But this does not mean that these are grounds on which to prefer the parasitism model. Perhaps the food quality data are characterized by a higher measurement error. Similarly, one model may be able to predict the dynamics best for a rather marginal value of one of the parameters. But again, we have no common metric to downgrade this model in relation to others. This means that not everything can be formalized, and some aspects of model performance will have to be left to the judgment of individual ecologists.

CHAPTER 2

Population Dynamics
from First Principles

2.1 INTRODUCTION

Ecologists rarely discuss the philosophical foundations of research into population dynamics. Foundational issues tend to work in the background shaping inquiry, and are rarely hauled out into the daylight to be closely examined (Cooper 2001). Several controversies in population ecology can be traced to a misunderstanding or a disagreement at the very basic level, for example, the density dependence debate (section 5.4).

My goal in this chapter is to explicitly discuss the logical foundations of population dynamics. The starting point is provided by a set of "self-evident truths" or postulates, from which much of the logical structure of the theory of population ecology can be derived. I will use these postulates to construct several foundational principles of ecological theory that we shall find particularly useful in explaining population oscillations. Furthermore, I will argue that one of the implications following from the principles is that there are three fundamental classes of population dynamics. This typology turns out to be of great practical utility in studying dynamics of specific populations. In particular, it provides a basis for classifying (1) qualitative kinds of dynamical behaviors and (2) the dynamical role of various ecological mechanisms. Additionally, the typology provides (3) the appropriate null hypotheses for testing ideas about population dynamics. In sum, the foundational principles and dynamical classes together serve as a general framework for investigating complex population dynamics.

Does Population Ecology Have "Laws"?   Like many scientists who are not physicists, ecologists are unable to resist unfavorable comparisons between their science and physics. Some argue that ecologists do not think like physicists, and that is why there is little progress in ecology (Murray 1992). Others reply that biologists should not think like physicists, because of the nature of biological science (Quenette and Gerard 1993; Aarsen 1997). On both sides of the debate, there is a widespread belief that ecology is different from physics because it lacks general laws, and is not a predictive (and, therefore, not a "hard") science. For example, Cherrett (1988) lamented that "there is unease that we still do not have an equivalent to the Newtonian Laws of Physics, or even a generally accepted classificatory framework" (see Kingsland 1995:222–223 for a commentary). Even eminent theoretical ecologists appear to subscribe to this view: ecology, apparently, is different from physics because one of its distinguishing features is the near absence of universal facts and theories (Roughgarden 1998:xi). As to ecology's ability to generate testable theories, Aarsen (1997:177) thinks that "on this scale, ecology admittedly has a weak record" (see also Weiner 1995). "Ecology was not and is not a predictive science" (McIntosh 1985).

  Much can be said to counter these arguments. For one, I doubt that many ecologists truly suffer from physics envy (as opposed to using it as a rhetorical technique to motivate colleagues to do better science). The ones I know would much rather chase butterflies in the field than become the one-hundredth member of a supercyclophasotron team! I also think that population dynamics is a much more exciting field, being (as I hope to demonstrate in this book) on the brink of a major synthesis, compared with high-energy particle physics, where most of the excitement is in the past. Furthermore, physics is not a monolithic science. In certain highly respectable fields, such as astrophysics, manipulative experiments are impossible. Does it mean that there is no progress in astrophysics? The number of articles on astrophysics in *Science* or *Nature* seems to belie that claim. Finally, it is a gross exaggeration to claim that physics is a predictive science in all its branches. Physicists assure us, on one hand, that they have a complete understanding of the laws of fluid dynamics that govern atmospheric movements. On the other hand, neither they nor anybody else can accurately predict weather more than five to seven days in advance.

I could go on, but I do not think that trying to counter each charge of the critics is necessary. A more productive approach is simply to do population ecology and eventually show that it is a vigorous, theoretical, and, yes, predictive science. I submit the synthesis part of this book (chapters 9–14) as evidence that population ecology is a mature science in this sense. Furthermore, I think that population dynamics has a set of foundational principles that are very similar, in spirit and in logic, to Newton's laws. Accordingly, my second agenda for this chapter is to highlight the similarities between the logical foundations of population dynamics and Newtonian mechanics.

The Concept of Population   The final task I need to address in this introduction is the definition of population, since that is the central object of study in population dynamics. Following Berryman (2001; see also Camus and Lima 2001), I define **population**[1] as a group of individuals of the same species that live together in an area of sufficient size to permit normal dispersal and migration behavior, and in which population changes are largely determined by birth and death processes. Groups of organisms living in smaller regions become *local populations*, whose dynamics are strongly affected by dispersal and migration (Berryman 2001). Because I cannot address spatial issues in this book, its central focus must be on global rather than local populations, that is, on populations for whom emigration/immigration terms can be neglected without a serious loss of predictability.

2.2   EXPONENTIAL GROWTH

Practically all textbooks start exposition of population ecology with the exponential law of population growth (Malthus 1798). The exponential law is a good candidate for the first principle of population dynamics (Ginzburg 1986; see also Brown 1997; Berryman

---

[1] Boldface type indicates terms whose definitions are in the glossary.

1999). Before discussing why exponential growth may qualify as a *law*, let us briefly review the logic underlying its derivation.

2.2.1   Derivation of the Exponential Model

I begin with two postulates. *Postulate 1* states that the number of organisms in a population can change only as a result of births, deaths, immigrations, and emigrations. If we use an area of sufficient size in the definition of population, then we can approximately set immigration and emigration terms to zero, so that the number of organisms would change only as a result of births and deaths.

Postulate 2 states that population mechanisms are individual-based. That is, all population processes affecting population change (births, deaths, movements) are a result of what happens to individuals. Thus, the expected number of new individuals appearing in a population per unit of time is obtained by summing the offspring produced by each adult in the population. Similarly, the rate at which population numbers are decreased is a summary result of the probabilities that any individual would die during the time interval. This postulate reflects the principle of methodological individualism, which provides the philosophical basis for the reductionist agenda in studying population dynamics (section 1.2.2).

From these two postulates, we can derive the first foundational principle of population dynamics, which states that a population will grow (or decline) exponentially as long as the environment experienced by all individuals in the population remains constant. *Environment* here refers to all environmental influences affecting vital rates of individuals: abiotic factors, the degree of interspecific crowding, and densities of all other species in the community that could interact with the focal species.

The derivation of the exponential law for the case when all individuals in the population are absolutely identical (in particular, there is no age, sex, size, or genetic structure) and reproduce continuously is very simple. We start by expressing population change as the balance between births and deaths (postulate 1), and then change to per capita rates using postulate 2:

$$\frac{dN}{dt} = B - D = bN - dN = (b - d)N = r_0 N \qquad (2.1)$$

where B and D are the total birth and death rates, b and d are the per capita rates, N is the total number of individuals in the population, and r_0 is the per capita rate of population change. There are no immigration/emigration terms, because I assumed that the population is closed. I have also glossed over the fact that individuals typically come in discrete packages, while N is a continuous variable. This is not a serious problem, since we can think of N as the *expected* number of individuals per unit of area in the population at time t. (Another approach is to think of N as biomass.)

This elementary derivation readily generalizes to more realistic settings:

1. For semelparous organisms (such as annual grasses or insects) we obtain the discrete form of the exponential law: $N_{t+1} = \lambda N_t$.

2. Adding age or stage structure is also relatively straightforward. However, we now have to wait awhile for the population to achieve a stable age distribution, after which all age classes (as well as total number of individuals) begin to grow according to the exponential law.

3. The general pattern of growth is still exponential when we consider finite populations and add demographic stochasticity. For example, Bartlett (1966) shows that the expected population size in a stochastic birth process is $N(t) = N(0)\exp[r_0 t]$, same as in the deterministic model. It is also possible to calculate the probability distribution of $N(t)$, including variance. The stochastic framework provides a natural way of handling discrete individuals. However, there is a caveat: when the number of individuals in a population becomes very low, we can observe deviations from the exponential law (Lande 1998).

4. The environment does not have to be constant. If the environment varies in such a way that the per capita rates b and d have stationary probability distributions, then we obtain a model of stochastic exponential growth/decline (Maynard Smith 1974:14–15). The expected population density is again $N(t) = N(0)\exp[r_0 t]$.

5. Finally, adding space and diffusive movements leads to a simple partial differential equation model, analyzed by

Fisher (1937) and Skellam (1951). In this model, the total
number of individuals continues to grow exponentially, even
as they diffuse out from the initial center.

In short, as long as the environmental influences do not change
in a systematic manner, we end up with one or another version of
the exponential law. In fact, we can formulate it even more generally
by substituting "stationary environment" for "constant environment"
in the definition given above. The exponential law is a very robust
statement.

2.2.2   Comparison with the Law of Inertia

But is it a *law*? Let us consider the nature of something about which
there is no argument regarding whether it is a law—Newton's first
law. Here is a very clear explanation, which incidentally comes from
a persistent critic of theoretical population ecology:

> In their survey of the development of the concepts of physics,
> Einstein and Infeld (1938) begin with the discovery by Galileo
> and Newton of a new way of thought, "scientific reasoning." The
> important lesson in this new thinking was that "intuitive conclu-
> sions based on immediate observations are not always trusted for
> they sometimes lead to the wrong clews." Fundamental problems,
> such as the motion of bodies, are often obscured by their com-
> plexity. For example, if we give a cart a push, it will move some
> distance before stopping. If we give it a bigger push, it will move
> farther before stopping. The intuitive interpretation is that of Aris-
> totle, a "moving body comes to a standstill when the force which
> pushes it along can no longer so act as to push it."
>
> Today we all "know" that Aristotle's concept is false. We can all
> recite Newton's law of inertia. "Every body continues in its state
> of rest, or of uniform motion in a right [= straight] line, unless
> it is compelled to change that state by forces impressed upon it"
> (Newton 1729). But, how did Newton achieve this insight?
>
> Einstein and Infeld (1938) return to the motion of the cart. If
> we give it a push, it will move a certain distance and stop. Sup-
> pose we oil the wheels and smooth the road? If we give the cart

the same push, it will move farther before stopping (because we have reduced the forces tending to stop the cart). Suppose we now remove all impediments to motion? If we give the cart any push at all, it will move at a constant speed in a straight line and never stop. But, we cannot do this experiment. We can only think it. As Einstein and Infeld (1938) state, "this law of inertia cannot be derived directly from experiment, but only by speculative thinking consistent with observation. The idealized experiment can never be actually performed, although it leads to a profound understanding of real experiments." The law allowed Newton to predict how far an apple would fall each second of free-fall near the earth's surface, the elliptical shape of planetary orbits, the flattening of the poles, and much else (Murray 1992:594).

The similarity between the exponential law and the law of inertia is striking (Ginzburg 1986). First, both statements specify the state of the system *in the absence of any "influences" acting on it.* The law of inertia says how a body will move in the absence of forces acting on it. The exponential law specifies how population numbers will change in the absence of systematic changes in the environment.

Second, the action of both laws is obscured by complexities characterizing real-life motions of bodies or population fluctuations. As a result, neither statement can be subjected to a direct empirical test. Just as we cannot observe a body on which no forces are acting, we cannot observe a population growing exponentially indefinitely. Inevitably, as a result of population growth, individuals will begin experiencing a higher degree of crowding, start running out of food, and suffer greater predation or disease. Thus, both laws have to be arrived at by speculative thinking, and only their consequences can be empirically tested.

Third, both statements are in some sense self-evident (at least, in retrospect!), so there is a suspicion that they are trivial, or tautological in some sense. However, we can also imagine an alternative universe in which different versions of first laws of population dynamics and classical mechanics would hold. The alternative to the law of inertia is Aristotle's concept, discussed by Murray in the quotation above. In fact, in pre-Galileo and Newton days, Aristotle's "first law" was widely believed.

Similarly, in pre-Pasteur days, many biologists believed that life could spontaneously generate from nonliving matter. During the Middle Ages it was thought that mice and flies spontaneously generated from dirty laundry and refuse, and that frogs fell from the skies with rain. If spontaneous generation were possible, then population dynamics theory would be completely different (just as classical dynamics based on Aristotle's first law would be a completely different science). In particular, the equivalent of equation (2.1) would be

$$\frac{dN}{dt} = S + B - D = S + bN - dN = S + r_0 N \qquad (2.2)$$

All notation is as in equation (2.1) except S, which is a constant rate of spontaneous generation. For N near 0, equation (2.2) can be approximated as

$$\frac{dN}{dt} \approx S, \quad \text{which solves to } N(t) = N(0) + St \qquad (2.3)$$

In other words, when population density is small, population will grow linearly with time. In contrast, population growing according to the exponential law exhibits a nonlinear, accelerating pattern of growth. Actually, equation (2.2) is not quite as ridiculous as it sounds. It can be used to model population dynamics in a sink habitat, dominated by immigration from some source habitat (S, then, would represent a constant flow of immigrants). Additionally, the plant regrowth equation in certain plant-herbivore models (see section 4.4.2) has the same relationship to equation (2.3) that the logistic has to the exponential. We shall see that regrowth is a strongly stabilizing feature in herbivory models, which allows us to conjecture that in the alternative universe where spontaneous generation occurs routinely, we would have much more stable population dynamics!

The fourth way in which the two "first laws" are similar is that they provide the basis for building predictive theories for population dynamics and for classical mechanics. Just as Newton was able to predict how far an apple would fall by using the law of inertia (plus several other laws, to be sure), population ecologists use the exponential law as the basis for modeling populations. This can be seen by rewriting equation (2.2) slightly, and making the per capita rate of population change a function of "all sorts of things":

$$\frac{dN}{N\,dt} \equiv r(t) = f(N, \text{ weather, food, predators, etc.}) \qquad (2.4)$$

Practically all population dynamics models have this form (or an equivalent if we use some other mathematical framework than ordinary differential equations). In fact, the exponential law is most profitably thought of as the null state in which any population would be if no forces (= environmental changes) were acting on it. It is a direct equivalent of the law of inertia, and is used in the same way, as a starting point to which all kinds of complications are added. Thus, the starting point in the analysis of time-series data is the discrete version of (2.4):

$$r_t \equiv \ln \frac{N_t}{N_{t-1}} = f\left(N_{t-1}, N_{t-2}, \ldots, U_t^1, U_t^2, \ldots, \epsilon_t\right) \qquad (2.5)$$

where r_t is the realized per capita rate of change and N_t is the density at time t. The lagged densities $N_{t-1}, N_{t-2}, \ldots$ represent endogenous feedbacks, U_t^i are known (measured) exogenous influences, and ϵ_t represents the unknown exogenous influences modeled as a random variable (section 7.2.1).

2.2.3 *"Laws": Postulates, Theorems, Empirical Generalizations?*

Note that up to now I have avoided defining just what exactly I mean by "law." I did this on purpose, because I wanted to avoid definitional wrangles. Instead, I adopted the approach of arguing by analogy with the law of inertia from classical mechanics. At this point, however, it is becoming clear that we have to think more carefully about the logical status of various "lawlike" statements. In particular, perhaps we should distinguish between elementary propositions that are taken without proof (postulates or axioms) and statements derived from a set of these postulates (theorems). Note, however, that this distinction is not absolute. For example, we can take exponential growth as a postulate. On the other hand, we can also derive it from more elementary principles (my postulates 1 and 2). In the second approach (which I favor), exponential growth is a theorem.

We may, therefore, have several kinds of lawlike statements in population ecology. Some have the logical status of axioms, and others theorems. Yet a third kind is an "empirical law" that arises as a generalization from some body of facts (e.g., many allometric laws, such as the relationship between body size and population cycle period; see Calder 1983). My preference is to reserve the label "law" for those foundational principles that guide theory building in population ecology. Laws, thus, provide the *hard core* of a research program (using Lakatosian terms). Therefore, a law, in the sense that I use, does not need to be directly substantiated by empirical tests (just as the law of inertia cannot be directly tested). Instead, we make judgments whether the whole research program is fruitful, or not. I shall return to this idea in section 2.6, but first let us consider whether there may be other foundational principles in population dynamics, in addition to the exponential law.

2.3   SELF-LIMITATION

2.3.1   Upper and Lower Density Bounds

One cannot predict the motion of planets with just the law of inertia. Similarly, we need more principles in addition to the exponential law to predict population dynamics, so that we can eventually subject the whole framework to empirical tests. The second foundational principle that I would like to propose is a formalization of the notion that population growth cannot go forever.

The notion of some upper bound beyond which population density cannot increase seems uncontroversial. There must be some absolute upper bound on density, simply because one can physically cram only so many organisms into a unit of area. Let us call the existence of an upper density bound *postulate 3*. The problem with this postulate is that it is not very useful in practical applications, because few organisms ever attain this upper limit in population density by completely filling their physical space. Thus, we shall need to come up with a better formalization of the notion of self-limitation (in the next section).

In contrast to the absolute upper bound, there could be no such hard lower bound, as some consideration of the issue shows. Most obviously, populations and whole species have been known to go extinct. It is true that on the ecologically relevant time scale the vast majority of species do not go extinct routinely, because at some low population density the rate of population growth becomes positive. However, we must qualify this statement with several important caveats. First, a population may need to spend some time at low density before ecological mechanisms acting with a delay would relax their downward pressure on the rate of population growth. Second, at extremely low densities, population growth may be negative as a result of an Allee effect. Thus, a population fluctuating around the mean level $\overline{N}$ may be prevented from extinction by positive growth rates around the $N_+ \ll \overline{N}$ level, which typically provides a lower bound on population fluctuations. However, should the population decrease below another threshold, $N_- \ll N_+$, where the growth rate becomes negative due to an Allee effect, then it would go extinct. Third, in a stochastic world it is not enough for persistence that the population growth rate is positive; it has to be positive enough to counteract the detrimental effect of environmental stochasticity (see Lande 1998; Holsinger 2000). Finally, individuals come in discrete packages. Models based on discrete-valued population state variables generally have an unhelpful property that eventual extinction occurs with probability 1 (Chesson 1981).

In short, I would argue that the notion of a lower limit on population fluctuations is not as obvious as one might think. Thus, it is not surprising that a very sophisticated mathematical apparatus is needed to handle it rigorously (Chesson 1978, 1982). Given these complexities, I feel that at this point in time it would be premature to attempt to codify the notion of a lower bound with some general principle. Fortunately, this issue is more important in conservation ecology, where the concern is with populations that are prone to extinction. For the main subject of this book, species that exhibit sustained oscillations, we can largely ignore these complications.

2.3.2 Formalizing the Notion of Self-Limitation

One way to formalize the notion of population self-limitation is to tie it to an upper density bound. Thus, we may require the rate of population change to become negative above some (possibly very high) density threshold: $r(t) < 0$ if $N > N_{upper}$, where the per capita rate of change $r(t) \equiv dN/(Ndt)$. The problem with this approach is that the value of N_{upper} is likely to change with time depending on environmental conditions, such as resource or natural enemy abundances. For example, individuals may defend smaller territories when food is abundant, and larger territories when it is scarce. It is desirable to separate the effects of self-limitation, understood as direct (or undelayed) density dependence, from population feedbacks involving time lags, such as depletion of food (when food is a slow dynamical variable), or increase in specialist natural enemies.

The alternative approach, thus, is to require that the partial derivative of $r(t)$ with respect to N is negative

$$\frac{\partial r(t)}{\partial N} < 0 \quad \text{for } N > \overline{N} \tag{2.6}$$

The biological meaning of this statement is that as we vary N, while keeping all other variables that affect $r(t)$ constant, increasing N leads to a decrease in $r(t)$, while decreasing N increases $r(t)$. In the example above of territories varying with food availability, we fix food density and then consider how increasing density will affect the per capita growth rate. Clearly, as N becomes large enough so that there are not enough territories for all individuals, $r(t)$ will decrease, and so equation (2.6) holds. We now also see why we have to hedge (2.6) with a condition that it should hold only for high enough densities. It is conceivable that $r(t)$ will not change with N when N is low. In fact, any relationship between $r(t)$ and N at low N is possible: negative if effects of density "percolate" all the way down to $N = 0$, positive if we have Allee effects, or none if density has to reach some high value before density-dependent effects begin operating. What is important for self-limitation is that there is a negative relationship between $r(t)$ and N at higher population densities, for $N > \overline{N}$, where $\overline{N}$ is related to some average level of density. To be maximally specific, let us define $\overline{N}$ as the long-term mean density. Now that we have a mathematical statement of this requirement—I will call it postulate 3—it is easy enough to check whether any particular ecological model includes self-limitation or not. We simply rewrite the equation for the focal species in per capita form, differentiate the right-hand side with respect to N, and check whether it is negative for N greater than the long-term mean density.

2.3.3   The Logistic Model

The existence of upper and lower bounds on population density, with all the caveats discussed above, clearly implies that most populations are in some sense regulated, and that the rate of population change must be density dependent (section 5.4). The simplest differential model of density-dependent population growth is the logistic equation, formulated by Verhulst in the nineteenth century, and promoted by Raymond Pearl during the 1920s (Pearl and Reed 1920; see Kingsland 1995 for the history of the ensuing debate).

There are two ways to view the logistic model: (1) the formulation as a differential model, and (2) the solution of the model, the infamous S-shaped curve. It was a great mistake, in my opinion, for Pearl to focus exclusively on the solution, rather than on the more mechanistic formulation as a differential equation. The S-shaped curve is the solution of the logistic for a rather special set of initial conditions, namely, when $N(0)$ is very small. Starting with an initial condition near the equilibrium, k, or above it, results in a J-shaped curve. Since most natural populations fluctuate within some typical bounds, we shall see an (even approximately) S-shaped trajectory only after a catastrophic decline, or after an invasion into a new habitat.

To make matters worse, when confronted with deviations from the symmetric S-shaped curve in applications, Pearl and Reed chose to modify the logistic curve in a completely phenomenological manner. In particular, using the logistic solution as the starting point,

$$N(t) = \frac{k}{1 + be^{at}}$$

they simply added extra terms to the exponent:

$$N(t) = \frac{k}{1 + be^{a_1 t + a_2 t^2 + \cdots + a_n t^n}}$$

(Kingsland 1995:70). In my opinion, they would have done much better were they to focus on the differential form—the dynamical rule underlying population change. The failings of the logistic model are well known: it assumes linear relationship between the realized per capita rate of population change and density, and there are no explicit consideration of effects of noise and, most important, no lags.

However, the diagnosis of these problems can be turned around as a prescription on how to modify the logistic equation in any specific case study. Thus, allowing for nonlinearities in density dependence results in an asymmetric pattern of growth. Adding noise changes the nature of the equilibrium from a stable point to a stationary distribution. Finally, time lags allow density to overshoot the equilibrium, potentially leading to cycles and chaos (e.g., in the discrete versions, such as the Ricker model).

My conclusion, therefore, is that the logistic model is a useful starting point for modeling density-dependent population dynamics. Its primary value is not in the specific equation but in the general framework it provides, both for including realistic features of single-species dynamics and for inclusion into models of multispecies interactions.

2.4 CONSUMER-RESOURCE OSCILLATIONS

Ecologists distinguish five general classes of pairwise species interactions, classified by the positive $(+)$, negative $(-)$, or no (0) effect of species on each other: interference competition $(-, -)$, mutualism $(+, +)$, commensalism $(+, 0)$, amensalism $(-, 0)$, and trophic $(+, -)$. Note that I consider the exploitative kind of interspecific competition as not a binary interaction but one that requires at least three species: two consumers and one resource. Although resource-consumer, or trophic, interaction is only one of five types, population ecologists have devoted a massive share of their attention to studying trophic interactions. This is not to say that other interactions, such as mutualisms, are unimportant. But unlike the trophic interactions, they do not seem to be of universal importance: mutualism could be the most important interaction in some specific population systems, but all organisms are consumers of something, and most are also a resource to some other species. Furthermore, the current state of empirical evidence suggests that population oscillations are primarily driven by trophic mechanisms (section 15.1). Thus, we need a thorough understanding of the logical foundations of consumer-resource models.

2.4.1   Three More Postulates

I propose that we need at least three postulates to begin developing general models of trophic interactions. *Postulate 4*, mass action, states that at low resource densities the number of resource individuals encountered and captured by a single consumer is proportional to resource density. In mathematical terms,

$$\text{the capture rate } = aN \text{ as } N \to 0 \tag{2.7}$$

where N is the population density of resources and a a constant of proportionality. This principle in an ecological setting was first formulated by Vito Volterra, who used it in his derivation of the Lotka-Volterra predator-prey model (although Volterra did not stress the important qualification that N needs to be low for the linear relationship to hold). In the modern ecological jargon, relationship (2.7) is called the linear **functional response**.

Postulate 5, biomass conversion, states that the amount of energy that an individual consumer can derive from captured resource, to be used for growth, maintenance, and reproduction, is a function of the amount of captured biomass (Maynard Smith 1974; Getz 1991; Ginzburg 1998). Most predator-prey models in ecology use a special case of this postulate: consumers will assimilate a constant proportion of energy in the resource biomass they capture. It is clear, however, that this version of the principle lacks generality, since nothing prevents consumers from eating a variable proportion of captured resource biomass. In fact, many predators indulge in what is known as "surplus killing," so we in fact would expect that at high resource density the proportion consumed would decrease. Thus, a most general version of postulate 5 should be prohibitive in nature (as are many laws of physics, including the law of energy conservation, which clearly underlies postulate 5): a consumer cannot derive more energy from captured resource than the energy contained in the resource multiplied by a maximum conversion rate characterizing the consumer species.

The last postulate that I would like to propose is complementary to postulate 4. *Postulate 6*, maximum consumption rate, states that no matter how high the resource density is, an individual consumer can ingest resource biomass no faster than some upper limit imposed by its physiology (e.g., the size of its gut and the speed with which food can be passed through the gut). Similarly, there is a maximum reproduction rate characterizing different species. Let us call this *postulate 6'* (maximum reproduction rate). This proposition may be thought of as a consequence of postulate 6, since a maximum rate of energy acquisition imposes a limit on maximum reproduction rate, although in practice the latter may be prevented by other mechanisms from reaching the energetically imposed maximum.

Postulates 4 and 6 specify the functional form of the relationship between resource density and the rate of consumption at two ends of the scale: very low N and very high N. Connecting the two limiting cases by a smooth curve, we obtain the familiar hyperbolic functional response (or Type II) curve (figure 2.1). This functional relationship is one of the best-supported generalizations in population ecology. Essentially whenever there is a situation in which a single consumer is foraging without interference for a single type of resource, its consumption rate will have the shape depicted in figure 2.1. Holling (1965) thought that hyperbolic functional response would characterize only invertebrate predators, while vertebrates would exhibit an S-shaped (Type III) response. However, it is now clear that the hyperbolic functional response is a characteristic of *specialist* predators, whether invertebrate or vertebrate. (The functional responses characterizing generalist predators will be considered in section 4.2.4). Hyperbolic functional responses are a general feature of all kinds of consumers, not only predators (narrowly understood). Thus, arthropod parasitoids exhibit hyperbolic responses (Hassell 1978). The ingestion rate of herbivores is related to the forage biomass density by a hyperbolic curve (Spalinger and Hobbs 1992). The relationship between consumption rate and resource density becomes complicated when there is more than one kind of resource, when consumers interfere with each other, and when resources are highly clumped.

The clear logical foundations and generality of the hyperbolic functional response make it a logical candidate for a general law of population dynamics. Additionally, it is one of the most important ingredients in population ecological models. For these reasons, it deserves to be added to the list of foundational principles.

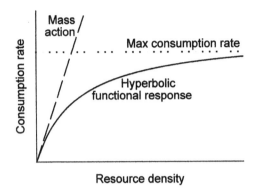

FIGURE 2.1. Constructing the hyperbolic functional response from postulates 4 and 6.

2.4.2 The Lotka-Volterra Predation Model

Putting postulate 4 (linear functional response) and a simplified version of postulate 5 (a constant proportion of prey biomass converted into new predators) together with assumptions that resource and consumer populations will grow or decline exponentially in the absence of the other species, we have the Lotka-Volterra model (Lotka 1925; Volterra 1926):

$$\frac{dN}{dt} = r_0 N - aNP$$

$$\frac{dP}{dt} = \chi aNP - \delta P$$

(2.8)

where N and P are population densities of resources and consumers, r_0 is the per capita rate of population growth of resources in the absence of predators, δ is the per capita rate of population decline of consumers in the absence of resources, a is the constant of proportionality from (2.7) (assuming that this relationship holds at all N), and χ is the constant of proportionality relating the number of consumed resources to the number of new predators produced per unit of time.

The Lotka-Volterra is not a very realistic model for real resource-consumer systems. To my knowledge, there has been no successful application of it to any field or laboratory population system. But this is all beside the point, because the Lotka-Volterra model seems to get at some extremely basic feature of trophic interactions: their inherent proneness to oscillation. In fact, the Lotka-Volterra model predicts a rather special kind of oscillation that has no fixed amplitude. Such dynamics, in which the amplitude of oscillation depends on initial conditions, and does not either decrease or increase with time (unless perturbed by an external force), are called **neutral oscillations**.

Let us rewrite model (2.8) using per capita growth rates (remember, this is the right thing to do, because it is what the first law of population dynamics tells us):

$$\frac{dN}{N\,dt} = r_0 - aP$$
$$\frac{dP}{P\,dt} = -\delta + \chi aN \tag{2.9}$$

Model (2.9) has one extremely interesting feature: the per capita rate of each species depends only on the density of the other species. For example, N does not appear on the right-hand side of the resource equation. Thus, there is no direct population feedback to resource density, although there is, of course, an indirect connection (since increase in prey density will eventually cause the predator to increase, which will in turn have a negative effect on prey per capita rate of change). Similarly, consumer dynamics depend directly only on resource density. A system in which per capita rates of change of both resource and consumer do not depend on their own density is a **pure resource-consumer system**. Also note that the right-hand sides in model (2.9) are linear functions. Thus, the Lotka-Volterra model is the simplest possible formulation of a pure resource-consumer system.

Two features of model (2.9), that it is a pure resource-consumer system, and that its dynamics are oscillatory, are clearly connected. This observation suggests that the Lotka-Volterra model has identified for us an important general principle, which may deserve the status of a law of population dynamics. Here is how I would formulate this consumer-resource oscillations principle: a **pure resource-consumer system** will inevitably exhibit unstable oscillations. By "unstable oscillations" I understand population oscillations that do not converge to a point equilibrium. Either they can be neutral, as in the Lotka-Volterra model, or they may actually diverge, getting away from the point equilibrium with each cycle, as in the Nicholson-Bailey model.

Does this statement depend on details of the Lotka-Volterra model? In particular, model (2.9) assumes that the right-hand sides are linear functions. However, the local stability of nonlinear generalizations of the Lotka-Volterra model will be determined by the stability of the linearized version. As long as the linearized version has the same signs in front of its coefficients (and lacks direct population feedbacks in both equations), we should obtain neutrally stable oscillations. (Of course, what happens when the oscillation gets away from the equilibrium will depend on the nonlinearities.) Interestingly, this is the argument by which Lotka accomplished his derivation of the Lotka-Volterra model. Unlike Volterra, who started by considering a specific mechanism of predators encountering prey, Lotka first wrote predator-prey equations in general form. He then considered the linearization of the general equations that leads to model (2.9).

We also need to check on how the general insight from model (2.9) depends on its mathematical formulation as a system of ordinary differential equations. One alternative framework is discrete difference equations. May (1973a) considered a discrete version of the Lotka-Volterra model, and showed that it is characterized by diverging oscillations for all values of parameters. The oscillations are not neutral, as in the continuous variant, because discretization introduces a lag in the responses of predators and prey to each other's densities, and lags are an inherently destabilizing feature in any model. By making the time step increasingly smaller, we can make the oscillations to diverge very slowly, and in the limit, when the time step is 0, we recover the neutral stability of the Lotka-Volterra model. May (1973a) further showed that the Nicholson-Bailey parasitoid-host model is equivalent in its stability properties to the discrete Lotka-Volterra model (despite different functional forms used by Nicholson and Bailey). In summary, it appears that the tendency of pure consumer-resource systems to show unstable oscillations does not depend sensitively on the specific assumptions of the Lotka-Volterra model.

There may be two objections to my proposal of consumer-resource oscillations as a general law of population dynamics. First, we know very well from experience that not all (and perhaps a minority) of real-life consumer-resource systems show persistent oscillations.

This objection, however, misunderstands the nature of a general law. Like the law of exponential growth, the law of consumer-resource oscillations is not meant to be tested directly. In real life, we never expect to encounter *pure* resource-consumer systems. Necessarily, the per capita growth rate of both resource and consumer populations would be affected by their densities (as postulate 3 tells us). Furthermore, there will be other species in the community. Consumers may be generalists. Resources may not be killed during the process of consumption, but only lose a part. There are refuges, spatial and temporal heterogeneity, and many other potentially stabilizing mechanisms known to ecologists. What the third law says, however, is that there is an inherent tendency for specialist consumer-resource systems to oscillate. This "signal" may or may not come through the "noise" of real-life complications.

2.5 PROCESS ORDER

In sections 2.2–2.4 I argued that the exponential, logistic, and Lotka-Volterra predation models illustrate three foundational concepts of population dynamics: exponential population growth, self-limitation, and trophic oscillations. The three elementary models are very simple, and usually we need to add various realistic features to use them in real-life applications. Yet, because of their simplicity, they lay bare some of the fundamental features of population dynamics, which we can capture with the concept of **process order**.

Order in Differential Models   Consider the general form of the three models (table 2.1). Taking the Lotka-Volterra model first, note that the per capita rate of population change, $r(t)$, depends on N indirectly, via some other dynamical variable, X (which in this case happens to be predator density, P). Thus, in order to describe the dynamics of this system fully, we need two equations, one for the rate of change of each state variable (N and P). Two-equation models such as the Lotka-Volterra are sometimes called **second-order** dynamical processes. In the logistic, by contrast, $r(t)$ depends only on N, leading to a one-dimensional system, or a **first-order** process. Finally, in the exponential model, the per capita rate of population change, $r(t)$,

TABLE 2.1. The concept of process order as illustrated by the three elementary models. $r(t) \equiv dN/(N\,dt)$

Model	Equations	General Form	Order
Exponential	$dN/dt = r_0 N$	$r(t) = r_0$	Zero
Logistic	$dN/dt = r_0 N(1 - N/k)$	$r(t) = f(N)$	First
Lotka-Volterra	$dN/dt = r_0 N - aNP$ $dP/dt = -dP + \chi aNP$	$r(t) = f(X)$ $dX/dt = g(N)$	Second

does not depend on N, and therefore, by extension, I will call this type of models a **zero-order** process.

Dynamical systems of different order are characterized by fundamentally different kinds of behaviors. Zero-order systems can only exhibit nonstationary behaviors—they tend to either increase to infinity or decline to zero. First-order systems, by contrast, are capable of stable equilibria, because they are characterized by population feedbacks. Finally, second-order systems add the ability to exhibit cycles to the spectrum of dynamical behaviors. However, such a strict mapping of process order to the qualitative type of dynamics occurs only in models formulated as ordinary differential equations. We need to make sure that the concept of order generalizes well enough to be useful to other mathematical frameworks. Thus, let us consider how we can define process order in the framework of difference equations.

Order in Difference Equations   The concept of order extends naturally to discrete-time population models. Note that in difference equations the per capita rate of population change $r_t \equiv \ln N_t/N_{t-1}$. For the discrete exponential model r_t is a constant, and therefore this model has order zero. First-order models have the general form $r_t = f(N_{t-1})$. An example of a first-order model in ecology is the Ricker equation:

$$N_t = N_{t-1} \exp[r_0(1 - N_{t-1}/k)]$$

A second-order model has the following general form:

$$N_t = F(N_{t-1}, X_{t-1})$$
$$X_t = G(N_{t-1}, X_{t-1})$$

$$(2.10)$$

where X_t is some other state variable. An example of a second-order ecological model is the Nicholson-Bailey model:

$$N_t = \lambda N_{t-1} \exp[-a P_{t-1}]$$
$$P_t = N_{t-1}(1 - \exp[-a P_{t-1}])$$

$$(2.11)$$

Here $X_t \equiv P_t$ stands for parasitoid density.

The general model (2.10) that involves two state variables that depend only on their first lags can be transformed into a model with a single variable and multiple lags, such as

$$N_t = f(N_{t-1}, N_{t-2}) \tag{2.12}$$

For example, solving for P_{t-1} in the first equation of the Nicholson-Bailey model (2.11), substituting it into the second equation, and rearranging terms, we have

$$N_t = \lambda N_{t-1} \exp\left[\frac{a}{\lambda} N_{t-1} - a N_{t-2}\right] \tag{2.13}$$

or, equivalently,

$$r_t = \ln \lambda + \frac{a}{\lambda} N_{t-1} - a N_{t-2} \tag{2.14}$$

In general, we should not expect to get closed-form expressions such as (2.13)—this "trick" works only in the simplest cases, to which the Nicholson-Bailey model belongs. We can see that it is a very simple model, because it actually implies a linear relationship between r_t and lagged densities. Furthermore, sometimes we will need more than two lags to represent a system with two state variables (more on this later in this section).

Finally, it is easy to do the reverse procedure, that is, to translate a model involving a single variable with multiple lags into a model of multiple variables with single lags. For example, taking model (2.12), let us equate N_{t-2} with X_{t-1} (and, therefore, $X_t \equiv N_{t-1}$). Then we can write an equivalent model,

$$N_t = f(N_{t-1}, X_{t-1})$$
$$X_t = N_{t-1}$$

$$(2.15)$$

which has the same form as model (2.10).

To summarize, the notion of order for discrete systems generalizes as follows:

$$r_t = r_0 \qquad \text{zero-order dynamics}$$

$$r_t = f_1(N_{t-1}) \qquad \text{first-order dynamics} \qquad (2.16)$$

$$r_t = f_2(N_{t-1}, N_{t-2}) \quad \text{second-order dynamics} \qquad (2.17)$$

Qualitative dynamics of discrete systems are in many ways similar to those of continuous systems. Thus, zero-order processes in both frameworks are capable only of nonstationary behaviors. Second-order discrete models are capable of producing "smooth" cycles, in which both the increase and decrease phase can take several time steps, which is very similar to limit cycles in continuous second-order systems. However, first-order systems in the discrete framework, unlike analogous **ODE** (ordinary differential equation) **models**, are capable of limit cycles, and even chaotic dynamics. Such **first-order oscillations**, nevertheless, are fundamentally different from second-order cycles. The typical period of first-order cycles is two, and thus unstable trajectories produced by first-order systems look saw-shaped. In contrast, periods of second-order cycles are typically longer. For example, in the generic second-order model of Ginzburg and Taneyhill (1994) the minimum period is 6 generations. Trajectories produced by second-order systems look much smoother.

First-order systems also can generate long periods. For example, the Ricker model goes through a series of period-doubling bifurcations. However, all resulting cycles of period 4, 8, and so on are dominated by 2-year periodicity. Adding a little stochastic noise usually makes such cycles indistinguishable from a 2-cycle (with noise). First-order systems can also generate 3-, 4-, and longer n-year cycles by the following mechanism: it takes $n - 1$ years for the trajectory to increase exponentially to the peak, and then collapse in one year to a very low value, after which the cycle repeats itself. However, such first-order cycles longer than 3 or 4 years are unusual, because one has to "tune" parameters very finely to get them. And, unlike second-order cycles, the population decline is accomplished in one time step.

Order in Other Mathematical Frameworks   How well does the concept of process order generalize to mathematical frameworks other than differential and difference equations? First, consider models with population structure, for example, age structure. Such models can be framed either as partial differential equations or as matrix models. These models have been particularly well understood for the linear case, in which there are no population feedbacks. As is well known, the typical behavior for such models is an exponential trend (increase or decline), around which there are oscillations with a period of roughly one generation. In other words, such models add another kind of behavior, which we can call **zero-order** cycles (or, using more established terminology, **generation cycles**). In density-independent models, generation cycles are usually a transitory feature of dynamics, because for most parameters and initial conditions the amplitude of generation cycles gradually decays with time, as stable age distribution establishes itself. However, generation cycles can be stabilized by population feedbacks, as some models for periodic insects and fish, such as salmon, show.

The second general framework that is increasing in popularity among theoretical ecologists is that of delayed differential equation models. As reviewed by Gurney and Nisbet (1998), these models can exhibit a large variety of dynamical behaviors, but we can classify them into the three general classes, according to the relationship between the period of oscillations and the value of developmental delay, τ. The first type of oscillations are with a period roughly the same as τ (these are generation cycles). In the second class, cycle periods are typically in the range of 2–4τ (these are first-order oscillations in my terminology). Finally, there are "Lotka-Volterra" type of cycles, with periods ranging from 6τ and greater, which correspond to second-order oscillations. It appears that the notion of order generalizes rather well from discrete models to delayed differential models.

A Somewhat More Formal Definition of Order   The brief review of theory in the preceding paragraphs leads me to propose the following definition of process order for ecological systems. This definition assumes that we can describe the behavior of the system by a general nonlinear autoregressive model,

$$r_t = f(N_{t-\tau}, N_{t-2\tau}, N_{t-p\tau}, \epsilon_t) \tag{2.18}$$

where r_t is defined in terms of the base lag τ: $r_t = \ln N_t / N_{t-\tau}$. Process order is the number of lagged densities, p, that we need to adequately describe the behavior of the focal population. It is clear that p depends on the temporal scale at which we choose to sample the dynamics, τ. If we use very small temporal steps, we shall need more lagged densities to capture the system's dynamics. Thus, any specific estimate of order must be conditioned on whatever value of τ was chosen. Therefore, it is a good idea to use the value of τ that is biologically relevant to the issues that we investigate. Usually, this means that τ should be closely related to generation time. An alternative, in certain cases, is to choose the periodicity of some external driver (e.g., seasonality). The issue of which base lag to choose will be further discussed in section 7.2.2.

Note that the concept of process order, as it is used in this book, is similar to, but not the same as, the **dynamical dimension**—the number of state variables employed in the model (and thus the dimensionality of the phase space). Dimension is well defined for mechanistic models such as

$$N_t = f(N_{t-1}, X_{t-1}, Y_{t-1}, \dots)$$

$$X_t = g(N_{t-1}, X_{t-1}, Y_{t-1}, \dots) \tag{2.19}$$

$$Y_t = h(N_{t-1}, X_{t-1}, Y_{t-1}, \dots)$$

and so on

in which all state variables $(N_t, X_t, Y_t, \dots)$ have some biological meaning. As we discussed earlier in this section, one can translate back and forth between single-lag multiequation formulations such as (2.19) and multiple-lag, single-equation ones such as (2.18). However, according to a mathematical theorem (Takens 1981), in order to properly "reconstruct" the dynamics of model (2.19) one may need twice as many (plus one) lags in (2.18) as there are equations in (2.19). This line of argument suggests that the estimated process order may be much larger than the actual dimension of the underlying mechanistic model.

On the other hand, suppose that we choose to model a multi-species community with a system of equations where each species is represented by a separate state variable. We may then find out that certain groups of species act as a **dynamical complex** with respect to the focal species. For example, several generalist predators feeding on the focal species may all act in a "first-order" manner, that is, in such a way that their influence is folded into the first lag, $N_{t-\tau}$ (a possible example is provided by the generalist predators of voles; see chapter 12). Furthermore, several specialist predators may oscillate in great synchrony with each other, so that their combined effect would be folded into the second lag (a possible example is the larch budmoth system; see chapter 9). Thus, at a mechanistic level, we would have a model with a large number of state variables, while at the level of population feedbacks we could adequately represent the dynamics with just second-order process. I suspect that this is precisely what happens with many real-life applications. This line of argument would suggest that the estimated process order could be much less than the actual dimension of the underlying mechanistic model. In other words, the description in terms of the autoregressive model could often be more parsimonious than the description in terms of the mechanistic model (if we insist on modeling each interacting species with a separate state variable).

My conclusion is that the process order and the dynamical dimension, while related, are logically separate concepts, and that there is no rigid connection between the two. Process order reflects the feedback structure of population dynamics, and can be estimated from time-series data on the focal species alone. It is independent of the specific mechanistic model we would construct and defend (and the dynamical dimension is the property of such a model). Estimated process order provides a very useful **probe** for choosing among alternative mechanistic models, but there is no reason why process order and the model dimension should always coincide. Thus, analytical approaches that mechanically equate order and dimension (e.g., Stenseth et al. 1997) are, in my opinion, methodologically flawed (at the very least, such attempts should establish by simulation that the postulated equivalence holds for the specific combination of models and data employed in the project).

An important addition to the preceeding discussion is that the equivalence between order and dimension begins to break down primarily for second- and higher-order systems. Zero-order and first-order systems are described by a single-dimensional equation (where N_t enters with a single lag). Thus, both the mechanistic model and the autoregressive model have the same general form, $N_t = f(N_{t-1})$. This is in contrast with a second-dimensional mechanistic model, to reconstruct which we may need an autoregressive model with order two, three, four, or even five (according to the Takens theorem). I suggest, therefore, that the primary utility of the concept of process order is in distinguishing between the crude classes of zero-, first-, and second- or higher-order dynamics. This classification is particularly appropriate, because as we go from zero-order to first-order, and also from first-order to second-order models, we see qualitative changes in the spectrum of dynamical behaviors. But as we go from second-order to third-order models (and further), we do not see such fundamental changes. For example, discrete second- and higher-order models are all capable of stability, limit cycles, quasiperiodicity, and different kinds of dynamical transitions to chaos (e.g., both period-doubling and quasiperiodic routes). This makes it both more difficult to estimate the precise order and less compelling to try.

To finish, I see the concept of process order as primarily of heuristic value. In practice, we shall often run into various kinds of difficulties in assigning any empirical system, or even model output, to one of the three discrete classes. These difficulties may have to do with paucity of data, or with trying to impose discrete boundaries on the continuous world. For example, two-dimensional difference equations, depending on specific functional forms and parameter values, are capable of all kinds of behaviors, including unbounded growth/decline, first-order stability, and second-order oscillations. If both first and second lagged densities affect the rate of population change negatively, then by varying the coefficients we can make the system exhibit cycles of all dominant periods, from two to infinity. It is not clear where we should stop calling the system first-order and start calling it second-order in this progression. Thus, the classification based on process order appeals to "ideal types" of dynamics, and some imprecision when trying to fit nature within the ideal typology is inevitable. Nevertheless, I believe that, with all the caveats, the concept of process order offers a very useful heuristic device for organizing information about complex population dynamics.

2.6 SYNTHESIS

Table 2.2 summarizes the postulates, foundational principles, and general classes of dynamics that I discussed in this chapter. Postulates are statements that are assumed to be true without proof (in other words, they are axioms). Statements about general dynamical classes, on the other hand, are logically derived from the postulates (in other words, they are theorems). In between are foundational principles, which are a mixture of axioms and theorems. In fact, there is some overlap between different categories (e.g., biomass conversion postulate and trophic coupling principle; or exponential growth, which is both a principle and a dynamical type). The reason for putting principles into a separate group is that they are the ones that are of immediate relevance to building theory. Of course, there is some unavoidable degree of arbitrariness in creating this group. For example, others may prefer to use mass action and maximum physiological rates postulates directly, rather than combine them in a hyperbolic functional response (see, e.g., the discussion of mass action law in Metz and Diekmann 1986:89–90). Additionally, the hyperbolic functional response could be criticized as not going far enough toward realism. My personal preference is to draw the line between the *hard core* and *auxiliary belt* (using Lakatosian terms) at the hyperbolic functional response, but I freely admit that others may do it differently, especially if they are addressing a different subject area (thus, the emphasis of Metz and Diekmann on the law of mass action in the context of their theory of physiologically structured populations).

   Revisiting once again the comparison between population dynamics and classical mechanics, my current thinking is as follows. The analogy between the exponential law and the law of inertia appears to be complete. Both laws describe the null situation that would obtain if no forces were acting on the object of study: an ecological population or a physical body, respectively. (Incidentally, another analogous principle is the Hardy-Weinberg equilibrium, which describes the null

TABLE 2.2. Three kinds of "laws" in population dynamics

Postulates	Principles	Dynamical Classes (order)
1. Conservation	Exponential growth	Exponential (0)
2. Method. individualism		
3. Upper density bound	Self-limitation	Bounded (1)
4. Mass action	Hyperbolic func. resp.	
5. Biomass conversion	Trophic coupling	Oscillatory (2)
6. Max. physiol. rates		

state of a population genetics system, obtaining in the absence of such forces as selection, assortative mating, and genetic drift.) Beyond the first laws, however, the analogy between population dynamics and classical mechanics cannot be pushed. Whereas the exponential growth principle describes population dynamics in the absence of forces, other population principles that I tentatively identified in this chapter really describe the nature of forces that may act on the population. I focused on two particular forces, self-limitation and trophic interaction. Conceivably, other principles may be advanced for other ecological forces. An example of another intrinsic mechanism (in addition to self-limitation) is cooperation, potentially leading to the Allee effect. In fact, Berryman (1999) chose to add cooperation to his proposed set of fundamental principles. Other interspecific interactions than resource-consumer may also require their own guiding principles, for example, mutualism and interference competition.

Here is an illustration of how principles of population dynamics are routinely used by ecologists in building mechanistically based theory. Consider the Lotka-Volterra model (2.8). As was discussed above, this model violates two of the four principles in table 2.2: it does not have any self-limitation terms, and it assumes a linear, rather than a hyperbolic, functional response. Substituting the hyperbolic response in place of the linear one, and adding self-limitation to prey in the simplest possible way (logistic), we have the following model:

$$\frac{dN}{dt} = r_0 N \left(1 - \frac{N}{k}\right) - \frac{cNP}{d + N}$$

$$\frac{dP}{dt} = \chi \frac{cNP}{d + N} - \delta P$$

This is of course the Rosenzweig-MacArthur model (1963), which has been gainfully employed in many real-life applications (in contrast to the Lotka-Volterra model). This is not to say, however, that this model cannot be improved. Notice, for example, that it "half-violates" the principle of self-limitation, because there is no direct density dependence in the predator equation. Additionally, the model accommodates the foundational principles with the simplest possible forms. For example, prey's per capita rate of change is affected by prey density in a linear manner. This, and other assumptions, may need to be modified in specific case studies (see chapter 4).

CHAPTER 3

Single-Species Populations

In this chapter I present an overview of mathematical models for single-species populations. The basic format is to show and explain model equations, and then to discuss the dynamical behaviors that the models can exhibit. As I stated in the preface, I will not discuss how the results are obtained, but simply provide the references to the appropriate literature. I will also not attempt to provide a comprehensive account of models of single-population dynamics. My primary focus is on models that can potentially exhibit complex dynamics, although I will review some models that are not capable of complex dynamics, but are useful as submodules in more complex models.

The survey begins with *unstructured* models, that is, models that have a single state variable: population density. Following the established conventions, I will denote it as $N(t)$ in continuous-time models (or simply N), and as N_t in discrete-time models. Both purely endogenous and mixed endogenous/exogenous models are discussed. Next, I review models with population structure, primarily focusing on age or stage dynamics. Finally, I discuss second-order models that incorporate some other population property as an extra state variable in addition to population density.

3.1 MODELS WITHOUT POPULATION STRUCTURE

There are two simple mathematical frameworks with which to approach modeling population dynamics: ordinary differential equations (ODE) and difference equations. ODE models, framed in continuous time, are a more natural starting point for building mechanistic population models than difference equations, because even individuals of annual species do not really perform all their actions—eating, reproducing, and dying—at one point in time once per generation. This observation is particularly true with respect to such time-distributed processes as foraging and mortality. For this reason, I begin the survey with ODE models (section 3.1.1).

The preceding is not an argument against using discrete-time models employing difference equations. Rather, I am saying that discrete models, by integrating over a period of time during which state variables may be changing, make a greater stride in abstracting away from individual-level processes than continuous time models. Therefore, to avoid excessive phenomenology, it is important to understand precisely what we assume when we make this abstraction. Thus, it is a good procedure to derive discrete models explicitly as approximations of continuous models (or, alternatively, directly as approximations of individual-based continuous-time processes). This is the approach that I follow in section 3.1.2.

Note that within the general class of continuous-time models, ODE present the simplest approach. The important assumption underlying ODE models is that all action and reaction is instantaneous. A more realistic framework that does not make this simplifying assumption is that of the delayed differential equations, to be discussed in section 3.1.3.

3.1.1   Continuous-Time Models

The simplest model for single-species dynamics that can exhibit long-term stationary dynamics is the **logistic**:

$$\frac{dN}{dt} = r_0 N \left(1 - \frac{N}{k}\right) \qquad (3.1)$$

The two parameters are the intrinsic rate of population growth, r_0, and the carrying capacity, k. As long as $r_0 > 0$ and $k > 0$, this model is always stable. The stable point equilibrium is k, and the population density approaches it monotonically.

The logistic equation is a very simple model, and its potential failings are well known: (1) the assumption that realized per capita rate of change, $r(t) = dN/(Ndt)$, is related to N linearly; (2) the rate of population change responds to variations in density instantaneously, without a time lag; (3) the model does not incorporate the effects of exogenous influences; and (4) it ignores effects of population structure. Despite these failings, the logistic model is taught by all ecology texts. The reason is that it provides a simple and powerful metaphor for a regulated population, and a reasonable starting point for modeling single-population dynamics, since it can be modified to address all four criticisms listed above. I now proceed to the discussion of how various realistic features can be added to the logistic model.

First, let us deal with the assumption of linearity. (Note, however, that although the realized per capita rate of change is a linear function of N, the logistic model itself is a nonlinear model.) This feature of the model arises during its derivation, when we make an assumption that per capita birth and death rates are linear functions of density (or, most generally, the difference between these two rates is a linear function of N). Readers wishing to refresh their recollection of how the logistic model is derived can consult a very clear explanation by Gotelli (1995). There is no particular biological reason why the relationship between vital rates and density should be linear; linear relationship is simply the most parsimonious functional form.

It is easy to modify the logistic model in a way that allows nonlinear relationship between the realized rate of population change and density. For example, the **theta-logistic model** is described by this equation:

$$\frac{dN}{dt} = r_0 N \left[1 - \left(\frac{N}{k} \right)^{\theta} \right] \qquad (3.2)$$

The exponent θ controls the shape of the relationship between $r(t)$ and N (figure 3.1a). On biological grounds, we may suspect that the case of $\theta > 1$ is the most realistic one. The argument goes like this: As N increases from 0, the per capita rate of change should stay nearly constant until N gets near k, where density dependence would finally kick in. As a result, the relationship between $r(t)$ and N should be convex: first flat, and then near k we should observe a rapid falloff. Although this argument seemingly makes sense, in practice we observe a variety of shapes for the $r(t)$ function (Turchin 1999), both concave and convex, and also approximately linear. To see why the above argument, based on biological intuition, is misleading, let us frame it in terms of log N instead of N. After all, initially a population

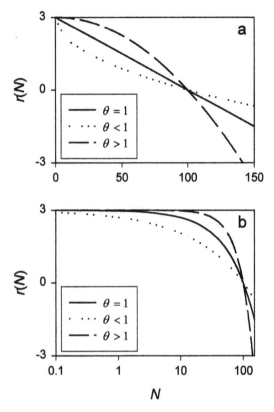

FIGURE 3.1. Relationship between the realized per capita rate of increase, $r(t) = dN/Ndt$, and population density in the theta-logistic model. Same three functions are plotted using (a) natural scale for N and (b) log-transformed N.

starting from N near 0 will increment itself in steps of constant $\log N$, rather than N. Plotting the several theta-logistic curves against log-transformed N, we observe that they all become much more convex (figure 3.1b). This point of view on the shape of $r(t)$, then, resolves the apparent paradox: when population starts growing from near 0, it will make such tiny steps in N that $r(t)$ will remain near r_0 for quite a while. The take-home message from this is that perhaps the linear approximation for $r(N)$ is not so bad, after all, and it certainly has a virtue of simplicity, since the alternatives require an extra parameter.

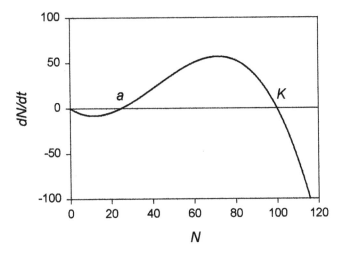

FIGURE 3.2. Allee-type population growth: $dN/dt = bN(N - a)(k - N)/k^2$ (parameters $b = 6$, $a = 25$, and $k = 100$).

The theta-logistic model belongs to a broader class of **logistic-like models** that are defined as those models in which the per capita rate of population change $r(t)$ is characterized by a maximum at $N = 0$, is positive for all $N < k$, and negative for all $N > k$. An important example of *not* logistic-like population growth is the **Allee effect**.[1] The simplest phenomenological form for per capita growth rate in an Allee population employs a quadratic polynomial (e.g., Lewis and Kareiva 1993):

$$r(t) = \frac{b(N - a)(k - N)}{k^2} \tag{3.3}$$

where k is the carrying capacity, a (satisfying the condition $0 < a < k$) is the density below which population growth is negative, and $b > 0$ a constant related to how fast the population will grow at its maximum possible growth rate. The rate of population change, dN/dt, is cubic and is illustrated in figure 3.2.

---

[1] All terms emphasized with boldface are defined in the glossary.

3.1.2   Discrete-Time Models

The discrete analogues of the logistic model are usually derived in a completely phenomenological fashion. Thus, we start with the discrete exponential model $N_{t+1} = \lambda N_t$, and then make λ a function of population density. Next, we attempt to figure out what the shape of the λ function should be like. A number of phenomenological relationships have been proposed (table 3.1). The problem with this approach, however, is that we do not really know how the postulated functional forms are related to assumptions about individual-based processes.

A better approach is to derive discrete models as approximations of continuous models. We know what assumptions we made when deriving, for example, the logistic equation; by going through an explicit derivation from the logistic to a discrete equivalent we shall have a good understanding of what assumptions we have to make about individuals in order to obtain the result. The point here is not that following this prescription will result in a more realistic model, but rather that we shall know better in what ways our model is unrealistic. Additionally, an explicit connection to individuals means that the resulting model's parameters are functions of quantities observable at the individual level. This feature makes parameter estimation much more rigorous.

There are at least three ways to "discretize" the logistic model known to theoretical ecologists (see, e.g., Gurney and Nisbet 1998:61–64). The naive (and flawed) approach is to discretize the derivative in the logistic model:

$$\frac{\Delta N}{\Delta t} \approx \frac{dN}{dt} = r_0 N \left(1 - \frac{N}{k} \right) \tag{3.4}$$

Next we set $\Delta t = 1$ (measured in generation units), add a subscript t to $N = N_t$ to emphasize that we are now dealing with discrete-time density, and replace ΔN with $N_{t+1} - N_t$. After some algebra, we obtain:

$$N_{t+1} = aN_t - bN_t^2 \tag{3.5}$$

TABLE 3.1. Some functional forms proposed for single-species discrete-time models of population growth ($\lambda_0 = \exp[r_0]$ is the intrinsic discrete or multiplicative rate of population increase, k is the carrying capacity, b some positive constant, and θ an exponent)

Label	Function
Exponential	$N_{t+1} = \lambda_0 N_t$
Quadratic map	$N_{t+1} = \lambda_0 N_t (1 - N_t/k)$
Ricker	$N_{t+1} = \lambda_0 N_t \exp[-bN_t]$
Gompertz	$N_{t+1} = \lambda_0 N_t^\theta$
Beverton-Holt	$N_{t+1} = \lambda_0 N_t/(1 + bN_t)$
Depensation	$N_{t+1} = \lambda_0 N_t^2/(1 + bN_t^2)$
Theta-Ricker	$N_{t+1} = \lambda_0 N_t \exp[-bN_t^\theta]$

where $a = 1 + r_0$ and $b = r_0/k$. This model is known as the discrete logistic, or better (to distinguish it from the Ricker model, which is also sometimes called discrete logistic) the "quadratic map." The reason model (3.5) is flawed for ecological applications is because if, for whatever reason, population density at time t happens to exceed $k(1 + r_0)/r_0$, then at time $t + 1$ population density becomes negative. For example, if $r_0 = 2$ (a rather typical value for biological populations; recollect that this value implies that population would increase by a factor of $e^2 \approx 7$ when N_t is near 0), then N_t would need to exceed only $1.5k$ in order for N_{t+1} to become negative. Clearly, we do not want to use a model that so easily produces nonsensical predictions.

The second derivation employs the following trick. We integrate the logistic model one time step forward, from t to $t + 1$, while pretending that the per capita growth rate, $r(t) = r_0(1 - N/k)$, stays constant during this interval of time (we fix it at the value obtaining at the start of the time interval, t). Since $r(t)$ is by assumption constant, we use the integrated solution of the exponential model:

$$N(t + 1) = N(t) \exp[r(t)]$$

Next, we substitute $r(t) = r_0(1 - N(t)/k)$ and change to subscripts, obtaining the well-known Ricker model:

$$N_{t+1} = N_t \exp\left[r_0\left(1 - \frac{N_t}{k}\right)\right] \tag{3.6}$$

Because $r(N)$ is inside the exponential, the Ricker model avoids the flaw of the quadratic map: for any $N_t > 0$ the model predicts a positive N_{t+1}.

The third derivation is also based on integrating the logistic model for one time step (Gurney and Nisbet 1998:63). The integrated solution of the logistic model is

$$N(t+1) = \frac{k}{1 + [(k - N(t)/N(t)]e^{-r_0}}$$

Rearranging the terms, substituting $\lambda_0 = \exp[r_0]$, and switching to subscripts, as usual, we have

$$N_{t+1} = \frac{\lambda_0 N_t}{1 + [(\lambda_0 - 1)/k]N_t} \tag{3.7}$$

It is readily apparent that equation (3.7) has the same functional form as the Beverton-Holt model (table 3.1).

In summary, I have reviewed three approaches to deriving a discrete equivalent of the continuous-time logistic model. The first one, based on direct discretization of the derivative, results in a model with pathological properties, namely, the tendency of the model to predict negative population density for certain initial conditions. This problem is of particular importance if we wish to add a stochastic component to the model, in which case sooner or later population density will go over the threshold and become negative. When simulating trajectories based on the quadratic map, it is certainly possible to prevent this from happening by, for example, setting population density to some small positive constant when a crash happens. However, this approach smacks of ad-hockery, and is not really necessary, since a well-behaved alternative exists (the Ricker model). The quadratic map, thus, is of primarily historical interest in population ecology (it happened to be the model using which Robert May discovered dynamical chaos; see May 1974a; Gleick 1988:69–73).

The other two approaches, leading respectively to the Ricker and Beverton-Holt models, are a different matter. Both of these approaches are very useful, but in different ways. If we want to obtain a discrete model that is the most faithful dynamical analogue of the continuous logistic, then we should opt for the Beverton-Holt. Like the logistic, the Beverton-Holt model is characterized by an exponential approach to a stable point equilibrium for all parameter values. In fact, the trajectories predicted by the Beverton-Holt are numerically very near the trajectories generated by the logistic model (Gurney and Nisbet 1998:63). On the other hand, it is well known that the Ricker model is capable not only of stability but also of limit cycles and chaos. The reason for the difference between these two models is clear from their different derivations. The derivation leading to the Ricker model makes explicit our assumption of imposing a generation-long lag on population dynamics: all vital rates depend only on the conditions at the beginning of the growth season. This feature leads to the possibility of a serious overshoot of equilibrium density. In the continuous logistic model, by contrast, vital rates are instantaneously adjusted to conditions throughout the growth season. By utilizing the integrated solution of the logistic model, the Beverton-Holt model also implicitly assumes no lag in the effect of the environment on vital rates. Adding lags in regulation is potentially a destabilizing feature in almost any model of population dynamics.

One advantage of the derivation leading to the Ricker model is that it is a general approach that can be used to discretize almost any continuous model of population dynamics. This is because the overwhelming majority of population dynamical models have the following form:

$$\frac{dN}{Ndt} = f(\text{all sorts of stuff}) \tag{3.8}$$

It is natural to approximate (3.8) with

$$N_{t+1} = N_t \exp[f(\text{all sorts of stuff})] \tag{3.9}$$

and in fact a number of discrete population models are obtained in just this way (e.g., see Berryman 1991). This procedure is certainly better than the one based on discretizing derivatives, but carries its own subtle dangers (Gurney and Nisbet 1998:64). Unfortunately, we can follow the third approach (exemplified by the derivation of the

Beverton-Holt model) in only few cases, since explicit solutions for nonlinear ecological models are rare. Therefore, in most cases, we are limited to the second procedure, but we should use it cautiously, and check the correspondence with the underlying continuous-time process by judiciously chosen simulations.

Although I have focused on deriving discrete-time models as approximations of continuous-time processes, I do not wish to imply that this is the only "correct" approach. A viable alternative is a direct derivation of a discrete model from postulates about how intraspecific competition affects individual reproduction and mortality. An example of this approach has been provided by Royama (1992:144–146). He considers the scenario where individuals possessing circular "areas of influence" are distributed randomly in space. As population density increases, so does the overlap between individual areas of influence. As a result, individuals tend to obtain fewer resources, and their mean reproductive rate is decreased. Royama show that this "geometric model" leads to the functional form of intraspecific competition that is identical to that assumed by the Ricker model. The subsequent section (Royama 1992:149–155) on the generalization of the discrete logistic theory is also well worth reading (even despite his less than flattering assessment of my multiple-lag extension of the Ricker model!).

3.1.3   Delayed Differential Models

The third mathematical framework, combining some features of both ordinary differential and difference equations, is the delayed differential equations (DDE). The simplest DDE model in population dynamics is the **delayed logistic** (Hutchinson 1948):

$$\frac{dN(t)}{dt} = r_0 N(t)\left[1 - \frac{N(t - \tau)}{k}\right] \qquad (3.10)$$

As far as I know, this model does not have a rigorous derivation, and is simply based on an argument that the self-regulatory mechanism may involve a time lag (Hutchinson 1948:237). Below I will consider DDE models that have a more explicit connection to individuals.

The dynamics of the delayed logistic model are well understood (e.g., May 1981). The qualitative type of dynamics is controlled by a single parameter combination, $r_0\tau$. Model (3.10) has a **monotonically damped stable point** for $0 < r_0\tau < e^{-1}$, an **oscillatory damped stable point** for $e^{-1} < r_0\tau < \pi/2$, and a **stable limit cycle** for $r_0\tau > \pi/2$. Once stable limit cycles arise, their period is approximately 4τ. Increasing $r_0\tau$ beyond $\pi/2$ simultaneously lengthens (slightly) the cycle period and increases the amplitude. An unpleasant feature of the delayed logistic is that the amplitude of cycles depends very sensitively on $r_0\tau$. For example, changing $r_0\tau$ from 2.4 to 2.5 increases the amplitude from approximately 1,000-fold to 3,000-fold (May 1981: table 2.1). At the same time, the period increases only slightly, from 5.11 to 5.36τ.

May (1981:23) suggested that the delayed logistic model provides a "detail-independent explanation of 'wildlife's 4-year cycle.'" In particular, he observed that a lag of $\tau = 9$ months in the delayed logistic model would produce a period intermediate between 3 and 4 years. Although May was careful to stress that this explanation is "independent of the biological mechanism(s) producing the time delay," empirical ecologists working on rodent cycles pursued his suggestion by searching for the putative 9-month delay in their data (e.g., Hörnfeldt 1994; Korpimäki 1994). This direction, in my opinion, is a blind alley, and it is worth discussing why this is so. One of the common themes in this book is that the most useful model for explaining population cycles has an intermediate complexity: it is neither too simple nor too complex. Well, the delayed logistic model, with its dynamics determined by a single parameter combination, is just too simple to be useful in investigating population oscillations. Worse, it has a very tenuous connection to mechanisms. Its only significance for an investigation of population cycles is that it provides further support to a well-known theoretical idea that time-delays promote the possibility of oscillations. Fitting this model to real population data does not get us any closer to an identification of ecological mechanisms that may be responsible for oscillations. There is no particular significance in the 9-month delay—this is a complete artifact of the simplistic structure of this model. As we shall see in chapter 12, predator-prey models explain the rodent cycle without any explicit time delays. The delay arises naturally as a result of population interaction between prey and their specialist predators.

A better, more mechanistic approach to deriving DDE models is to separate the birth and death rates, and consider separately how these vital rates may be affected by time delays (May 1981; Nisbet and Gurney 1982). Death rate should be related to the current population density, but birth rate most plausibly depends on population density some time ago, as a result of developmental delays. These assumptions lead to an equation of the form (Nisbet and Gurney 1982:41)

$$\frac{dN}{dt} = B(N(t - \tau)) - D(N(t))$$ (3.11)

Nisbet and Gurney (1982) used this general model in their investigation of population cycles in Nicholson's blowflies. This model with biologically plausible functions and parameter values predicted population cycles that were very similar to those observed by Nicholson, even capturing the "double-peak" shape of the observed cycles (see Nisbet and Gurney 1982:285–308).

3.2   EXOGENOUS DRIVERS

Real-world populations are complex, even messy, systems. They are affected by a multitude of different factors, some of which we model explicitly (as reviewed in this and the next chapters), while others we have no choice but to leave out of the model. Nevertheless, these **exogenous factors**—fluctuations in weather, immigration events, erratic fluctuations in food resources or natural enemies, and so on—continuously affect population change. Although we cannot model such factors explicitly, we need to include their effect somehow in the model. The reason is that even small effects of a multitude of exogenous factors add up, and can potentially change the nature of dynamics. A very useful approach in such situations is to model the collective effect of such influences with random variables.

"Randomness" in models can potentially represent three kinds of processes. The first one is "true randomness," a process that is, as far as we know, irreducible to any deterministic explanation. Physicists tell us that the behavior of subatomic particles is random in this sense. It is not clear whether the behavior of such micro-level entities is relevant to the macro world inhabited by ecological entities. One hypothesis is that population dynamics is insulated against effects at the subatomic level by all the intervening levels of physical and biological organization. However, now that we understand the nature of dynamical chaos better, we know that small random influences at the micro level can potentially be amplified (rather than damped, as the insulation hypothesis would have it) and have macro consequences for nonlinear dynamical systems.

The second meaning of "randomness" is a reflection of our ignorance. For example, suppose we are studying a population of mice who are eaten by owls. Owl numbers may fluctuate in response to dynamics of another rodent, for example, a vole species. After a vole outbreak and collapse there will be lots of hungry owls severely impacting the population of mice we are studying. However, we have no data on voles, and do not even suspect that owl numbers are actually predictable. To us it would appear that owl numbers fluctuate randomly, and therefore the best model, under the circumstance, would be to model owl numbers as a stochastic variable with a certain mean and variance, and perhaps some autocorrelation structure. Ignorance is probably the most common reason why we need to use random variables in population dynamics models.

The third kind of randomness arises when we know about certain processes affecting population change, but decide not to incorporate them explicitly in the model, modeling them instead as random variables. This situation can arise only in well-studied and -understood systems. We may know about several processes that each have a minor effect on population rate of change. We may find out that explicitly modeling these processes doubles the model complexity (as measured, e.g., by the number of parameters), but increases our ability to predict dynamics (as measured by R^2_{pred}) by only a couple of percentage points. In such situations, we may decide that the extra prediction ability is simply not worth substantially increasing model complexity. Alternatively, we may find that adding extra state variables to the model *decreases* its prediction accuracy. This may happen because parameters are never known precisely. Additionally, extra state variables may be measured with a substantial observation error. Thus, we may be in a seemingly paradoxical situation that we know perfectly well that some process is operating, but we cannot capitalize on it for prediction purposes. I argue that in such situations we should ruthlessly expunge these processes from the model. We should keep in mind, however, that the optimal mix of explanatory mechanisms to be included in the model may change depending on the purpose of the model.

3.2.1 Stochastic Variation

As I discussed above, the usual reason that we need to include stochasticity is our ignorance. By its nature, then, a stochastic factor has to be modeled phenomenologically. The question is how best to include it in the model. The most direct approach is to add noise to the population rate of change:

$$r_{t+1} = f(\mathbf{Z}_t) + \epsilon_t \qquad (3.12)$$

Here $r_{t+1} = \ln(N_{t+1}/N_t)$, $\mathbf{Z}_t$ is the vector of state variables (i.e., processes that are modeled explicitly), and ϵ_t is a random variable characterized by some probability distribution; for example, it could be normally distributed with mean $= 0$ and variance σ^2. Once every time step, we choose an ϵ_t from the specified distribution, add it to r_{t+1}, and calculate the new N_{t+1}. Note that noise is added to r_{t+1} in an additive manner (which means that N_{t+1} is affected by noise multiplicatively). The justification for it is that environmental fluctuations are likely to affect per capita death and birth rates, and those two rates are combined additively in determining r_{t+1}.

An alternative approach is to add a random number directly to N_t. This approach is rarely used, because it does not have as ready justification as the first one. One possible mechanism is random immigration events, but we would still expect that environmental fluctuations would affect vital rates, so it seems better to simply perturb r_{t+1}. Additionally, if adding or subtracting from N_t, we need to ensure that we do not inadvertently decrease N_t below zero.

The third approach is to randomly vary parameters of the model. This approach is perhaps the most mechanistic, especially if we have some information about where in the population process the environmental influences are the most important. An example is provided by the wildebeest model in Hilborn and Mangel (1997), in which the carrying capacity was assumed to be a function of annual rainfall.

So far I have been focusing on including noise in discrete models, where there is a natural time step, and therefore it makes sense to add noise once per step (e.g., a generation). In continuous-time models exogenous factors should affect population dynamics continuously. This creates a problem of how such continuous effect should be modeled. Of course, continuous models have to be discretized in order to be simulated on the computer, but it is a bad idea to add a random variable to $dN/(Ndt)$ once per iteration step. The problem is that the effect of noise becomes affected by the step length. For example, if we solve an ODE model and use a time step of 0.1 yr, then we "kick" population trajectory 10 times per year, but if use a time step of 0.01, then we kick it 100 times per year. Clearly, in the second case, we add more noise to the trajectory, even though we are drawing ϵ from the same distribution. The situation is even worse if we are employing some variable-step ODE-solving routine. Mathematically rigorous approaches to simulating stochastic differential equations exist, but a simpler solution, and in my opinion perfectly adequate for most population applications, is to add noise once per some natural time unit, rather than discretization step.

3.2.2 Deterministic Exogenous Factors

Not all exogenous factors should be modeled with random variables. Effects of seasonality, for example, are clearly exogenous (since population density does not affect the change of seasons), but at the same time highly predictable. Another class of nonstochastic exogenous factors is systematic environmental trends, for example, long-term successional change of the community within which the studied population is embedded. Yet another class is exogenous factors that may change erratically from year to year, but for which we have measurements, so that we can treat them as "fixed." For example, a mouse population may be strongly affected by fluctuations of their food supply caused by tree masting. If we have measured the amount of seeds produced over a long period of time, then one approach to modeling the effect of masting on mouse dynamics could be simply to include the measured mast time-series into the model in some mechanistic fashion.

In this section I focus on seasonality as one of the most ubiquitous deterministic exogenous factors that affect population dynamics.

Let us consider how we can model seasonality, using the logistic model as an example. It is clear that seasonality could affect both birth and death rates. During the unfavorable season, such as winter in temperate communities, individuals will be forced to reduce or even completely curtail their reproduction. Additionally, we might expect that death rates would be higher due to inclement weather, reduced food, or higher vulnerability to predation. A naive approach to including such an effect into the logistic model is to make the intrinsic rate of population growth, r_0, a function of time while keeping carrying capacity constant. In the simplest case, the seasonality function would take one of two values, $r_0(t) = r_w$ during winter and $r_0(t) = r_s$ during summer (other seasonal functions will be considered later). The reason why this modeling approach is not satisfactory becomes clear when we consider what would happen if $r_w < 0$. Such a situation is perfectly possible, in fact expected in those cases in which reproduction ceases in winter while mortality factors continue to operate. However, allowing a negative r_w leads to nonsensical prediction (figure 3.3). The line labeled "summer" in figure 3.3 depicts the usual configuration assumed by the logistic model. If $r_w > 0$, we have a similar relationship in winter, although the slope (or strength of density dependence) is decreased compared with the summer situation. However, if $r_w < 0$, then we observe a positive slope, or inverse density dependence! Clearly, varying r_0 seasonally while keeping k constant does not make sense. We run into a similar problem if we attempt to change k seasonally while keeping r_0 constant. Thus, simple assumptions about how seasonality might affect logistic growth fail to yield meaningful models.

There is actually a general lesson here. Parameters of the logistic model combine individual-level parameters in subtle ways, and it is dangerous to modify them in a phenomenological fashion. If we wish to build models that behave in ways consistent with biology of organisms, than it is a better procedure to derive models from first principles (which means basing the derivation on what individuals do). To give a simple example, let us assume that per capita birth rate is density independent, and varies seasonally, so that $b(t) = b_s$ in summer and $b(t) = 0$ in winter. Per capita death rate, on the other

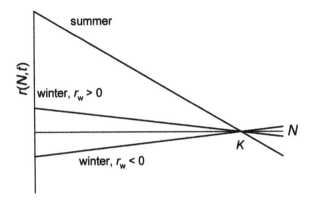

FIGURE 3.3. Relationship between the per capita rate of population change and population density: when seasonality affects r_0, but not k.

hand, does not change seasonally, but is affected by density, $d(N) = d_0 + d_1 N$. Putting these assumptions together, we have

$$r(N, t) = r_0(t) - d_1 N \tag{3.13}$$

where $r_0(t) = b_s - d_0$ in summer and $r_0(t) = -d_0$ in winter. In other words, the intrinsic rate of population growth changes with season, while the slope of density dependence is kept constant. This argument suggests that if we want to model seasonal effects within the logistic framework, and wish to do it in the simplest possible way, by adding a single parameter, than the way to do it is by varying $r_0(t)$ while keeping the slope of density dependence constant. A detail-independent equation for seasonal logistic model takes, then, the following form:

$$\frac{dN}{dt} = r_{avg}[1 - e\Sigma(t)]N - \frac{r_{avg}}{k}N^2 \tag{3.14}$$

where $\Sigma(t)$ is some periodic function with period of 1 year, varying between -1 and 1 (for example, a sine function). The parameter r_{avg} is the intrinsic rate of population increase averaged over all seasonal values, and k is the carrying capacity, similarly averaged over all seasonal values. Equation (3.14) is illustrated in figure 3.4. Note that if $e > r_{avg}$, then both the intrinsic rate of increase and carrying capacity during the worst season will be negative. As discussed above, this is the situation that should be characteristic of most organisms inhabiting temperate environments, where there is no possibility of

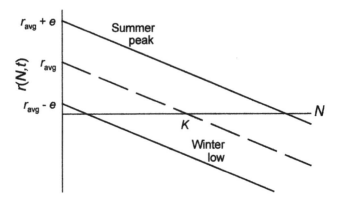

FIGURE 3.4. Relationship between the per capita rate of population change and population density: varying r_0 while keeping the slope of density dependence constant.

reproduction during winter, nor enough resources to indefinitely support the population. As a result, these populations are always decreasing during winter, and denser populations decrease faster. Organisms are prevented from going extinct only by periodic recurrence of the favorable season, summer.

3.3   AGE- AND STAGE-STRUCTURED MODELS

Whereas the first part of this chapter focused on models that lumped all different kinds of individuals in the population into a single state variable, N, in this section I review models that explicitly address population structure. Whether to include population structure in a model of a particular dynamical system is a very important issue. The advantages of structured models are twofold (Caswell et al. 1997). First, population structure may have important consequences for population dynamics. A population responds to environmental changes with time lags that reflect individual development. These lags may result in oscillatory dynamics, which would not appear in an unstructured variant of the model (although note that delayed differential models provide an approach of intermediate complexity to handle these kinds of phenomena). Second, because structured models are written in terms of the vital rates of individuals, their parameters often have a clear operational definition, and so are amenable to direct measurement (Caswell et al. 1997). The third reason that I would add to this list is that sometimes we are explicitly interested in issues that can be addressed only with a structured model, for example, oscillations in age structure.

On the other hand, adding structure to population models inevitably increases their complexity. The important question that needs to be resolved by the model builder is whether this added complexity is warranted: whether the gain in accuracy is worth the price of extra parameters. In this book, I advocate a hierarchical approach to model building that yields a series of models of variable complexity. Determination of optimal complexity, thus, can be put on an empirical basis. More specifically, the question of whether to include population structure or not can be resolved by constructing and contrasting with data two versions of the model: one that explicitly incorporates individual variability, and the other that averages over this variability with judiciously chosen functional forms. Additionally, as one gains experience with various kinds of models, it becomes possible to make an educated guess as to whether population structure may or may not be important. For example, if we are trying to understand population oscillations of southern pine beetles (chapter 10), we may note that the cycle period is around 6–9 years, and that there are about 6 generations per year. This means that cycle period is on the order of 40–50 generations. As we shall see below, structured population models without trophic interactions produce cycles of much shorter period. This suggests that an attempt at explaining cycles in this beetle with a single-species structured population model is highly unlikely to succeed.

3.3.1   *Mathematical Frameworks*

There are three general mathematical frameworks for modeling population structure: matrix models, systems of delayed differential equations (DDE), and partial differential equations (PDE) (Caswell et al. 1997). These frameworks differ in how they represent time and population structure. As Gurney and Nisbet (1998:237) point out, discrete-time formalism yields conceptually simple models that are especially well adapted to numerical simulation, but analytical treatment beyond determination of steady states poses considerable mathematical challenge. By contrast, continuous-time models are much less adapted to numerical realization, but they are often easier to treat analytically. PDE models, in which both time and structure are continuous variables, are thus a very powerful framework for obtaining theoretical insights. However, they can boast few empirical applications, probably because their numerical solution is a highly technical field (see the discussion in Gurney and Nisbet 1998:242–245), it is difficult to add stochasticity to these models, and parameterizing them is conceptually not straightforward.

Matrix population models are formulated within the discrete-time, discrete-structure framework. These models, therefore, assume that each individual can be assigned to one of a set of discrete age (or stage, size, etc.) classes. Matrix models are formally the same as a system of difference equations, such as

$$N_{t+1}^1 = f_1(N_t^1, N_t^2, \ldots, N_t^m)$$

$$N_{t+1}^2 = f_2(N_t^1, N_t^2, \ldots, N_t^m)$$

$$\ldots$$

$$N_{t+1}^m = f_m(N_t^1, N_t^2, \ldots, N_t^m)$$

where m is the number of classes, N_t^i is the number (or density) of individuals in class i, and f_i are functions relating the past densities to the future ones. The same model can be also written in the matrix form

$$\mathbf{N}_{t+1} = \mathbf{A}\mathbf{N}_t \tag{3.15}$$

where $\mathbf{N}_t$ is the vector of N_t^i, and $\mathbf{A}$ is a matrix whose ij coefficient specifies how many individuals of class i appear at time $t + 1$ per individual of stage j at time t (Caswell et al. 1997). Matrix models may be linear or nonlinear ($\mathbf{A}$ is constant or not), and either deterministic or stochastic (elements of $\mathbf{A}$ change randomly over time).

Matrix models enjoy a great amount of popularity in ecological applications. Their main advantage is that they are easy to construct using life-table data. Additionally, since they are framed as discrete-time models, they are easy to simulate on the computer. Another advantage of a discrete-time framework is that adding stochasticity is quite straightforward (section 3.2.1). The only disadvantage of matrix models is that they require discrete stages, which may cause difficulties if natural stages do not exist (Caswell et al. 1997, but see Easterling et al. 2000).

The third popular class of models, delayed differential equations (DDE), employs the continuous-time but discrete-structure framework. We have already encountered an example of a DDE model in the context of models without population structure (section 3.1.3). Adding structure simply means that we need multiple equations, one for each of the population classes. For example, if we have two stages, larvae and adults, than we might write the following model

$$\frac{dL}{dt} = \beta A - \delta_1 L - M(t) \tag{3.16}$$

$$\frac{dA}{dt} = M(t) - \delta_2 A \tag{3.17}$$

where L and A are densities of larvae and adults. Parameter β is the per capita fecundity rate, while δ_1 and δ_2 are per capita death rates of larvae and adults, respectively. The most interesting part of the model is $M(t)$, the maturation rate of larvae into adults. One specific form investigated by Gurney and Nisbet (1998:253) is as follows:

$$M(t) = \beta \exp[-\delta_1 t] A(t - \tau) \frac{L(t - \tau)}{L(t)} \tag{3.18}$$

where the new quantity τ is the larval developmental time. This formulation assumes that larvae compete among themselves for a single limiting resource supplied at a constant rate (for details, see Gurney and Nisbet 1998).

The literature on structured population models in ecology is voluminous, and I cannot review it in this book. Readers comfortable with mathematics should consult Metz and Diekmann (1986). A good survey is Tuljapurkar and Caswell (1997; see in particular chapters by Caswell and Nisbet). The definitive book on matrix models in ecology is Caswell (2000). Finally, an excellent, and mathematically not-too-demanding introduction to structured models is chapter 8 of Gurney and Nisbet (1998).

3.3.2   An Example: Flour Beetle Dynamics

Flour beetles in the genus *Tribolium* have been a subject of population dynamic investigations since the "golden age" of population ecology (Chapman 1928; Park 1948). Recently, laboratory studies showed that it is possible to demonstrate "chaos in a bottle" using this organism (Costantino et al. 1997), although rather heroic measures had to be taken in order to get flour beetle populations to oscillate chaotically. As part of their theoretical/empirical investigation, Costantino, Desharnais, Dennis, and Cushing constructed and parameterized a model of *Tribolium* population dynamics, which provides a good illustration of the discrete-time, discrete-stage framework.

The formulation of the model capitalizes on the fact that the maturation interval of the feeding larvae, 2 weeks, is about the same as the cumulative time spent in the prepupal, pupal, and callow adult stage (Dennis et al. 1995; Costantino et al. 1995; Costantino et al. 1997). Thus, the natural time step is 2 weeks, and model equations are written as

$$L_{t+1} = \beta A_t \exp[-c_1 A_t - c_2 L_t]$$

$$P_{t+1} = L_t(1 - \delta_1)$$

$$A_{t+1} = P_t \exp[-c_3 A_t] + A_t(1 - \delta_2)$$

The state variables in this model are L_t, the number of larvae, P_t, the number of all nonfeeding stages (prepupae, pupae, and callow adults), and A_t, the number of adults. The model, thus, is known as the "LPA model" (larvae-pupae-adults). Parameter β is the fecundity rate, or the number of larval recruits per mature adult per unit of time (two weeks). Parameter δ_1 is the fraction of larvae dying of causes other than cannibalism, while δ_2 is the death rate of adults per unit of time. Finally, c_1 and c_2 are cannibalistic rates by larvae and adults on eggs. Thus, the fraction $\exp[-c_1 A_t]$ is the probability that an egg is not eaten during the 2-week period in the presence of A_t adults, and similarly for larval cannibalism. Cannibalism of adults on pupae is modeled analogously, with parameter c_3. Note that the functional form of cannibalism is the same as Nicholson-Bailey parasitism. Stochasticity can be added to the model by multiplying

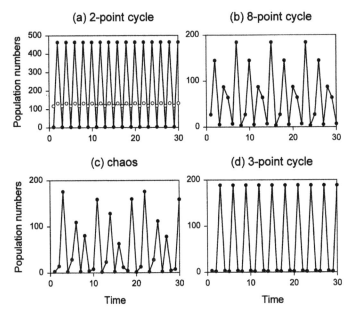

FIGURE 3.5. Dynamics of the LPA model. Solid circles: larvae. Open circles in (a): adults. In (a) parameters are $\beta = 11.7$, $\delta_1 = 0.51$, $\delta_2 = 0.11$, $c_1 = 0.011$, $c_2 = 0.009$, $c_3 = 0.018$. In (b–d) common parameters are $\beta = 6.598$, $\delta_1 = 0.2055$, $\delta_2 = 0.96$, $c_1 = 0.01209$, $c_2 = 0.0155$, and c_3 is varied: (b) 0.05; (c) 0.35; (d) 0.50.

each state variable with a lognormally distributed random variable (Dennis et al. 1995; Costantino et al. 1995; Costantino et al. 1997).

The LPA model is capable of a usual variety of dynamics typical for this class of discrete-time models: stable point equilibria, limit cycles, quasiperiodicity, and chaos (figure 3.5). There are two interesting observations. First, for parameter values that obtain in unmanipulated populations, the dynamics are typically characterized by a 2-point cycle (figure 3.5a) or a stable-point equilibrium with oscillatory convergence, also with 2-point periodicity (not shown). Adult survival over a 2-week period is high, and, as a result, adult numbers fluctuate very little (figure 3.5a: open circles). In other words, the cycles are primarily in the stage structure of the model, not the total numbers. Second, reducing adult survival and increasing cannibalism rate on pupae (Costantino et al. 1997) results in a variety of nonlinear oscillatory dynamics depicted in figure 3.5b–d. However, in every case the dominant oscillation period is between 2 and 3. By "dominant period" I mean the average length of a typical excursion from peak to peak, or from trough to trough. Note that even in the 8-point cycle, each "mathematical" cycle consists of two oscillations of 3 time steps each, and one oscillation of 2 steps. In the presence of noise, this fine structure would be largely washed out, leaving an average period of 2–3 steps. The reason that this observation is interesting is that the LPA model appears to fit the general mold of single-species models. Single-species models tend to produce oscillations with periods of 2–4 τ, where τ is the developmental delay (section 3.1.3). The LPA model produces oscillation periods in the same range.

3.4   SECOND-ORDER MODELS

Single-species second-order models have the following structure:

$$N_{t+1} = N_t f(N_t, X_t)$$
$$X_{t+1} = g(N_t, X_t)$$

$$(3.19)$$

where X_t is some **intrinsic** dynamical variable, that is, some characteristic of the focal population that changes with time. These models are slightly suspect because of the phenomenology resulting from reducing the distribution of some variable, for example, population quality, to dynamics involving only its average. However, some models of population interactions, for example, herbivore–plant quality (section 4.4.4), suffer from the same problem. Here I review two second-order models of intrinsic hypotheses that were proposed for explanation of population cycles.

3.4.1   Maternal Effect Hypothesis

Ginzburg and Taneyhill (1994) proposed the following general model for dynamics of populations affected by maternal effects:

$$N_{t+1} = N_t f(X_t)$$
$$X_{t+1} = g(N_{t+1}, X_t)$$

$$(3.20)$$

The N_t equation in this model has no direct density dependence. Thus, population density grows exponentially, but with the intrinsic rate of change being a function of average quality, X_t. In the quality equation, note that X_{t+1} depends on N with the subscript of $t + 1$, reflecting the assumption that quality is affected by density in the current generation (as a result of intraspecific competition). (Despite this feature, the model is still a special case of equations 3.19, as can be seen by substituting N_{t+1} from the first equation into the second one.)

Ginzburg and Taneyhill argued that f should be a monotonically increasing function of X_t (the higher the average quality, the greater the per capita rate of population increase). Function g decreases with N_{t+1}, and increases with X_t. The last assumption is why we call this model "maternal effects," because the quality of offspring is positively correlated with the quality of mothers. Ginzburg and Taneyhill proposed the following specific functions that conform to the postulated shapes:

$$N_{t+1} = \lambda_0 N_t \frac{X_t}{b + X_t}$$

$$X_{t+1} = X_t \frac{\mu_0}{1 + dN_{t+1}}$$

(3.21)

The relationship between the realized discrete rate of increase and quality is hyperbolic, with b the half-saturation quality, and λ_0 the maximum reproductive rate (realized when $X_t \to \infty$). The second equation is analogous to the first, but the hyperbolic function is of the decreasing, rather than increasing, kind. Parameter μ_0 is the maximum (proportional) rate of increase in quality, occurring when N_{t+1} is near zero. As density increases, the rate of proportional change declines below 1. Thus, when N_{t+1} exceeds $(\mu_0 - 1)/d$, quality begins to decline.

The analysis of this model by Ginzburg and Taneyhill indicated that it produces undamped oscillations when parameters λ_0 and μ_0 are greater than 1. Thus, if the population persists, then it inevitably undergoes cycles. Furthermore, these cycles are *neutrally stable*. In other words, this model is analogous to the Lotka-Volterra predation model, which also exhibits neutral cycles for all parameter values. In fact, if we rewrite equations 3.21 in the per capita form, we see that this is a **pure second-order** model (similar to the

Lotka-Volterra predation, which is a pure resource-consumer model; see section 2.4.2), although formulated in a discrete-time framework:

$$\frac{N_{t+1}}{N_t} = f(X_t) = \frac{\lambda_0 X_t}{b + X_t}$$

$$\frac{X_{t+1}}{X_t} = g(N_{t+1}) = \frac{\mu_0}{1 + dN_{t+1}}$$

It appears, therefore, that the neutrally stable oscillatory dynamics in this model are due to lack of direct feedback either in density or in quality.

Another interesting feature of this model is that the period of oscillations depends only on parameters λ_0 and μ_0 (this is another analogous feature to the Lotka-Volterra model, in which the period is a function of the exponential growth/decline rates of predator and prey). Low values of λ_0 and μ_0 lead to longer oscillations. This is understandable: the longer it takes for density, for example, to increase above the threshold where quality begins to be negatively affected, the longer the cycle length will be. What is of particular interest, however, is that even for high values of λ_0 and μ_0 the minimum cycle length is 6 time steps. Thus, the range of periods that can be generated by this pure second-order model, 6–∞, is clearly differentiated from periods in first-order models (2–4).

3.4.2   Kin Favoritism Model

The basic idea that interactions between kin may play an important role driving a population cycle was proposed by Charnov and Finerty (1980). This idea was extended and applied to red grouse dynamics by Robert Moss, Adam Watson, and their coworkers. As Moss and Watson (2000) explain in their recent review of tetraonid population dynamics, young red grouse cocks tend to settle near their fathers, thus forming spatial clusters of related territory owners. It is possible that population change can be related to the size of such kin clusters (Mountford et al. 1990), as a result of the following mechanism. Bigger kin clusters may facilitate a higher recruitment rate, which in turn results in bigger kin clusters, setting a positive feedback loop during the increase phase of a cycle. At peak densities, crowding halts this positive feedback, and kin clusters decay because they are not replenished by new recruits. Smaller kin clusters result in lower recruitment, setting the stage for the decline phase. When low enough densities are reached, kin clusters can start forming again, and the cycle repeats itself (Moss and Watson 2000). This proposed mechanism for population cycles has been modeled with a spatially explicit, individual-based simulation (Hendry et al. 1997) and with a simple analytically tractable model (Matthiopoulos et al. 1998). Here I discuss the simpler model developed by Matthiopoulos et al. (1998) (for the general assessment of the "kin favoritism" hypothesis in the context of red grouse population cycles, see chapter 11).

The kin favoritism model of Matthiopoulos et al. (1998) has the basic form of a stage-structured model, and has two state variables: O_t, the number of old males that have survived and reproduced for at least one year, and Y_t, the number of young males that were recruited last year. The model keeps track only of male numbers, because it is assumed that only territorial birds reproduce, and the number of territories is determined by the interactions between males only. That is, there are always enough females to populate all male territories.

Dynamics of red grouse numbers are driven by territorial interactions between males. In the absence of differential treatment of kin versus nonkin, the model is

$$N_{t+1} = N_t \left(s + \frac{b}{1 + cN_t} \right) \qquad (3.22)$$

where $N_t = O_t + Y_t$ is the total number (or density) of territorial males. The first term in the parentheses, s, reflects the yearly survival rate of territorial males. The second term reflects recruitment, and has two components: b is the number of new (male) recruits reared by each territory, and $1/(1 + cN_t)$ is the probability that a recruit will establish a territory. Note that the only way density dependence enters in this model is via probability of acquiring a territory: it is 1 when N_t is near zero, and declines monotonically with increased N_t. Parameter c is proportional to the minimum territory size (see Matthiopoulos et al. 1998 for derivation details). Checking with table 3.1, we see that the recruitment function assumed by Matthiopoulos et al. is in the Beverton-Holt form. Thus, it is not surprising that model (3.22) can have only one kind of stationary dynamical behavior, stable point with monotonic damping (Matthiopoulos et al. 1998).

To add the effect of kin favoritism, Matthiopoulos et al. made the key assumption that the minimum territory size required for establishment of males is proportional to the ratio (index of crowding)/(index of kinship). "Crowding" reflects the number of kin clusters that are competing for space. If cocks can recognize only their fathers and brothers, then, as Matthiopoulos and colleagues argue, the density of old birds, O_t, can serve as the crowding index. "Kinship" should be related to the cluster size. Matthiopoulos et al. propose that the kinship index should be proportional to Y_t/O_t. When $Y_t = 0$ all clusters consist of a single old male, and the kinship index is zero. When many young birds are added, the ratio Y_t/O_t will be large, as will the kinship index. Putting together these assumption leads to the following functional form for the parameter c in equation (3.22):

$$c \sim \frac{\text{index of crowding}}{\text{index of kinship}} \sim \frac{O_t}{Y_t/O_t} = \frac{O_t^2}{Y_t} \tag{3.23}$$

The model resulting from these assumptions is

$$O_{t+1} = s(O_t + Y_t)$$

$$Y_{t+1} = \frac{b(O_t + Y_t)}{1 + kO_t^2(O_t + Y_t)/Y_t} \tag{3.24}$$

where k is the constant of proportionality between c and O_t^2/Y_t.

Model (3.24) is derived by assuming the dynamics of kinship clusters, but the final product is expressed in terms of state variables of a stage-structured population dynamics. This makes it difficult to see what really is going on in the model's guts. Thus, I am going to rewrite this model in terms of different state variables, the total population density of males, $N_t = O_t + Y_t$, and the kinship index, as defined by Matthiopoulos and colleagues, $X_t = Y_t/O_t$. Furthermore, to reduce the number of parameters, I will scale the density as follows: $N_t' = \sqrt{k}N_t$. After some algebra, I obtain the following model:

$$N_{t+1} = N_t\left\{s + b\left[\frac{X_t(X_t + 1)}{X_t(X_t + 1) + N_t^2}\right]\right\}$$

$$X_{t+1} = \frac{b}{s}\left[\frac{X_t(X_t + 1)}{X_t(X_t + 1) + N_t^2}\right] \tag{3.25}$$

(where the primes associated with N_t have been dropped). We see that the N_t equation is a modification of model (3.22), in which the recruitment probability (the quantity in square brackets) is now a function of both density and kinship. The recruitment probability has a general form of $a^2/(a^2 + b^2)$. It is S-shaped in X_t for fixed N_t, and vice versa. The same factor (note that the quantities in square brackets are identical in both equations) enters the X_t equation. Note that kinship at time $t + 1$ is simply a linear function of recruitment at time t.

The model appears to work as follows. For low to medium N_t values (below the flex point of the S-curve), the factor N_t^2 is very small. Thus, recruitment at time t is near maximum, 1. Correspondingly, kinship at time $t + 1$ is also near its maximum (b/s). As a result, population grows at the maximum rate $(s + b)$. When N_t passes through the flex point, recruitment collapses, causing a decline in both density, N_{t+1}, and kinship, X_{t+1}. At the next time step, density continues to decline because kinship (and, therefore, recruitment) is very low. Only after kinship comes back to near its maximum can density start increasing again. As a result, the model exhibits a typical second-order cycle (several time steps of increase, followed by at least two steps of decrease).

The kin favoritism model is an interesting attempt to include the influence of behavioral interactions between territorial animals on their population dynamics. Because this attempt opens up a largely unexplored territory, it should not be judged too harshly. However, I have two major criticisms of the approach taken by Matthiopoulos and colleagues. First, I am bothered by the formulation of the model in terms of age-structured dynamics. This approach required the authors to translate their assumptions about behavioral interactions first into demographic dynamics, and only secondly and indirectly into population dynamics. For example, I do not really understand why the relationship between the density-dependent parameter (related to the minimum territory size) and stage class densities should be as derived by Matthiopoulos et al.:

$$C \sim \frac{O_t^2}{Y_t} \tag{3.26}$$

Currently Matthiopolous et al. (2000a, b) are attempting to remedy this problem, but only the future will show whether more mechanistically explicit models of kin favoritism would yield oscillatory dynamics for biologically plausible parameter values.

The second serious problem with model (3.25) is that it has only two parameters affecting its qualitative dynamics, s and b, neither of which measures the strength of the mechanism that is postulated to drive the oscillations. As Matthiopoulos et al. show, model (3.25) will cycle whenever values of b and s are large enough (their figure 2B). What it means is that the kin favoritism effect has been *hardwired* in the model. Generally, this is not a good theoretical approach, because we usually wish to investigate the following question: "how strong should the effect A be in relation to effects B, C, ... in order for population to cycle (or to be stable, etc.)?"

3.5   SYNTHESIS

The survey of theory for single-species population dynamics suggests certain recurrent themes. In particular, oscillations predicted by models appear to fall within three general classes, corresponding to zero-, first-, and second-order dynamics (section 2.5). Zero-order cycles with periods approximating a generation time arise in the simplest models of age-structured populations (e.g., in linear matrix models). Stochastic exogenous influences (e.g., an episodic massive die-off of a particular age class) may help perpetuate generation cycles in situations where deterministically the system would tend to a stable age distribution. Furthermore, interesting dynamics may result when periodic exogenous influences resonate with the periodicity due to age-structure oscillations. It should not be forgotten that annual populations, which we often model within a discrete-time framework, have a basic generation cycle underlying all other kinds of dynamics.

Another type of dynamics that arises naturally in first-order discrete models, such as the Ricker equation, is stable cycles with the typical period of about 2–3 time steps. Unlike generation cycles, whose origin is due to population inertia, first-order cycles are a result of negative feedback (strong enough to cause an overshoot of equilibrium density). First-order cycles also arise in delayed differential equations, and their periods are typically between 2 and 4τ, where τ is the developmental delay. Note that τ is typically less than the generation time. Whereas τ can be interpreted as the age of first reproduction, the generation time, T_c, is usually defined as the mean age of reproduction. Thus, typically $T_c \geq \tau$, with equality obtaining when all reproduction is concentrated in one pulse. Furthermore, Ginzburg

(1970) showed that in PDE models of age-structured populations, the dominant period of oscillations, $T \leq 2B$, where B is the last possible age of reproduction. In sum, the typical period of oscillations that arise as a result of feedbacks acting directly on the population growth rate is roughly 2 generations.

The final generic class is second-order oscillations, which arise when population feedback operates indirectly, via some slow dynamical variable other than population density. *Pure second-order models* (those that do not include first-order feedback) generate longer periods than those characterizing generation or first-order cycles. For example, the minimum period in the Ginzburg and Taneyhill model is 6 time steps. However, models that include both first- and second-order feedbacks are capable of the complete spectrum of cycle periods, depending on parameter values.

To summarize, even if at risk of oversimplifying, periods of cycles characterizing single-population models fall into the trichotomy of "one-two-many" generation times, corresponding to generation, first-order, and second-order cycles. I believe that this trichotomy offers a useful heuristic device for imposing some structure on the chaos of observed variety of empirical population dynamics. It is important, however, not too approach this classificatory scheme too dogmatically. First, in both real-world and more complex models, there is a great degree of overlap in periods predicted by various population processes, depending on the specific mix of mechanisms and parameter values. Second, dynamical systems can be affected by more than one cycle, "nested" within each other. As a simple example, take the Ricker model for parameter values that generate 2-year cycles. There is actually another, generation, cycle occurring at the same time, although we typically ignore it by sampling the population only once a year.

CHAPTER 4

Trophic Interactions

Consumer-resource interactions are inherently prone to oscillations and are, therefore, the obvious suspect to investigate as a potential mechanism of a population cycle. However, not all models of trophic interactions exhibit cycles. The purpose of this chapter is to survey the theory of consumer-resource dynamics, and ask two major questions: how is the propensity to cycle affected by (1) structural assumptions of the model, and (2) parameter values? Theoretical literature on resource-consumer interactions is enormous, and even answering this narrow question could take a whole book in itself. To make the task more manageable, I shall primarily focus on models that have been invoked by authors in discussions of real-life case studies of complex population dynamics.

Consumer-resource interactions can be classified along two independent axes: *intimacy* (the closeness and duration of the relationship between the individual consumer and the organism it consumes) and *lethality* (the probability that a trophic interaction results in the death of the organism being consumed) (e.g., Stiling 1999). Thus, *predators* are high on lethality and low on intimacy, *parasitoids* are high on both scales, *parasites* are high on intimacy and low on lethality, and, finally, *grazers* are low on both scales. This functional classification affects the form of trophic interaction models. I start this chapter with predation as the paradigmatic trophic interaction, and then discuss the other three functional classes of consumers.

Throughout this chapter I discuss a multitude of equations, which use the same symbols over and over again. In order to avoid unnecessary repetition, I chose to explain the symbols only when they are first encountered. Additionally, all symbol definitions are given in the list of mathematical symbols at the front of the book.

4.1 RESPONSES OF PREDATORS TO FLUCTUATIONS IN PREY DENSITY

Changes in prey density affect predators at both individual and population levels. Temporal variation in prey density affects the rate at which prey are killed by predators (**functional response**). Predators often respond to spatial variation in prey density by moving in ways that result in their aggregation in areas where prey are abundant (**aggregative response**). Functional and aggregation responses are a direct result of individual behaviors, and as a rule occur on a fast timescale. Variation in prey density also affects predator population numbers, via its effect on predator reproduction and death rates. Such a **numerical response** typically occurs on a slower timescale. For example, many predators reproduce only once a year. Thus, if prey density increases right after a reproduction period, it may take another year before predators could respond by increasing their population numbers. Even more important, it often takes an appreciable time for predators to increase numerically to the point where they impact prey survival rates. In general, therefore, functional and aggregative responses, occurring on a fast behavioral timescale, tend not to introduce time lags into population dynamics of predator-prey interaction, while numerical responses, occurring on a slower timescale, often do introduce time delays. This tendency is not absolute, however. Aggregation responses may be very slow if predator movement rate is slow compared to the scale of prey patchiness. For example, Abrams (2000) suggests that aggregation responses may introduce enough of a lag to have major effects on population dynamics. Vice versa, some natural enemies, most notably pathogenic microorganisms, are characterized by very fast numerical responses.

4.1.1 Functional Response

Functional response is defined as the temporal rate at which an individual predator kills prey. Note that functional response is a *double* rate: it is the average number of prey killed *per* individual predator *per* unit of time. Thus, units of functional responses are [prey] [predator]$^{-1}$ [time]$^{-1}$. Alternatively, the quantity of prey killed may be measured as biomass, rather than individuals (this is particularly appropriate for plant-herbivore models). Finally, note that I define functional response in terms of the number of prey individuals *killed*, rather than *eaten*. The reason for this is that functional responses are part of the prey equation, and from the point of view of prey the important factor is how many prey are removed from the population, rather than how many are actually consumed. The consumption part (translated into increased predator survival and reproduction) belongs to the predator equation, where it becomes an important input into the predator numerical response. For plants, similarly, functional response refers to the amount of biomass removed by a herbivore, which includes both what is consumed and what is "wasted": cut off and discarded, trampled, etc.

In the following, I provide an overview of most commonly used functional responses, starting from the simplest ones. The formulas that I discuss are listed together in table 4.1 where they can easily be compared.

Linear Response   The basic classification of functional responses (Types I, II, and III) was proposed by Holling (1959). This was an extremely important conceptual breakthrough, which continues to serve as the basis of modern theory of predator-prey interactions. Unfortunately, Holling muddied the waters by defining Type I functional response as a linearly increasing function up to a point where it abruptly becomes a flat horizontal line. There are several problems with this particular definition. First, it leaves the basic functional response used in the Lotka-Volterra model unnamed (after all, their equations assume no ceiling for the functional response). Second, it is my opinion that the difference between Type I as defined by Holling and Type II is rather minor. In fact, both of them satisfy the postulates of mass action and maximum physiological rate (see chapter 2), and consequently are characterized by two parameters. Both have similar dynamical consequences (destabilization of the neutrally stable Lotka-Volterra cycle). Finally, I dislike the sharp corner in the functional response Type I—nature seems to abhor sharp corners, smoothing them by averaging over event stochasticity and individual heterogeneity. In short, Type I does the same job as Type II, but more poorly.

TABLE 4.1. Some functional responses. Variables: N, prey density; P, predator density. Parameters: c, maximum killing rate; a, predator searching rate; h, handling time; w, wasted time; b, maximum searching rate; g, parameter regulating how fast search rate saturates with prey density; d, half-saturation constant; θ, an exponent

Label	Functional Form
Constant	c
Linear	aN
Hyperbolic	$\dfrac{aN}{1 + ahN}$
Exponential (Ivlev)	$c(1 - \exp[-N/a])$
Sigmoid	$\dfrac{cN^2}{d^2 + N^2}$
Sigmoid (mechanistic)	$\dfrac{bN^2}{1 + gN + bhN^2}$
θ-sigmoid	$\dfrac{cN^\theta}{d^\theta + N^\theta}$
Predator interference (mechanistic)	$\dfrac{aN}{1 + awP}$
Predator interference (phenom.)	$aNP^{-\theta}$
Beddington[1]	$\dfrac{aN}{1 + awP + ahN}$
Hyperbolic ratio dependent	$\dfrac{cN}{dP + N}$
Linear ratio dependent	$c\dfrac{N}{P}$

[1]Combines the hyperbolic response with predator interference.

Adding to the terminological confusion, many authors define Type I as a purely linear functional response. Thus, I decided to abandon the usage of "Types I, II, and III," switching instead to descriptive labels. Accordingly, the **linear functional response** is $f(N) = aN$. It embodies the mass action principle in its purest form, and is the component of the Lotka-Volterra predation model.

Hyperbolic Response   Two functional forms have been proposed for the saturating response (Holling's Type II): exponential (Ivlev 1961), which is now rarely used, and hyperbolic:

$$f(N) = \frac{aN}{1 + ahN} = \frac{cN}{d + N} \tag{4.1}$$

The hyperbolic form is solidly based on mechanisms at the individual level (Holling 1965:8; for a more gentle exposition, see Gotelli 1995:150). The first parameterization in equation (4.1) is based on the searching rate, a, and the handling time, h. The second parameterization employs $c = h^{-1}$, the maximum killing rate, and $d = (ah)^{-1}$, the half-saturation constant (prey density at which the killing rate is half of the maximum).

Recall that the hyperbolic functional response simultaneously satisfies the principles of mass action and maximum physiological rates. Thus, it has two limiting cases, each embodying one of the two principles: the linear and constant functional responses. The latter is rarely used in the theoretical literature, but I nevertheless list it in table 4.1 for completeness.

Sigmoid Response   Sigmoid (Holling's Type III) functional responses are often used in ecological theory without much thought given to their mechanistic underpinnings. The phenomenological form

$$f(n) = \frac{cN^\theta}{d^\theta + N^\theta} \tag{4.2}$$

as far as I know, has no mechanistic derivation. Originally, it was thought (e.g., Holling 1965) that hyperbolic functional responses were characteristic of invertebrate predators, while sigmoid responses were characteristic of vertebrate predators. Since then, it has become clear that the distinction is functional rather than taxonomic: specialist predators should be characterized by the hyperbolic response, while generalists are expected to exhibit a sigmoid response, if they are characterized by switching behavior. Generalist predators, by definition, kill several kinds of prey, including the focal species (the one we study). Accordingly, when the density of the focal species is low, generalist predators should focus on other prey species. When the density of the focal prey is high, predators will switch to hunting it (Murdoch 1969), because it becomes profitable for them to do so. Perhaps the most likely mechanism for such switching behavior is habitat choice. For example, a house cat (an ultimate generalist predator!) will hunt voles in the grassy patch when voles are abundant there. When voles become sparse, cats will shift their attention to birds coming to the feeder. Meanwhile, cat numbers will not respond numerically to changes in either vole or bird abundance, since they are regulated by the amount of cat food provided by their doting owners.

A more mechanistic form of the sigmoid response can be derived by starting with the hyperbolic response and assuming that predators will search more actively as prey density rises (Hassell 1978:38). Specifically, let us suppose that the search rate, a, is an increasing function of prey density (Hassell 1978:43):

$$a(N) = \frac{bN}{1 + gN} \tag{4.3}$$

The saturating functional form assumed by equation (4.3) reflects the obvious biological constraint that there is a maximum search rate above which $a(N)$ cannot increase no matter how high prey density becomes. Substituting this form into the equation for the hyperbolic response, we have

$$f(N) = \frac{a(N)N}{1 + a(N)hN} = \frac{bN^2}{1 + gN + bhN^2} \tag{4.4}$$

Although this form of the sigmoid functional response has a clear derivation from first principles, it is rarely used in population models. A much more common form to be found in current theoretical literature is the following one

$$f(N) = \frac{cN^2}{d^2 + N^2} \tag{4.5}$$

This form is a special case of (4.4), which assumes $a(N) = bN$; that is, we set $g = 0$ in equation (4.3). New parameters are $c = 1/h$ and $d^2 = 1/(bh)$.

If we replace the second power in (4.5) with an exponent parameter θ, we obtain equation (4.2). This allows us to regulate the nonlinearity of the transition between low predation rate at N near 0, and the saturated level of predation at $N \to \infty$.

Although equation (4.5) has obvious limitations, I believe that it strikes the right balance between detailed mechanistic and detail-free phenomenological approaches to generalist predation for most case studies where we do not desire to model generalist dynamics explicitly. Its main practical advantage is the parsimony (only two parameters). And it does have a derivation from first principles, so we know exactly in what ways we oversimplify nature when we employ equation (4.5). This form has certainly been useful in both theoretical investigations (e.g., Yodzis 1989:84–104) and practical applications (e.g., Turchin and Hanski 1997). If more detailed description of the process of generalist predation is desired, then perhaps we need to model the switching process explicitly.

Predator Interference   So far my survey has dealt only with prey-dependent functional responses; that is, the assumption is that the killing rate by a predator, $f(\cdot)$, is a function of prey density only $f(\cdot) = f(N)$. This may be a tenable assumption for an experimental situation where a single predator searches for prey in an arena, but in real life it is likely that individual predators will interact with each other. Leaving aside the issue of predator cooperation in hunting and subduing prey, it is likely that predator encounters will lead to antagonistic interactions. Intraspecific competitive interactions between individual predators can affect their birth and death rates (this will be discussed later in the context of predator numerical responses). Antagonistic interactions may also affect predator efficiency in finding and killing prey, that is, predator functional response. It seems most natural to model predator interference using the same logic as that used to derive the hyperbolic functional response. Accordingly, let us assume that each encounter between predators results in wasted time, w' (this is analogous to handling time, h). If predators encounter each other at rate b (analogous to the encounter rate, a, with prey), then these assumptions lead to the following formula (Beddington 1975):

$$f(N', P') = \frac{aN'}{1 + bw'(P' - 1)}$$

Beddington derived his functional response in terms of prey and predator numbers, N' and P', so I am using primes here to distinguish from densities N and P, as used throughout my book. The term $(P' - 1)$ in Beddington's formulation arises because he accounts for the fact that an individual predator cannot encounter itself. Expressing this term in terms of densities (i.e., $P = P'/A$, where A is the area occupied by the population), we have $P - 1/A$. Further, neglecting the $1/A$, which is likely to be close to 0, we have

$$f(N, P) = \frac{aN}{1 + bw'P}$$

Next we note that parameters b and w' enter the above formula as a single combination. This suggests that we do not actually need two separate parameters. Predator encounter rate, b, is likely to be similar to the search rate, a (differing only if predators detect each other at distances much different from those at which they detect prey). This observation suggests the following reparameterization: $w = w'b/a$, leading to

$$f(N, P) = \frac{aN}{1 + awP} \tag{4.6}$$

This form is clearly analogous to the hyperbolic functional response, which should not be surprising, since it was derived following the same logic.

Equation (4.6) is what I call the mechanistic form of the functional response with predator interference because it was derived from first principles. There is also a more phenomenological form (see table 4.1). This form was proposed by Hassell and Varley (1969), but later Hassell (1978:84) pointed out that it has problems, and suggested that the more mechanistic form is more appropriate for modeling host-parasitoid systems. Unfortunately, the phenomenological form continues to be used. I strongly recommend the more mechanistic form: it has the same number of parameters, it has a clear derivation from first principles, its parameters are interpretable in terms of individual behavior, and it does not make a biologically implausible assumption that predator efficiency would increase indefinitely as predator density declines to zero (Hassell 1978:84). In short, the contrast between the two forms once again illustrates the value of theory derived from first principles.

Yet another advantage of the form (4.6) is that it allows us a natural method for including predator interference in the hyperbolic functional response. The logic is that predators "waste" time both handling prey and dealing with other predators. Thus, both handling time and wasted time are added together to reduce the amount of time left for search. This logic leads to the following form for combined functional response:

$$f(N, P) = \frac{aN}{1 + awP + ahN} \tag{4.7}$$

This functional form was first proposed by Beddington (1975) and independently (although without a derivation from first principles) by DeAngelis et al. (1975).

Ratio-Dependent Responses Perhaps the most direct way to derive a ratio-dependent functional response is to start with the interference response, equation (4.7), and simplify it by assuming that "1" in the denominator is small relative to $awP + ahN$. We might expect that this would not be a bad approximation in situations where predator interference is very strong ($awP \gg 1$). Following this logic, we have

$$f(N, P) = \frac{aN}{1 + awP + ahN} \approx \frac{aN}{awP + ahN} = \frac{cN}{dP + N} = \frac{cN/P}{d + N/P} \tag{4.8}$$

which is the hyperbolic functional response, but with the ratio N/P replacing prey density, N. The advantage of equation (4.8) over (4.7) is that it has one fewer parameters.

Ratio-dependent predation has been a subject of intense controversy (see, e.g., Arditi and Ginzburg 1989; Abrams 1994; Akcakaya et al. 1995). Recently, two of the most vocal opponents in this debate collaborated on a very useful summary that clearly delineated the areas of agreement and disagreement (Abrams and Ginzburg 2000). Surprisingly, Abrams and Ginzburg agreed on many more issues than they disagreed on. Most important, it seems clear that predator density should have a strong effect on predator functional response in nature. Many biological processes can produce such predator dependence: (1) group hunting by the predator, (2) facultative and costly antipredator defense by the prey, (3) density-dependent and time-consuming social interactions between predators, (4) aggressive interactions between searching predators that encounter each other, and (5) a limited number of high-quality sites where predators capture prey rapidly (for references, see Abrams and Ginzburg 2000:339). However, including such realistic processes into the functional response $f(N, P)$ typically leads to complex, multiparameter functions. Prey-dependent, $f(N)$, and ratio-dependent, $f(N/P)$, responses offer simpler, less parameter-rich formulations. Such functions, therefore, are better starting points for modeling real predator-prey systems, since starting with simple equations and building in complexity only where necessary is a much better methodology than jumping into complexity right away. Of course, the cost of simplicity is lack of realism. Precise prey- or ratio-dependence should be rare. In particular, prey dependence must break down for sufficiently high predator densities, while ratio dependence must break down for sufficiently low predator densities (Abrams and Ginzburg 2000:338–339).

All these issues are ones on which Abrams and Ginzburg agree. The disagreement is on which of the alternative simpler approaches provides a better starting point. Ginzburg feels that averaging of functional responses (e.g., over reproductive intervals) shifts trophic functions toward ratio dependence, and that such averaged responses should be the basis of both differential and difference equation models because these averages determine long-term population dynamics. By contrast, Abrams favors prey-dependent models as the basic building blocks for theory because they are based on a single well-defined set of assumptions; this makes it clear how to modify them when the assumptions are known to be deficient (Abrams and Ginzburg 2000:341).

My philosophy of modeling inclines me to agree with Abrams. I feel that the hyperbolic (i.e., prey-dependent) functional response provides the appropriate starting point, especially if we formulate models within the differential equations framework. Predator interference can be added transparently, leading to the Beddington response. If

reproduction and predation occur on different timescales (e.g., reproduction is seasonal), reproduction can be appropriately modeled in a
pulsed (or some other time-varying) fashion. Logically, I do not see
the need to use any averaging, as suggested by Ginzburg; the differential equations framework allows us to model both predation and reproduction processes explicitly and mechanistically. Matters are different
when we employ discrete-time (difference) equations, where some
temporal averaging is inevitable (see section 4.3.1). Furthermore, it is
not necessary to be dogmatic about this choice. For example, if the
ratio-dependent response fits data better than the prey-dependent one
with the same number of parameters, then by all means we should
consider basing our model on the ratio-dependent form.

As a final comment, I wish that ecologists would write more papers
like Abrams and Ginzburg (2000). If we did so, we might find that
our disagreements are not as deep and wide as they may appear. Furthermore, clearly stating the remaining areas of disagreement allows
the protagonists in the controversy to advance the empirical research
agenda, as Abrams and Ginzburg did.

4.1.2 Aggregative Response

Aggregation is an explicitly spatial process and consequently does not
fit easily within the dichotomy of functional and numerical responses
postulated by the standard (aspatial) theory of predator-prey interactions. Most textbooks do not even mention aggregative responses,
and there is some confusion in the literature about appropriate ways
of treating aggregation. For example, some authors consider it as a
kind of a numerical response. Indeed, aggregation results in a local
increase of predator numbers. However, in most cases it is inap-
propriate to confuse aggregative and numerical responses, because
the numerical response results from births and deaths, and typically
occurs on a much slower timescale than behavior-based aggregative
(and functional) responses. In reality, a full treatment of aggregative
responses requires an explicitly spatial approach, and therefore cannot be pursued here (interested readers may wish to consult my book
on movement and spatial dynamics, Turchin 1998). There are some

circumstances, however, when aggregative responses can be approximated with simple, spatially *implicit* functional forms.

Consider a landscape of many patches with very variable prey density in each. Suppose that predators can move freely over the whole landscape. The simplest possible scenario for predator aggregation is provided by the ideal free distribution, which suggests that predators should forage in a patch only if prey density there is greater than a certain threshold, say h. The number of predators in low-prey patches will be zero, and in high-prey patches it will be P_{high}, equal to the total number of predators within the landscape, divided by the number of high-prey patches. This is not a terribly realistic foraging model, but let us see how far it will take us. Let us make a further simplification and assume that each predator will kill prey at a constant rate, c. The reason for this assumption is that in high-prey patches, where all predators are, their functional responses are likely to be saturated. These assumptions imply that in a patch with prey density N

$$\text{total killing rate} = \begin{cases} 0 & \text{if } N < h \\ cP_{high} & \text{otherwise} \end{cases}$$

Because both c and P_{high} are constants, we can reparameterize their product as g. Thus, the total killing rate will be 0 if $N < h$ and g if $N > h$. It is not very likely, however, that killing rate will change in such a discontinuous, steplike fashion at density threshold h. For example, predators will probably need to forage in a patch for some time before they can obtain enough information to decide whether to leave it or stay. This and other considerations suggest that the sharp corners of the step function that we derived above should in reality be blurred. We can represent this blurring with the phenomenological sigmoid form:

$$\text{total killing rate} = \frac{gN^{\theta}}{h^{\theta} + N^{\theta}} \tag{4.9}$$

The exponent θ controls the degree of sharpness, so that as $\theta \to \infty$ equation (4.9) approaches the step function.

The argument leading to equation (4.9) rides roughshod over all kinds of biological sensibilities, some of which were touched upon and others were not. This is clearly an extremely crude approximation, but the alternative is an explicitly spatial approach, which would require collecting much more data than are typically available in real-life applications. As often is the case, the question of applicability of equation (4.9) becomes empirical—how bad an approximation is it in any particular application?

One interesting feature of equation (4.9) is that it is clearly related to the sigmoid functional response (table 4.1). However, it is very important to note that equation (4.9) refers to the *total* killing rate, rather than killing rate *per predator*, which is the functional response.

4.1.3 Numerical Response

Numerical response refers to the rate of change of predator population as a function of prey and predator densities. There are three major components of numerical response (Beddington et al. 1976b): growth of individual predators, predator reproduction, and predator death. All of these processes require energy derived from consumed prey. Individual growth and reproduction can be thought of as two aspects of a single process, increase in predator biomass. The probability of death, on the other hand, should be related to the amount of energy needed for maintenance. Recollect that postulate 5 connects the amount of energy derived from consumed prey to that available for growth/reproduction and maintenance. The simplest assumption is that energy available for growth, reproduction, and maintenance is a linear function of food intake—the *linear conversion* rule (Ginzburg 1998). Translating this assumption into per capita rate of growth of predator population, we have

$$\frac{dP}{P\,dt} = \chi(I - \mu) \tag{4.10}$$

The easiest way to understand equation (4.10) is by casting it in energetic terms. Accordingly, let P be the energy contained in predator biomass, and I the rate at which energy in prey biomass is ingested by an individual predator. Then, χ is the assimilation efficiency, and χI is the rate at which prey energy is assimilated by an individual predator. A part of assimilated energy, $\chi\mu$, is used for maintenance (or respiration), and the rest, $\chi(I - \mu)$, is allocated to growth and reproduction (or secondary production). Finally, if the rate of ingestion is zero, biomass of an individual predator will decrease at the rate $\chi\mu$.

Recasting equation (4.10) in population-level terms, let P be predator density, rather than biomass. Parameter μ is the ZPG (zero population growth) consumption rate, because it is the consumption rate at which an individual predator just manages to satisfy its maintenance requirements and replace itself. Parameter χ is the rate at which ingested prey in excess of the replacement requirement is translated into predator population increase. Finally, $\chi\mu$ is now seen to be the death rate of predators in the absence of prey.

Numerical response based on equation (4.10) provides a very useful starting point for modeling predator-prey interactions. However, because it is the simplest implementation of postulate 5, it may need to be modified in real-life applications. For example, some theoretical ecologists argued that the assumption of constant death rate of predators in the absence of prey is unrealistic (Ginzburg 1998).

The next step is to link equation (4.10) to quantities appearing in the prey equation. The simplest assumption is that ingestion rate equals the killing rate, or functional response. Thus, if predators are characterized by a hyperbolic functional response, then equation (4.10) becomes

$$\frac{dP}{P\,dt} = \chi\left(\frac{cN}{d+N} - \mu\right) \tag{4.11}$$

This relationship between the per capita rate of predator increase and prey density is depicted in figure 4.1. Note that the relationship saturates at high prey density, so that predator's $r_0 = \chi(c - \mu)$. In other words, equation (4.11) does not violate the principle of maximum physiological rates (postulate 6).

How likely is the assumption that ingestion rate equals functional response? Some predators, such as predatory fish, swallow their prey whole. Other predators, such as many mammalian ones, always leave some parts of prey unconsumed (hooves, horns, large bones, etc.). If the proportion of prey biomass that is not consumed does not depend on prey density, then no special handling is required. We simply fold the proportion of biomass that is edible into the constant χ. However, if predators consume a smaller proportion of biomass at higher prey density, or even indulge in surplus killing , then we need to modify the predator numerical response accordingly.

How should the proportion consumed vary with prey density? When prey density is low, it should be at the maximum. Let us set it

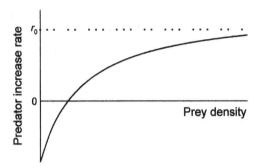

FIGURE 4.1. The relationship between the per capita rate of predator change and prey density, resulting from the hyperbolic functional response and the assumption of equation (4.10).

to 1, without any loss of generality (essentially I am assuming that we are talking about the proportion of *edible* biomass). As prey density increases, this proportion should decrease, approaching either 0 or some small positive number. Here is a possible approach to modeling this process. Let us assume that a predator spends h_s units of time subduing prey (this includes the whole process of pursuing, subduing, and killing prey, and perhaps resting before starting the search again). On the consumption side, let h_d be the time needed to digest one prey item. Because digestion is a slower process than subduing prey, $h_d > h_s$. These assumptions imply that both the functional response, $f(N)$, and the consumption rate, $c(N)$, are hyperbolic functions, and that the proportion consumed is the ratio $c(N)/f(N)$:

$$f(N) = \frac{aN}{1 + ah_s N}$$

$$c(N) = \frac{aN}{1 + ah_d N}$$

$$\frac{c(N)}{f(N)} = \frac{1 + ah_s N}{1 + ah_d N}$$

We see that this simple modeling of proportion consumed satisfies the constraints we set on it a priori. In particular, as $N \to 0$, $c(N)/f(N) \to 1$, and as $N \to \infty$, $c(N)/f(N) \to h_s/h_d < 1$, since $h_s < h_d$. Furthermore, the proportion consumed would be 0 only if the subduing time, h_s, were 0, that is, if the functional response is linear.

So far I discussed only how predator consumption of prey affects predator numerical response. Predators also interact directly in a variety of competitive and cooperative ways, and these interaction terms affect their numerical dynamics. Such mechanisms, however, are best considered within the context of dynamical consequences of various assumptions about functional and numerical responses, which is the next topic to be discussed.

4.2 CONTINUOUS-TIME MODELS

Predator-prey theory is one of the best-developed areas in population ecology, so that modeling any particular predator-prey system is increasingly an application of craft rather than art. The most popular framework for modeling specialist predator-prey interactions has the following structure:

$$
\frac{dN}{dt} = + \boxed{\begin{array}{c} \text{prey growth in} \\ \text{the absence of} \\ \text{predators} \end{array}} - \boxed{\begin{array}{c} \text{total killing rate} \\ \text{by predators} \end{array}}
$$

$$
\frac{dP}{dt} = - \boxed{\begin{array}{c} \text{predator growth} \\ \text{(decline) in the} \\ \text{absence of prey} \end{array}} + \boxed{\begin{array}{c} \text{conversion of} \\ \text{eaten prey into} \\ \text{new predators} \end{array}}
$$

where N and P are densities of prey and predators, respectively. The majority of ecological predator-prey models make a further assumption that the rate at which eaten prey is converted into new predators is directly proportional to the killing rate, yielding

$$
\frac{dN}{dt} = r(N)N - f(N, P)P
$$
$$
\frac{dP}{dt} = \chi f(N, P) - \delta(P)P
$$

(4.12)

Here $r(N)$ is the density-dependent per capita rate of prey growth in the absence of predators, $\delta(P)$ is the per capita decline rate of predators in the absence of prey, $f(N, P)$ is the predator functional response, and χ the conversion rate of eaten prey into new predators. The origins of equations (4.12) go clearly back to the Lotka-Volterra model, which is the simplest possible example of it (since it assumes exponential growth/decline terms and the linear functional response). Much of the standard predator-prey theory, therefore, is an elaboration of the Lotka-Volterra model that substitutes more sophisticated assumptions in one box or another of the framework depicted at the beginning of this section. In fact the state of the theory is such that one can simply take modules appropriate to the specific empirical system "off the shelf," fit them within the general framework, and usually have a reasonable starting point for modeling the system.

In the following review of predator-prey models, I start with those that conform to the generalized Lotka-Volterra form (4.12) and then consider models that fall outside this framework. At the end of the section I also discuss generalist predation. My primary concern is what various assumptions mean for real-life applications, and the qualitative dynamics that the resulting models are capable of. To keep track of different models, I label them with the name(s) of people who first proposed or analyzed them (to the best of my knowledge; I apologize to any authors whom I inadvertently slight by not giving them proper credit).

4.2.1 Generalized Lotka-Volterra Models

The Volterra Model   The first obvious place to add some realism to the Lotka-Volterra model is by relaxing the assumption of density-independent prey growth. Volterra (1931) proposed a model in which prey grows logistically in the absence of predator:

$$\frac{dN}{dt} = r_0 N(1 - N/k) - aNP$$
$$\frac{dP}{dt} = \chi aNP - \delta_0 P$$

(4.13)

The effect of assuming density-dependence in prey growth is to rotate the prey isocline clockwise (see figure 4.2). A good rule of thumb in evaluating the effect of structural assumptions on the stability of a predator-prey model is to check whether the modifications rotate the isoclines clockwise (this yields more stable dynamics) or counterclockwise (this tends to destabilize the system). The Volterra model, accordingly, should be more stable than the Lotka-Volterra model. In

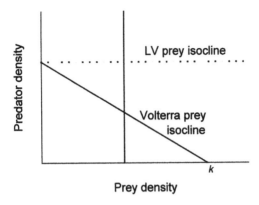

FIGURE 4.2. Clockwise rotation of the prey isocline in the Volterra model, as a result of adding the logistic self-limitation term to the Lotka-Volterra model.

fact, the Volterra model is characterized by a globally stable equilibrium point: any trajectory starting at positive prey and predator density will be attracted to this equilibrium. This should not be surprising, since we expect that any density dependence should contribute to the stability of the system.

The Rosenzweig-MacArthur Model The Volterra model is a step in the right direction (and it gets us away from the pathological neutral stability of the Lotka-Volterra model), but it does not go far enough. We also need to do something about the assumption of the linear functional response. Rosenzweig and MacArthur (1963) are generally credited with adding the assumption of the hyperbolic functional response to the Volterra model:

$$\frac{dN}{dt} = r_0 N \left(1 - \frac{N}{k} \right) - \frac{cNP}{d+N}$$

$$\frac{dP}{dt} = \chi \frac{cNP}{d+N} - \delta_0 P$$

(4.14)

This is perhaps the simplest model that can actually be applied to real-life systems. As a result, it has become something of a standard predator-prey model for resource-consumer interactions in theoretical ecology. Sometimes a different parameterization of the predator equation is used:

$$\frac{dP}{P\,dt} = \chi\left(\frac{cN}{d+N} - \mu\right) \qquad (4.15)$$

The ZPG consumption rate, μ, is the rate of prey consumption that an individual predator needs to survive and replace itself (see section 4.1.3). This parameterization is often useful in relating the Rosenzweig-MacArthur model to the data, since it is often easier to estimate μ from bioenergetic considerations than the predator death rate in the absence of prey, δ_0.

This basic model is capable of two nontrivial kinds of dynamic behaviors: a stable equilibrium and a limit cycle. Cycles occur for parameter values satisfying the inequality $d/k < (c - \mu)/(c + \mu)$. In general, decreasing the d/k ratio destabilizes dynamics (via the well-known mechanism of the **paradox of enrichment**). In addition, the model becomes more prone to cycle when μ is decreased in relation to c. Parameter r_0 does not affect qualitative stability, but affects quantitative aspects of dynamics: low values of r_0 result in long cycles with high amplitude, while high values produce shorter, less extreme oscillations.

The Rosenzweig-MacArthur model succeeds as the paradigmatic model for predator-prey interactions for two reasons. First, it fixes the worst failings of the Lotka-Volterra model. It does not satisfy all the postulates proposed in chapter 2 (there is no density dependence in the predator equation), but this can be considered a minor problem in applications where predators do not directly affect each other's performance, so that predator regulation occurs solely as a result of their running out of food. Second, the Rosenzweig-MacArthur model is capable of the complete spectrum of dynamical behaviors that can, in principle, characterize this kind of model (a system of two ordinary differential equations): a stable point equilibrium and stable limit cycles (and extinction of predator or both species, but these are not interesting outcomes from the point of view of population dynamics).

The Yodzis Model   If, instead of hyperbolic response, predators are characterized by a sigmoid functional response, then we have the following model

$$\frac{dN}{dt} = r_0 N \left(1 - \frac{N}{k}\right) - \frac{cN^2}{h^2 + N^2} P$$

$$\frac{dP}{dt} = \chi \frac{cN^2 P}{d^2 + N^2} - \delta_0 P \qquad (4.16)$$

I will name this model after Peter Yodzis, who extensively analyzed its dynamics (Yodzis 1989:84–99; this is a good nontechnical introduction to the mathematical analysis of predator-prey models), although he probably was not the first to propose it. Despite their rich diapason of behaviors, however, I do not recommend using equations (4.16) in practical applications. My main problem with this model is its assumption of the sigmoid functional response. Equations (4.16) model *specialist* predator-prey interaction, since in the absence of its primary prey ($N \to 0$) the predator would rapidly die out. However, it does not make sense for specialist predators to reduce their searching effort at low prey density, because they do not have an alternative prey to switch to. One might suggest that predators go into hibernation to wait for better times, but that is not what model (4.16) implies: although at low prey density predators stop hunting, they continue to die off at the maximum rate δ_0. To conclude, it is my opinion that sigmoid functional responses should be used only for generalist predators (whose numerical dynamics, therefore, should not change in response to variations in prey density).

The DeAngelis Model The next step in our consideration of the effect of various functional responses on predator-prey dynamics is the Beddington form that combines hyperbolic response with predator mutual interference:

$$\frac{dN}{dt} = r_0 N \left(1 - \frac{N}{k}\right) - \frac{cNP}{d + bP + N}$$

$$\frac{dP}{dt} = \chi \frac{cNP}{d + bP + N} - \delta_0 P \qquad (4.17)$$

This model was analyzed by DeAngelis et al. (1975; their equations 5 and 6 include density dependence in predator growth rate, but in the actual analysis they set this density dependence to zero). The DeAngelis model is dynamically very similar to the Rosenzweig-MacArthur model (to which it collapses as $b \to 0$), and has a similar spectrum of dynamics. However, mutual interference between consumers reduces predator killing efficiency at high predator densities, therefore imposing a positive feedback on predator density. This is a stabilizing influence, and in general the chance of oscillations in the DeAngelis model tends to be reduced (DeAngelis et al. 1975:888).

Setting parameter $d = 0$ in the DeAngelis model leads to the model analyzed by Arditi and Ginzburg (1989). Thus, the Arditi-Ginzburg model results from an incorporation of a ratio-dependent functional response into the Lotka-Volterra framework.

The Bazykin Model   Returning to the assumption of the hyperbolic functional response, let us consider the effect of adding self-limitation terms in the predator equation. Assuming linear density dependence, we have

$$\frac{dN}{dt} = r_0 N \left(1 - \frac{N}{k} \right) - \frac{cNP}{d + N}$$
$$\frac{dP}{dt} = \chi \frac{cNP}{d + N} - \delta_0 P - \delta_1 P^2 \tag{4.18}$$

This model was proposed by Bazykin (1974). The assumption of linear self-limitation is really the logistic term in disguise. To see this, let us assume that prey density is at the level where predator functional response is saturated. In other words, we are removing prey dependence from the second equation in (4.18) to highlight the self-limitation term:

$$\frac{dp}{dt} = \chi cP - \delta_0 P - \delta_1 P^2 = s_0 P \left(1 - \frac{P}{\kappa} \right)$$

We see that we end up with a logistic model, in which $s_0 = \chi c - \delta_0$ and $\kappa = (\chi c - \delta_0)/\delta_1$. One interpretation of κ may be as the greatest number of territories that can be fitted into a unit of space.

The Bazykin model is inherently more stable than the Rosenzweig-MacArthur model. In order for it to cycle, its parameters must satisfy all the conditions for cycles in the Rosenzweig-MacArthur model, and in addition the self-limitation coefficient, δ_1, must be weak enough.

Variable Territory Model   One possible biological mechanism that could lead to the self-limitation term in the Bazykin model is territoriality. Note, however, that the Bazykin model assumes a fixed territory size; in particular, it is independent of food availability. In situations where the size of territories is affected by food availability, we would need to use a somewhat different model, as the following argument shows. Assume that dominant individuals will defend an area that is sufficient to fulfill their energetic requirements. If q is the minimum amount of prey biomass required by an individual predator, then the territory size that needs to be defended for a particular value of prey biomass, N, is $T = q/N$. The predator carrying capacity is the number of territories that can be fitted within 1 unit of area, in other words, $\kappa = 1/T$. Thus, this argument suggests that predator carrying capacity should be directly proportional to current prey biomass, $\kappa = N/q$. Rewriting the Bazykin model in terms of s_0 and κ, and then substituting N/q, we obtain the following equation for consumer rate of change (Turchin and Batzli 2001):

$$\frac{dN}{dt} = r_0 N \left(1 - \frac{N}{k}\right) - \frac{cNP}{d + N}$$

$$\frac{dP}{dt} = \chi \frac{cNP}{d + N} - \delta_0 P - \frac{s_0 q}{N} P^2$$

$$(4.19)$$

I call this modification of the Bazykin model a *variable territory* model, because it is derived by assuming that the territory size changes in response to food availability.

4.2.2 Models Not Conforming to the LV Framework

Logistic Predation Abandoning the general LV framework, I now consider two models based on a direct generalization of the logistic equation. The first model was proposed by Leslie (1948):

$$\frac{dN}{dt} = r_0 N \left(1 - \frac{N}{k}\right) - aNP$$

$$\frac{dP}{dt} = s_0 P \left(1 - q \frac{P}{N}\right)$$

$$(4.20)$$

It retains the prey equation from the Volterra model, but for predators it assumes a logistic-like term, in which predator carrying capacity is directly proportional to prey density. Parameter s_0 is the intrinsic rate of predator increase. May (1974b) modified the Leslie model by assuming the hyperbolic functional response in the prey equation

$$
\begin{aligned}
\frac{dN}{dt} &= r_0 N \left(1 - \frac{N}{k} \right) - \frac{cNP}{d + N} \\
\frac{dP}{dt} &= s_0 P \left(1 - q\frac{P}{N} \right)
\end{aligned}
\tag{4.21}
$$

This model was later used by Tanner (1975) to investigate dynamics of mammalian predator-prey systems, and by Hanski et al. (1991) to model vole-weasel dynamics (see chapter 12).

One problem with the Leslie and May models is that they violate postulate 5, trophic coupling (Ginzburg 1998). This leads to some anomalies in their predictions. For example, the May model predicts that even at very low prey density, when the killing rate by an individual predator is essentially zero, predator populations can nevertheless increase, if predator/prey ratio is very small (that is, predator population is even smaller than prey). Clearly, this feature of the model violates the energetic principle—how can predator populations increase when individual predators are starving?

It is instructive to consider whether the logistic predation term may be derived as an approximation of some other, more mechanistic model. For example, the May model appears to be related to the variable territory model: if we approximate $cNP/(d + N)$ term in the predator equation of model (4.19) with cP (this is a reasonable approximation as long as $N \gg d$), then we obtain the May model. In other words, the May model can be justified as an approximation of the variable territory variant of the Bazykin model. We now see why the May model breaks down for low prey densities: the approximation $cN/(d + N) \approx c$ cannot hold.

The logic that connects the May and the variable territory models can be pushed one step further. Recollect that the May model can be derived by assuming that predator consumption rate is located within the saturated region; in other words, it does not depend on prey density. The functional response in the May model, however, is still modeled by the hyperbolic function. To harmonize the killing and consumption rate, therefore, we might consider the model in which the functional response is constant:

$$\frac{dN}{dt} = r_0 N \left(1 - \frac{N}{k} \right) - cP$$

$$\frac{dP}{dt} = s_0 P \left(1 - q \frac{P}{N} \right)$$

(4.22)

Here the predator intrinsic rate of increase, s_0, and the (constant) consumption rate, c, are connected: $s_0 = \chi c - \delta_0$, where δ_0 is the predator density-independent death rate in the absence of prey consumption, and χ is the conversion efficiency, as usual. I will call this the *Eberhardt model* since a discretized version of equations (4.22) was used by Eberhardt (1997) in an analysis of wolf-ungulate dynamics. This is not a good model for investigating population cycles, because prey density would periodically achieve low values, where approximations on which the model is based cannot hold. On the other hand, it may be a reasonable model for small-scale fluctuations in the vicinity of the equilibrium (providing that $d \ll k$). It certainly has the advantage of great simplicity, and its parameters can be approximated from regularly available life-history information. In particular, one great advantage of model (4.22) is that it does not require estimation of the search rate, a (or, alternatively, the half-saturation constant, d).

Prey-Dependent Consumption The final model that I consider is the one based on the idea explored in section 4.1.3 that predators may consume an increasingly smaller proportion of killed prey as prey density increases. To investigate the effect of this assumption on model stability, I use the extreme form of prey dependence in predator consumption rate that leads to the linear functional response and hyperbolic numerical response:

$$\frac{dN}{dt} = r_0 N \left(1 - \frac{N}{k} \right) - aNP$$

$$\frac{dP}{dt} = \chi \frac{aNP}{1 + ahN} - \delta_0 P$$

(4.23)

A little algebra shows that this model has the same isocline structure as the Volterra model. Thus, this model is characterized by a globally stable point equilibrium for all values of parameters. Because this model is intermediate between the Volterra and the Rosenzweig-MacArthur, this interesting result suggests that the destabilizing feature in the RM model is not the limited ability of predators to increase numerically, but their limited ability to kill prey. This conclusion follows from the observation that leaving linear functional response in the prey equation prevents the model from being able to exhibit stable limit cycles.

4.2.3   Anatomy of a Predator-Prey Cycle

The population interaction between a specialist predator and its prey has certain stereotypical features that affect the topology of the predator-prey cycle. More specifically, predator and prey cyclic peaks tend to be characterized by different shapes—"sharp" versus "blunt," and this feature can be a useful diagnostic in time-series analysis (see section 7.1.3). In many trophic systems, prey reproduce faster than predators. Faster intrinsic rate of increase of prey means that during the initial phase of the cycle, when both prey and predators are at low densities, prey numbers easily outgrow predator numbers. Prey numbers then approach some population "ceiling," typically imposed either by food availability or by prey social interactions (or both). Meanwhile predators, who have plenty of food, increase more slowly. As a result, prey populations remain at peak densities for some extended time, depending primarily on predator intrinsic population growth rate. Prey peaks, thus, have a blunt shape.

By contrast, predator peaks are sharp. When predators eventually increase to the point where they begin to affect prey density, the whole system enters rather abruptly a new regime, in which there are too many predators chasing too few prey. As prey begin collapsing, predator density also starts declining. The decline phase continues until predator densities drop to the point where they cannot affect prey anymore. Thus, from the point of view of a predator in an oscillatory trophic system, there are only two cyclic phases: one of prey abundance and few competitors, and the other of prey scarcity and many competitors. The transitions between the two phases are quite abrupt, and the resulting dynamics are of the saw-toothed pattern.

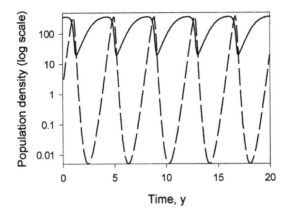

FIGURE 4.3. Peak shapes in a predator-prey system. Solid curve: prey density (N). Broken curve: predator density (P).

From the prey point of view, a cycle has three phases. First, there is plenty of food and no predators. This phase smoothly grades into the second one, when predators are still absent, but there is high competition for food or space. Finally, there is an abrupt transition to the phase of high predation mortality. The transition between the third and first phases is also fairly abrupt. The typical topology of predator-prey cycles is exemplified by numerical solutions of the Rosenzweig-MacArthur model (figure 4.3).

The topological pattern in figure 4.3 does not depend on the details of the Rosenzweig-MacArthur model, and is exhibited by other predator-prey models. In practice, however, it is not always going to be easy to detect that prey peaks are blunter than predator peaks. First, oscillations must be of high enough amplitude to reveal the pattern. Second, predators must have a lower r_0 than prey, so that the predator increase phase is long enough for prey to reach the plateau imposed by first-order regulatory factors (competition for food or space). If predators increase too fast, then the plateau phase in the prey trajectory may be hard to detect. Finally, population trajectories must be sampled finely enough. Obviously, if we have only two points per oscillation, all we would see would be an alternation between high and low values. Thus, a saw-toothed pattern may be consistent with either prey or predator dynamics in a trophic interactions model. The reverse is not true, however: in a two-species oscillatory system, predators cannot have blunt peaks. An imposition of a first-order regulatory mechanism strong enough to slow the approach of predator density to the peak, necessary for the blunt-peak pattern, has a side effect of stabilizing the predator-prey system.

4.2.4   Generalist Predators

As I discussed in section 4.1.2, there are two main approaches to modeling the effect of generalist predators. The more realistic, but also more complicated, approach is to model explicitly densities of alternative prey species and behavioral mechanisms that generalists use to switch between them. The second, much simpler, approach is to assume that generalist predators exhibit no numerical response to variations in prey density (i.e., predators are an exogenous factor in the system). This is the approach that I review here.

Since the numerical response is set to zero, and by convention in this book I do not consider explicitly spatial aggregation responses, the only choice we need to make in the generalist predation model is the form of the functional response. There are two sensible options: hyperbolic versus sigmoid responses (section 4.1). A simple model employing the hyperbolic response and assuming that prey population growth in the absence of predators is logistic is

$$\frac{dN}{dt} = rN\left(1 - \frac{N}{k}\right) - \frac{gN}{d + N} \qquad (4.24)$$

Note that the parameter g is the total killing rate by generalist predators. That is, $g = cP$, where c is the saturation killing rate by individual predator and P is the density of generalist predators. Since both c and P are assumed to be constant, so is g. There is no equation for predator density, since it is constant (at least in the deterministic version of the model; we can make parameter g a random function of time to model stochastic variation in generalist predator density).

The effect of generalist predators on prey dynamics can be easily understood by plotting together two per capita rates: that of prey growth and that of prey death due to predators (figure 4.4a). Prey density will increase if the dashed curve (predation rate) is below the solid line (growth rate in the absence of predators), and decline

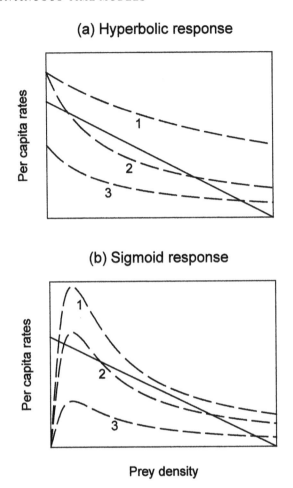

FIGURE 4.4. Effect of generalist predators on prey dynamics: (a) hyperbolic functional response; (b) sigmoid functional response. Solid lines: per capita growth rates of prey population in the absence of predators. Dashed curves: per capita death rate of prey as a result of predation. Numbers correspond to various cases discussed in the text.

otherwise. Equilibrium points occur where the two curves intersect. There are three possible configurations. In case 1, prey death rate is always greater than its reproduction rate, and therefore prey population will go extinct. In case 2, there are two equilibria. The upper one is stable, the lower one unstable. Thus, if prey density starts high enough (above the lower unstable equilibrium), then it will approach the upper equilibrium. However, if for some reason prey density decreases below the lower equilibrium, prey goes extinct. In case 3, only one equilibrium (upper and stable) is present. In sum, there are two generic outcomes: either generalist predators drive prey extinct, or prey exists at an upper stable equilibrium, where predator impact is slight. In this model, predators cannot control prey at a low stable density.

Model (4.24), unlike the Ludwig model to be considered next, is not often utilized by population ecologists, probably for two reasons. First, we typically associate the sigmoid response with generalist predators, rather than the hyperbolic one as assumed by this model (but see a note on this in the next paragraph). Second, the model predicts either a slight effect of predation on prey equilibrium densities, or an unstable (and therefore nonstationary) situation in which prey go extinct (or both). Neither of these outcomes is terribly interesting in practical applications. However, model (4.24) is important from the theoretical point of view, because it elucidates the population-level consequences of generalist predation with the hyperbolic functional response.

The Ludwig Model   The assumption of the sigmoid functional response leads to the following model:

$$\frac{dN}{dt} = rN\left(1 - \frac{N}{k}\right) - \frac{gN^2}{h^2 + N^2} \qquad (4.25)$$

This model, analyzed by Ludwig et al. (1978), can also be characterized by three possible configurations of per capita growth and death rates (figure 4.4b). In case 1, only the lower equilibrium is present, which is stable unlike in model (4.24). In case 2, we have the **metastable** situation, in which prey density may be attracted to either the upper or the lower stable equilibrium, depending on the initial conditions. The intermediate equilibrium is unstable, and serves as the separatrix between the basins of attraction of the two stable equilibria. In case 3, only the upper (stable) equilibrium is present. Thus, the qualitative difference between models (4.24) and (4.25) is that in the latter predators can control prey at a low density without driving prey to extinction. On biological grounds, we would expect that the model with the sigmoid response would be more likely to describe any particular generalist predator-prey system, because there is no reason for generalists to waste time searching for very rare prey. Unlike specialist predators, they are likely to have other prey species to switch to. This is the basis for associating the hyperbolic response with specialists and the sigmoid one with generalists. There is, however, one important exception to this rule. It occurs when alternative prey are found in the same spatial locations and habitats as focal prey. In this case, generalist predators will continue encountering focal prey even if those prey are very rare, simply as a side effect of searching for other more common prey species. Assuming that predators do not form a search image, then encounters are likely to result in prey death. In this case, the predator's functional response is likely to be of the hyperbolic kind. The population consequence of this biological feature is that prey will be driven to extinction (at least, locally) if they become rare.

The Hanski Model Finally, it is very straightforward to combine generalist and specialist predators in the same model, assuming that these predators do not interfere with each other. For example, let us assume that specialist predator-prey interaction is described by the May model, while the generalist predators are characterized by the sigmoid functional response:

$$
\begin{aligned}
\frac{dN}{dt} &= r_0 N \left(1 - \frac{N}{k} \right) - \frac{cNP}{d + N} - \frac{gN^2}{h^2 + N^2} \\
\frac{dP}{dt} &= s_0 P \left(1 - q \frac{P}{N} \right)
\end{aligned}
\tag{4.26}
$$

The conceptual basis for this model was developed by Hanski et al. (1991) in the context of small rodent cycles. This model presents a very interesting paradigm for population dynamics of prey attacked by a community of specialist and generalist predators. In a community where generalist predators are rare, the dynamics of the Hanski model will be essentially the same as those predicted by the May model: either stable limit cycles or a stable point equilibrium. Let us assume that the parameters of the interaction between specialists and prey are such that the dynamics of this model are fairly deep in the cyclic region, and examine the effect of "cranking up" generalist predation pressure. At first, as generalist predators increase, they will depress the prey equilibrium in the absence of specialists only slightly (this is the situation associated with case 3 in figure 4.4b). Then, for some (rather narrow) range of generalist predator densities, we obtain case 2 with its three equilibria. Increasing generalist numbers even further, we move to case 1, with its low stable equilibrium. Recollect that the May model belongs to the class of models that are characterized by the paradox of enrichment. In other words, as prey equilibrium density in the absence of specialist predators decreases, the model dynamics switch from stable limit cycles to a stable point equilibrium. Thus, in our scenario of both generalist and specialist predators present, case 3 corresponds to cycles, while case 1 is likely to lead to stability (observe that prey equilibrium is at a very low prey density: this is likely to lead to stable dynamics). What is particularly remarkable about this biological mechanism for the stabilization of specialist predator-prey dynamics is that rather minor changes in generalist predator densities are needed to transit between cases 1 and 3 (figure 4.4b). Note that this is a generic result, and a very similar scenario would obtain if we add generalist predators to the Rosenzweig-MacArthur model (or, in fact, to any model that enjoys the paradox of enrichment). Thus, smooth geographic gradients in generalist predation pressure may translate into very abrupt changes in population dynamics. This idea will be pursued within the context of a specific case study, field voles in Fennoscandia (see chapter 12).

4.3 DISCRETE-TIME MODELS: PARASITOIDS

In most ways parasitoid-host models are very similar to predator-prey ones. However, parasitoid models are usually formulated as discrete-time equations, while models of predation often utilize the continuous-time framework. The focus of this section (in contrast to the preceeding one), thus, will be on discrete-time models.

4.3.1 Functional and Numerical Responses

Functional responses of parasitoids tend to be modeled using the same logic as that reviewed in section 4.1.1. The most important difference is that parasitoid models are discrete, and we need to replace the instantaneous rate at which prey is killed by an integral of this rate over the period of time during which parasitoids search for prey. Derivation of various functional responses for discrete models is given by Hassell (1978: appendix I); here I simply give the resulting formulas. The discrete-time form of the linear functional response is

$$N_{\text{att}} = N_t(1 - \exp[-aP_t])$$ (4.27)

Here N_{att} is the number of hosts attacked (killed), N_t and P_t are the numbers of hosts and parasitoids in generation t, and a is the attack rate (Hassell's formulation also involves T, the total time available for parasitoids to search for and kill hosts, but we can set T to 1 without any loss of generality).

The expression for the hyperbolic functional response is derived using the same logic:

$$N_{\text{att}} = N_t\left(1 - \exp\left[-\frac{aP_t}{1 + ahN_t}\right]\right)$$ (4.28)

The perceptive reader will by now realize that equations (4.27) and (4.28) are derived using the second approach for discretizing continuous models in section 3.1.2. Recollect that this approach works by assuming that the various conditions stay approximately constant during the period corresponding to each time step. In terms of equation (4.28) it means that we have to assume that parasitoid and host numbers do not change during the time step (one generation), and are approximately P_t and H_t, the numbers at the beginning of the time step. This is clearly a great oversimplification. Parasitoid numbers should change during the time step as a result of death (not all parasitoids will survive to the end of the season). However, if parasitoid death rate is independent of host numbers, then we can include it by multiplying a by some constant (less than 1). Host densities will also decline during the course of the season, particularly as a result of parasitism. However, if parasitized hosts continue to be available to repeated parasitism, then we can assume that N_t is approximately constant. This assumption, however, really breaks down if we consider predators instead of parasitoids, because predators remove prey (by consuming it). Thus, prey density is constantly declining as a result of predation. Hassell (1978: appendix I) shows that if prey are removed as a result of parasitism or predation, then the expression for the discrete version of the hyperbolic response is

$$N_{att} = N_t \left\{ 1 - \exp\left[-aP_t\left(1 - h\frac{N_{att}}{P_t} \right) \right] \right\} \qquad (4.29)$$

The problem is that the quantity that we need, N_{att}, enters this formula in an implicit way, and to determine it we have to solve equation (4.29) numerically. Thus, two alternative approaches are often followed: either use the parasitism version (4.28) because of its simplicity, and hope for the best, or construct a two-scale model. At the within-generation timescale, the process of predator search is modeled using standard continuous-time predator-prey models (excluding prey and predator reproduction terms). At the between-generations timescale, the model uses the number of prey killed to calculate the prey and predator numbers in the next generation.

Turning now to the numerical response, we note that the connection between prey killed and the production of new offspring tends to be more direct in parasitoids. While predators typically have to use some part of consumed prey biomass to sustain their life processes, and only surplus goes to production of new predators, in many parasitoids the number of new parasitoids equals the number of hosts parasitized. This is true for the so-called solitary parasitoids, or the ones that deposit a single egg onto a host (and if more than one egg is deposited, only one completes development). In "gregarious" parasitoids that lay multiple eggs on a single host, the number of new parasitoids will be directly proportional to the number of hosts parasitized (although the constant of proportionality could conceivably vary in time, if the average host size changes and the number of parasitoids that can hatch is a function of host size). A final complication is host feeding by parasitoids. In the final analysis, therefore, the general numerical response in parasitoid-host models will be of the following form:

$$P_{t+1} = \chi(N_{att} - N_{hf}) \qquad (4.30)$$

where P_{t+1} is the number of parasitoids in the next generation, N_{att} is the number of hosts killed (attacked), N_{hf} those hosts that were eaten by the adult parasitoid (host feeding), and χ is the average number of new parasitoids hatching from a single host. Comparing this model with equation (4.10) in section 4.1.3, we see that they follow essentially the same logic.

4.3.2 Dynamical Models

The basis of parasitoid-host models is provided by the Nicholson-Bailey model:

$$N_{t+1} = N_t \exp[r_0 - aP_t]$$
$$P_{t+1} = N_t(1 - \exp[-aP_t])$$

(4.31)

(a is the parasitoid search rate). As is well known, the Nicholson-Bailey model is characterized by unstable diverging oscillations (although not a very realistic feature for field applications, it is apparently possible to mimic such dynamics in the lab; see Burnett 1958). Whole books have been written on the topic of how including greater biological realism can stabilize the Nicholson-Bailey model (e.g., Hassell 1978). Perhaps the easiest way to do so, and certainly the most obvious from the point of view of chapter 2, is to add self-limitation to the prey equation. Beddington et al. (1976a) combined the Nicholson-Bailey and Ricker models, yielding

$$N_{t+1} = N_t \exp\left[r_0\left(1 - \frac{N_t}{k}\right) - aP_t\right]$$
$$P_{t+1} = N_t(1 - \exp[-aP_t])$$

(4.32)

This model, depending on parameter values, can generate a great variety of dynamical behaviors, including stable points, stable limit cycles, quasiperiodicity, and chaos (Beddington et al. 1976a). This is one of the reasons that make the Beddington model a good paradigm for host-parasitoid interactions.

Adding other realistic features to the basic host-parasitoid model is straightforward. Instead of running through the whole sequence of models that include a variety of possible functional responses, I give here the equations for the Beddington model with the Beddington-type functional response (yielding the "Beddington²" model, if I may be forgiven this bit of creative nomenclature):

$$N_{t+1} = N_t \exp\left[r_0\left(1 - \frac{N_t}{k}\right) - \frac{aP_t}{1 + ahN_t + awP_t} \right]$$

$$P_{t+1} = \chi N_t \left(1 - \exp\left[-\frac{aP_t}{1 + ahN_t + awP_t}\right]\right)$$

(4.33)

(h and w are handling and wasted times defined in section 4.1.1). Parameter χ is the average number of next-generation parasitoids produced per parasitized host. Additionally, the probability of survival through winter can be folded into this parameter (so that χ conceivably could be less than 1). Models with a pure hyperbolic or predator-interference functional responses may be obtained from (4.33) simply by setting w or h to 0. The dynamical consequences of these various assumptions are largely analogous to continuous-time models (host self-limitation and parasitoid interference are stabilizing, while hyperbolic functional response is a destabilizing feature).

4.4   GRAZING SYSTEMS

According to the functional classification of trophic interactions (see the chapter opening preceding section 4.1), a *grazer* is a consumer that scores low on both *intimacy* and *lethality* scales. In other words, grazers rarely kill the resource individuals on which they feed, and throughout their life grazers will take a bite from many resource individuals. Most grazers are *herbivores*, that is, consumers of primary producers, usually plants. However, not all herbivores are grazers. Most insect herbivores—for example, aphids—are functional parasites (because they score high on the intimacy scale). As we shall see, both the nature of grazing and the nature of autotroph growth affect how we should model grazer-vegetation systems. This section reviews models of grazer-vegetation interaction (because grazing carnivores are relatively rare, I do not devote space to them here). My main focus in sections 4.4.1–4.4.3 is on the interplay between herbivory and *quantity* of vegetation, assuming that plant *quality* does not change dynamically. In the last section (4.4.4), I shift gears and consider models in which plant quantity remains the same but quality is a dynamical variable.

4.4.1 Grazer's Functional Response

Whereas predators tend to kill their prey, and thus it is natural to measure a predator's functional response in the units of prey individuals, grazers consume only part of a resource individual. Accordingly, a grazer's functional response is typically measured in units of resource biomass removed. The second important difference between predators and grazers is that grazers often specialize on particular organs or tissues of their resources. The distinction between edible versus inedible biomass is especially stark where inedible biomass is simply inaccessible to the herbivore. For example, aboveground and belowground plant biomasses serve as resource bases for two completely separate grazing communities. Furthermore, different plant tissues are characterized by widely different nutrient content, digestibility, temporal availability, and antiherbivore defenses, so there are typically different herbivore guilds that specialize as leaf chewers, sap suckers, twig browsers, stem borers, and so on. The importance of this specialization for grazing functional responses is that we need to measure the available vegetation biomass, V, appropriately: only that which constitutes the true resource for the herbivore. This often creates great practical difficulties in defining what the appropriate biomass is. The task is further complicated because two seemingly similar patches of vegetation may differ in nutritional quality, and structural or spatial arrangement of plants, and therefore constitute two different food availabilities for herbivores. As a result, quantifying functional responses of herbivores is a much more challenging task than doing the same for predators.

It may also seem that the hyperbolic functional response, paradigmatic of predators' killing rate, may not apply to herbivores. The world is green, and therefore herbivores should have no difficulty filling their guts. This argument would suggest that herbivores should be characterized by constant functional responses. In fact, numerous studies show that herbivores are characterized by the hyperbolic response, and that herbivore intake rate can decline well below the saturation level at the lower end of the natural variation in food availability (e.g., Spalinger and Hobbs 1992: figure 2).

Theoretical arguments also suggest that the hyperbolic response should be a reasonable approximation of grazer foraging, although, as we shall see, food availability may have to be measured appropriately. Herbivore foraging is affected by food distribution at several spatial scales (Spalinger and Hobbs 1992). At the smallest scale, food comes in different bite sizes. Bites are typically arranged in patches. Finally, patches are distributed within a landscape. Spalinger and Hobbs (1992) consider three scenarios. In the first case, bites are dispersed within a patch, requiring that the herbivore spend some time traveling between them. Additionally, bites are not readily apparent, so at low bite density the herbivore's encounter rate with them is directly proportional to the bite density within patch. Clearly, this description closely matches the typical derivation of the predator's functional response, and leads to the hyperbolic response, as Spalinger and Hobbs (1992) show. In the second case, bites are also dispersed within the patch, but are very apparent, so that herbivores travel directly from one bite to the next nearest one. Spalinger and Hobbs (1992) show that the functional response of the herbivore in this scenario will have the following form:

$$f(V) = \frac{a\sqrt{V}}{b + \sqrt{V}} \tag{4.34}$$

where V is the density of bites and a and b are some parameters. Equation (4.34) is similar to the hyperbolic form, except that vegetation density appears in it in the square-root transformed form. Spalinger and Hobbs note that this model cannot be a good description of the foraging process at very low V, when average distance between food items is much larger than the perception range of the herbivore. They propose a piecewise form that is hyperbolic at low V and then shifts to the form (4.34) for medium V, and finally hits the constant "ceiling" (figure 4.5). In my opinion, in dynamical models of plant-herbivore interactions, we might simply approximate this complex curve with the hyperbolic functional response.

The third case considered by Spalinger and Hobbs is the one where bites are concentrated and apparent. In this situation, herbivores will consume bites at a constant rate, determined by their ability to process them. Herbivore food intake, then, will be determined not by

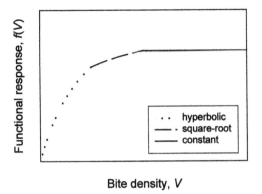

Bite density, *V*

FIGURE 4.5. The composite functional response proposed by Spalinger and Hobbs (1992: figure 1C).

bite density (which is effectively assumed to be infinite), but by bite size. Spalinger and Hobbs derive the following formula for herbivore intake:

$$f(S) = \frac{aS}{b + S} \qquad (4.35)$$

where S is the bite size and a and b are some constants. We see that we again obtain a hyperbolic response, albeit a function of a different independent variable.

The consequences of this scenario for long-term population dynamics of herbivores and vegetation will depend, in my opinion, on the landscape-level arrangement of patches. Two extreme cases can be distinguished. In the first case, the landscape consists of a number of discrete patches of the same size, characterized by the same bite size. While in the food patch, an individual herbivore feeds at the maximum rate (determined by the bite size in the patch), and leaves the patch once it is depleted. The important variable in this case is the density of patches within the landscape, since that will affect the travel time between patches. Thus, the natural approach to modeling herbivore functional response in such a situation is to treat each patch as an individual resource item. In effect, we shift to a larger spatiotemporal scale: measuring resource availability at the landscape level (patch density) and herbivore consumption over long enough periods that cover consumption of multiple patches, as well as travel between them. It is clear that this idealized situation should again lead to the usual hyperbolic functional response: as patch density within the landscape increases, herbivores spend less time traveling, and their long-term intake rate will eventually approach the maximum possible. On the other hand, the fewer patches there are, the more time a herbivore needs to travel between them. Thus, at the lower end of the patch density spectrum, long-term herbivore intake will be directly proportional to patch density. In turn, patch density will be determined by the balance between some process creating new patches, and herbivores destroying the existing ones.

The other extreme is when plant biomass is evenly spread through the landscape. In effect, the whole landscape is one huge patch. In this case, the relevant variable is how much biomass can be taken by a herbivore in a single bite. For example, consider a grazer within a grass sward. If the density of grass blades is constant, then the amount of grass biomass per unit of area will be determined by the average length of grass blades. The bite size will also be directly proportional to blade length. Thus, at very low biomass density, the larger the biomass, the greater the bite that a herbivore can take, suggesting that the functional response will be linear. As we increase grass length, however, at some point herbivores will not be limited by the bite size, and will be feeding at the maximum (saturation) rate. Grass length (and therefore bite size) will be determined by the balance between plant regrowth rate, and herbivore cropping rate.

The point of the preceding discussion is that the hyperbolic functional response provides a reasonable starting point for modeling grazer foraging. Spatial arrangements of plants can influence herbivores in a variety of complex ways, but we can reduce this complexity to two limiting situations. In the first, food comes in discrete packages—which could be single bites or whole patches—and grazer foraging is not conceptually different from what predators do. In the other limit, the relevant variable is the average bite size, and how it may be decreased by herbivore foraging. In both cases, the relevant variable is average plant biomass density, although in the first case it is related to the average density of bites or patches, while in the second case it is related to the average bite size. And, finally, in both cases herbivore functional response is hyperbolic. This is, perhaps, as much as one can say generally, and more complex functional response curves would depend on the details of the empirical system that is modeled.

4.4.2 Dynamics of Vegetation Regrowth

As I mentioned above, one distinctive feature that distinguishes grazers from predators is the tendency of grazers to specialize on attacking only some portion of vegetation biomass. Simple theoretical models of herbivore-plant interactions (Caughley and Lawton 1981; Crawley 1983) implicitly ignore this feature, by employing the logistic model for vegetation regrowth after herbivory:

$$\frac{dV}{dt} = v_0 V \left(1 - \frac{V}{m}\right) \tag{4.36}$$

where V is the vegetation biomass (per unit of area), and v_0 and m are the intrinsic (per capita) rate of plant growth and the maximum biomass approached in the absence of herbivory, "carrying capacity." The logistic implies that when vegetation biomass V is near 0, its growth rate is an accelerating function of V that reaches its maximum at $m/2$, and then slows to 0 as V approaches m (figure 4.6). The logic underlying the logistic model is that the more plant biomass is present, the more solar energy it can fix, and the faster it will grow (until it starts approaching the limit, m). The problem with this logic, when applied to herbivore-vegetation systems, can be illustrated with the following example. Suppose a grass-eating grazer reduces grass biomass in a savanna to practically zero. Do we expect grass to regrow logistically, with an initial acceleration phase? No, because the total biomass of grass has been hardly affected. Typically, at least 80% (and usually close to 90%) of graminoid—grasses and sedges—biomass is underground (Wielgolaski 1975), where it is protected from aboveground herbivory. What is likely to happen, therefore, is that plants will mobilize nutrients from belowground storage to fuel aboveground growth. If plants allocate a constant amount of energy/nutrients for regrowth, then the initial regrowth pattern will be linear, eventually saturating to the maximum standing biomass, m (figure 4.6).

This informal argument can be made more precise with the following simple model. Let A be the aboveground biomass density, and B the corresponding belowground biomass. Belowground biomass is

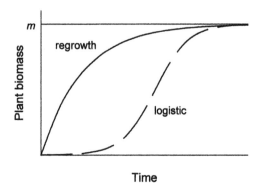

FIGURE 4.6. Temporal dynamics of vegetation obeying the logistic (broken curve) and regrowth (solid curve) equations.

increased by moving some portion of photosynthates down. Assuming that photosynthesis rate is directly proportional to the amount of aboveground biomass, and that a constant proportion of fixed energy is transferred belowground, the rate at which B increases is therefore sA, where s is a constant of proportionality. Similarly, some proportion of belowground energy is mobilized for the growth of aboveground biomass. Thus, when A is near 0, it will grow at the rate cB. However, as A approaches the maximum possible standing biomass, m, less energy should be mobilized for growth from belowground. Accordingly, I assume that A increases as $cB(1 - A/m)$. In addition, aboveground biomass will increase because a portion of energy fixed aboveground will be allocated to growth. This term is $rA(1 - A/m)$, where r reflects the rate at which energy is fixed by A and the proportion of the fixed energy that is allocated to growth when A is near 0. The term $(1 - A/m)$ governs how the proportion allocated to aboveground growth declines as A approaches m. Finally, I assume that belowground biomass is degraded at the rate d. The resulting model is

$$\frac{dB}{dt} = sA - cB\left(1 - \frac{A}{m}\right) - dB$$

$$\frac{dA}{dt} = (cB + rA)\left(1 - \frac{A}{m}\right)$$

(4.37)

At the equilibrium, $\widehat{A} = m$ and $\widehat{B} = sm/d$; that is, the aboveground biomass approaches the vegetation carrying capacity, and the belowground biomass reaches the equilibrium determined by the balance of decay rate and transport of fixed energy from above ground.

To simplify this model even further, let us suppose that B is well buffered against fluctuations of aboveground biomass, so that we can set it to some constant. Furthermore, since $B \gg A$, the term $(cB + rA) \approx cB = u_0$, a constant ($u_0$ is interpreted as the initial regrowth rate, when A is near 0). Then, the second equation in (4.37) simplifies to

$$\frac{dA}{dt} = u_0 \left(1 - \frac{A}{m} \right) \qquad (4.38)$$

where I replaced A with V (since we are not keeping track of the belowground biomass anymore). This model was proposed by Turchin and Batzli (2001), who called it the *linear initial regrowth* model, or the *regrowth model* in short. This equation has been previously used in models of nutrient dynamics in a chemostat (see, e.g., Edelstein-Keshet 1988:121), as well as in theoretical treatments of species competing for "abiotically" growing resources (MacArthur 1972; Schoener 1976; Abrams 1977; Gurney and Nisbet 1998). In contrast to the logistic, equation (4.38) implies no acceleration period; instead, when V is low it increases linearly at the maximum rate, u_0, and gradually slows to 0 as V approaches m (figure 4.6).

The logistic and regrowth models are oversimplifications of reality, and it is best to think of them as ideal cases, rather than representing the growth dynamics of actual plants. In fact, the logistic model is the limiting case of (4.37) for $B \to 0$. As noted above, theoretical population ecologists interested in the dynamics of herbivory have historically represented plant dynamics with the logistic model. By contrast, both empirical and theoretical ecosystem ecologists interested in vegetation dynamics developed models that imply the regrowth equation (see equation 14 in Parton et al. 1993; equations 8.1 and 9.5 in Ågren and Bosatta 1996). For example, in the Century model of grassland primary productivity, aboveground production rate in the beginning of the season is not affected by accumulating biomass, and foliage initially grows linearly (Parton et al. 1993). As the season progresses, growth slows down and eventually stops as a result of several processes (e.g., increased shading, shoot death, and depletion of nutrients in the soil). Clearly, the implied growth of aboveground biomass is of the regrowth type, with m a phenomenological parameter reflecting the combined action of several mechanistic processes.

Depending on the characteristics of the empirical system, plant growth can be represented with one or the other simple model. For example, in the arctic systems where Norwegian or brown lemmings (*Lemmus* spp.) are the dominant herbivore, different types of vegetation are characterized by different types of regrowth dynamics: green mosses may be better described by the logistic because nearly all of their living biomass is accessible to herbivores, while many graminoids (grasses and sedges), in which 80–90% of biomass is underground, may be better described by the regrowth equation. On the other hand, even graminoid dynamics may be better described by the logistic if there is extensive damage to their root systems resulting from a herbivore outbreak. For example, at high population densities during the spring thaw brown lemmings grub for rhizomes (Pitelka 1957), and root voles eat rhizomes of graminoids during winter (Tast 1974).

4.4.3   Dynamics of Grazer-Vegetation Interactions

We expect that the primary productivity of the plant community would be one of the most important factors affecting grazer-vegetation dynamics. As we shall see in a minute, this is indeed the case. However, the qualitative choice of the model for vegetation growth dynamics (logistic versus regrowth) has very profound effects on the resulting dynamics. Taking first the case of logistically growing vegetation, combining it with the hyperbolic functional response by grazers as the simple but reasonable assumption (section 4.4.1), and assuming no density-dependent interactions between grazers, we obtain the *Rosenzweig-MacArthur model* (section 4.2.1):

$$\frac{dV}{dt} = v_0 V \left(1 - \frac{V}{m} \right) - \frac{aVN}{b + V}$$

$$\frac{dN}{dt} = \xi N \left(\frac{cV}{d + V} - \eta \right)$$

(4.39)

where dynamic variables V and N are biomass density of plants and herbivore density, respectively. Recollect that the stability of this model is determined by the relationship between b/m and $(a - \eta)/(a + \eta)$. The second quantity is completely determined by the characteristics of the grazer. Plant productivity affects only the first quantity, via m, the equilibrium standing crop biomass approached in the absence of herbivory. Thus, the dynamics of herbivore-plant systems obeying the Rosenzweig-MacArthur model will suffer from the "paradox of enrichment": as plant standing biomass is increased, the dynamics of the system become increasingly less stable. In fact, the Rosenzweig-MacArthur model is extremely easy to destabilize, and it is likely to be in the oscillatory region for almost any parameter values that would characterize rodent-vegetation systems (Turchin and Batzli 2001), or indeed any mammalian grazing system. The logic underlying this insight is readily apparent in the condition of stability for the Rosenzweig-MacArthur model $b/m > (a - \eta)/(a + \eta)$. In mammalian grazers, η, the consumption rate needed to sustain and replace one herbivore, is likely to be not much less than a, the maximum rate of vegetation consumption, because by far the greatest proportion of ingested energy is used by mammals for thermoregulation (mammalian secondary productivity rarely exceeds a few percentage points). Taking a rather low value of $\eta = a/2$, we observe that the ratio $(a - \eta)/(a + \eta)$ should be at least $1/3$. The ratio b/m, on the other hand, is likely to be less than $1/3$. Remember that b is the half-saturation constant, that is, vegetation biomass density at which herbivore consumption rate is half of the maximum. Unless we deal with a highly unproductive system (e.g., arctic desert), we would expect that herbivore functional response would saturate well before vegetation density reaches m, implying that b is an order of magnitude less than m. Assuming that $b = m/10$, we see that the stability condition is easily violated for mammalian grazing systems: $b/m = 1/10 < 1/3 = (a - \eta)/(a + \eta)$.

The inescapable conclusion is that were real-world mammalian grazing system to obey the Rosenzweig-MacArthur model, then all but the least productive systems would exhibit oscillations. Furthermore, these oscillations would likely be of very great amplitude, because once the Rosenzweig-MacArthur system enters the oscillatory regime, the parameter range for which oscillations are characterized by "reasonable" amplitude (e.g., less than three orders of magnitude) is exceedingly narrow. Of course, the Rosenzweig-MacArthur model is not a reasonable description of most herbivore-vegetation systems, as we shall see in part III (see particularly chapters 12–14).

As we saw in section 4.2.1, adding consumer self-limitation terms (e.g., the Bazykin or variable-territory models) or employing a consumer-dependent functional response tends to result in models characterized by greater stability than the Rosenzweig-MacArthur one. It then becomes an empirical question, to be answered in any specific case study, whether herbivore self-limitation (or herbivore interference in the functional response) is strong enough to stabilize the inherently oscillatory dynamics of these generalized Lotka-Volterra models.

The models based on the Lotka-Volterra framework utilize the logistic equation for vegetation growth dynamics. However, as I argued in section 4.4.2, in most situations where only a part of vegetation biomass is accessible to herbivores, a more appropriate simplification is not the logistic but the regrowth equation. Replacing the logistic growth in the vegetation equation of the Rosenzweig-MacArthur model with the regrowth term has a profound effect on model dynamics. Model equations for this *herbivory-regrowth model* are (Turchin and Batzli 2001)

$$
\begin{aligned}
\frac{dV}{dt} &= u_0 \left(1 - \frac{V}{m} \right) - \frac{aVN}{b+V} \\
\frac{dN}{dt} &= \xi N \left(\frac{aV}{b+V} - \eta \right)
\end{aligned}
\tag{4.40}
$$

Here V is the vegetation biomass density, N is the population density of grazers, u_0 is the (linear) regrowth rate of vegetation at $V = 0$, and other parameters are as in the Rosenzweig-MacArthur model. It turns out that model (4.40) is globally stable for all values of its parameters. The key difference between this model and the Rosenzweig-MacArthur model is that logistic growth has an inherent lagtime built into it—the more vegetation is depleted by herbivory, the longer it takes to grow back. Thus, logistically growing vegetation consumed down to 0.01% of its maximum standing crop will take a much longer time to grow back compared with vegetation decreased to 1% of m. By contrast, regrowing vegetation will need essentially the same time to get back to m whether it starts from 1%, 0.01%, or even 0% of m. We might expect that in most temperate and tropical ecosystems, regrowth-type vegetation will essentially come back within one growing season. As a result, it acts as a fast dynamical variable, explaining why model (4.40) behaves as a quasi-first-order dynamical system.

Another way to think about model (4.40) is to consider it a model of consumer-resource interaction in which resources possess an absolute refuge. In this interpretation, the belowground plant biomass, inaccessible to herbivores, is a refuge, and there is movement of biomass from the refuge to the vulnerable, aboveground biomass, represented by V. An absolute, or "constant-number" refuge should exert a powerful stabilizing influence on dynamics in consumer-resource models (Maynard Smith 1974; see McNair 1986 for the distinction between "constant-number" and "constant-proportion" refuges, and the potentially contrasting effects of these refuge types on stability).

4.4.4 Plant Quality

The preceding discussion of herbivore-plant systems has focused exclusively on plant *quantity* as the relevant variable. However, another characteristic that distinguishes grazing from predation systems is that dynamical changes in resource *quality* are much more likely. There are two main sources of changes in average plant quality. First, by preferentially consuming better-quality plant individuals or tissues, herbivores may depress the average quality of what vegetation remains. Second, upon experiencing herbivory, plants may increase the degree to which they defend their remaining biomass, or their biomass newly produced to replace losses due to herbivory.

The theory of herbivore–plant quality dynamics has been largely neglected by mathematical ecologists. An exception is the work by Edelstein-Keshet (1984; Edelstein-Keshet and Rausher 1989), which I discuss next. The theory formulated by Edelstein-Keshet was framed in terms of continuous-time partial and ordinary differential equations, and it would be useful to adopt it for discretely reproducing organisms, such as forest insects. Thus, after reviewing the Edelstein-Keshet model, I discuss a discrete-time model of herbivore–plant quality interaction.

The Edelstein-Keshet Model Edelstein-Keshet assumed that the state of vegetation at any particular point in time could be represented by a frequency distribution of plant qualities. Herbivory modifies this distribution. Typically, increased herbivory tends to depress the average plant quality. In the absence of herbivory, average plant quality tends to increase. Based on these principles, Edelstein-Keshet derived a general model of herbivore–plant quality dynamics, framed in terms of partial differential equations (PDE), which I do not give here. Edelstein-Keshet further considered the conditions under which the full PDE model could be reduced to mathematically simpler equations. One particular case she discussed was herbivores that "integrate" over plant quality distribution by virtue of their high mobility. She showed that this assumption resulted in a model framed as a system of ordinary differential equations. For the specific functional forms suggested by Edelstein-Keshet, the following model results:

$$
\frac{dQ}{dt} = k - cQN(N - d)
$$
$$
\frac{dN}{dt} = r_0 N \left(1 - q\frac{N}{Q} \right)
$$

(4.41)

Here Q is the average plant quality and N is the herbivore density. The first equation assumes that in the absence of herbivores Q will increase at a constant rate k. The second term, $cQN(N - d)$, reflects the influence of herbivores. Thus, as long as herbivore density is small, $N < d$, plant quality will be further increased. When herbivore density is above the threshold d, plant quality will be decreased by herbivory. The effect of herbivory is magnified by high values of Q. This means that when herbivores are abundant, the high-quality plants will gradually dwindle in quality, while the lower-quality plants will be initially ignored. The second equation is simply the logistic, but assumes that herbivore carrying capacity is directly proportional to average plant quality.

The dynamics of model (4.41) are stable (Edelstein-Keshet and Rauscher 1989). Typically, herbivore attack will lead to decaying oscillations in both average plant quality and herbivore density which will eventually settle to an equilibrium unless constantly perturbed by some exogenous factors. Edelstein-Keshet further investigated the dynamics of the full PDE model, using numerical methods, and showed that if herbivores preferentially attack plants at the high-quality end of the spectrum (certainly a reasonable supposition), then plant quality will tend to become more uniform with time. This central tendency results from lower-quality plants becoming better, since they are largely spared herbivory, while higher-quality plants are being reduced in quality, as a result of herbivore attack. Meanwhile, the whole quality distribution oscillates toward an equilibrium, as predicted analytically by the simplified model (4.41).

The important theoretical insight from the work by Edelstein-Keshet is that, under certain conditions, a simplified model that tracks only average plant quality is capable of capturing the essence of the dynamical interaction between herbivores and plant quality. Building on this insight, we can extend the approach of Edelstein-Keshet to modeling plant-herbivore systems in discrete time. For example, it has been theorized that the interaction between inducible plant defenses and herbivory may drive population cycles in forest insects (Haukioja et al. 1987). Here is one possible approach of modeling such a hypothesis.

A Discrete-Time Plant Quality Model To make derivation of the model more transparent, let us start by formulating it in terms of induced plant defenses, D_t, which have a negative effect on herbivore population growth, and later switch to plant quality, defined as $Q_t = 1 - D_t$. Let us start with the discrete logistic (Ricker) model for the insect population, and further postulate that D_t affects the intrinsic rate of population growth, rather than carrying capacity:

$$N_{t+1} = N_t \exp\left[r_0 \left(1 - D_t - \frac{N_t}{k} \right) \right] \qquad (4.42)$$

When D_t is 0, insect population grows according to the Ricker model. It is prevented from expanding by direct competition for food *quantity* (this imposes the carrying capacity, k). Next year, the same quantity of vegetation is available for herbivores; thus vegetation quantity acts as a first-order check on population density. As D_t increases, the intrinsic rate of insect population growth declines, and can even become negative should D_t increase beyond 1. The dynamics of D_t should be governed by two processes. First, there is induction by herbivory. I will assume that the increase in plant defenses is hyperbolic in form. That is, at low N_t, D_t is incremented by an amount directly proportional to N_t. However, at very high N_t, the increment approaches a constant ceiling. This would suggest the following equation: $D_{t+1} = cN_t/(d + N_t)$. This equation is not a complete model, because there is the second process that affects D_t: carryover from one year to the next. Suppose that herbivore density crashes. It is not likely, then, that D_t next year will immediately decrease to 0. Instead, it is likely to decrease gradually. A simple model for this mechanism is a linear autoregressive process: $D_{t+1} = \alpha D_t$. Putting the two processes together, we have the equation for D_t:

$$D_{t+1} = \alpha D_t + \frac{cN_t}{d + N_t} \qquad (4.43)$$

This model suggests that D_t will fluctuate in the range between 0 and $c/(1 - \alpha)$. The minimum is the equilibrium that D_t approaches in the absence of herbivory. The rate with which D_t decays to 0 is determined by α (the smaller the value of α, the faster D_t approaches 0). The upper end of the range is the equilibrium for $N_t = \infty$, which is determined by the balance of induction (at the maximum rate, c) and decay toward 0 (governed by α).

The final step is to translate the model into terms of plant quality, defined as $Q_t = 1 - D_t$. That is, the maximum of Q_t is 1, when the herbivore population enjoys the highest intrinsic rate of increase, r_0. The lowest that Q_t can get is $1 - c/(1 - \alpha)$, which, depending on parameters c and α, can be negative. The complete model is

$$N_{t+1} = N_t \exp\left[r_0\left(Q_t - \frac{N_t}{k}\right)\right]$$

$$Q_{t+1} = (1 - \alpha) + \alpha Q_t - \frac{cN_t}{d + N_t} \qquad (4.44)$$

The application of this model to the larch budmoth system will be pursued in section 9.3.1.

4.5 PATHOGENS AND PARASITES

There are two general kinds of pathogen-host models. One class, generally applied to *macroparasites* such as parasitic worms, is broadly similar to usual trophic models in that we keep track of both resource and consumer densities. In the second class, applied to *microparasites* such as bacteria and viruses, we do not track pathogen numbers directly, instead focusing on the numbers of healthy versus infected hosts. This simplification results in more tractable models, and is justified in cases where within-host pathogen dynamics are rather stereotypical.

In this section I review the form and dynamics of pathogen-host models. As usual, I start by discussing dynamical coupling terms for these trophic systems (in this case, the disease transmission process). Then I review models of microparasite-host interactions. Finally, I discuss the somewhat more complex models of macroparasitism.

4.5.1 Transmission Rate

Dynamics of disease transmission is the key process in host-pathogen interaction. The starting point for modeling transmission rate is the *mass action assumption*, which takes the following form in microparasitism models:

$$\text{transmission rate} = \beta S I \qquad (4.45)$$

where S is the density of susceptible (pathogen-free) hosts, I is the density of infected hosts, and β is a constant of proportionality. The logic underlying the mass action transmission rate is the same as that for the linear functional response. It is assumed that the encounter rate between susceptibles and infectives is proportional to their densities (assuming homogeneous space and complete mixing). In other words, this is another example of the mass action principle (section 2.4.1). The constant β is analogous to the search rate in predator-prey models. Transmission in macroparasitism systems is modeled analogously, but instead of I in (4.45) we substitute the density of infective parasite stages.

The assumption of linear functional response in predator-prey models is difficult to defend, because high prey density must saturate the predator killing rate (section 2.4.1). In the microparasitism models, by contrast, there is no saturation of transmission rate at high host density, because it is assumed that each infected host has an effectively infinite supply of pathogenic spores (or whatever the actual transmission agent is). In macroparasitism models, once an infective parasite stage has encountered and parasitized a host, it is immediately removed from the population of free-living parasites. In effect, the parasitism rate of free-living infective stages is saturated by a single host, thus not requiring any additional modeling (as in the hyperbolic response with its extra parameter of handling time).

Mass action transmission has been the standard assumption in host-pathogen models since they have been developed at the beginning of the twentieth century (McCallum et al. 2001). Alternative functional forms, proposed in the theoretical literature, are listed in table 1 of McCallum et al. (2001). One alternative is *frequency-dependent transmission*, $\beta SI/N$, where N is the total population size. This mode of transmission is often assumed in models of sexually transmitted diseases, because the number of sexual partners of an individual usually depends on the mating system and is weakly related to host density. Another alternative is the negative binomial form, arising when hosts differ in their susceptibility to infection (more on this in section 4.5.3). Field studies reviewed by McCallum et al. generally suggest that mass action transmission is not an adequate model. Unfortunately, a clear alternative that could serve as a starting point for modeling host-pathogen interactions has not yet emerged.

4.5.2   Microparasitism Models

The approach to constructing simple models for microparasitic diseases proceeds by dividing all hosts into discrete classes with respect to their disease status. For example, for viral diseases such as measles, three classes of hosts are typically distinguished. *Susceptible* hosts are those that can be infected (their density is S), *infectious* are those that have already contracted disease and can now infect susceptibles (density I), and *recovered*, those that are immune to the disease, and

cannot be infected (density R). Sometimes, another class is added—
exposed (density E), or those that have already been infected, but are
not infectives themselves (are in a latent period). Similarly to the LPA
(larvae-pupae-adults) models (section 3.3.2), these models are known
by the acronyms referring to their state variables, SIR or SEIR. A
simple SEIR model proposed for measles epidemics in human populations is (e.g., Anderson and May 1991):

$$\frac{dS}{dt} = \mu N - \beta SI - \mu S$$

$$\frac{dE}{dt} = \beta SI - \sigma E - \mu E \qquad (4.46)$$

$$\frac{dI}{dt} = \sigma E - \chi I - \mu I$$

$N = S + E + I + R$ is the total number, assumed to be constant, since
the primary interest is on relative frequencies of people in different
classes (this assumption is justified because numerical dynamics of
human populations are slow in relation to dynamics of measle epidemics; also, it is assumed that there is no pathogen-induced mortality). To make sure that N does not change, the model forces birth
rate to be equal to death rate (both parameterized with μ). Thus, μN
models births (who are added to the class of susceptibles), and μS,
μE, and μI are density-independent (as well as disease-independent)
death rates of susceptibles though infectives, respectively. Note that
there is no separate equation for R, because $R = N - (S + E + I)$
is uniquely determined by the three equations of model (4.47). Rate
parameters σ and χ govern the speed with which latent exposeds
become infectives, and infectives turn into recovered. The key part
of the model is the mass action term βSI, which governs the rate
at which susceptibles contract the disease. Note that this is the only
nonlinear term in this very simple model.

Model (4.47) looks eerily similar to the Lotka-Volterra equations.
To accentuate this similarity, let us further simplify it by assuming
that infected individuals become infectives without a lag. This allows
us to eliminate variable E:

$$\frac{dS}{dt} = \mu(N - S) - \beta SI$$

$$\frac{dI}{dt} = \beta SI - (\gamma + \mu)I \qquad (4.47)$$

Apart from the term $\mu(N - S)$, this model has the same structure as the Lotka-Volterra model, with susceptibles playing the role of "prey" and infectives playing the role of "predators." The only difference between this and the Lotka-Volterra model is in the "prey" birth rate: in the epidemic model the birth rate is proportional to $N - S = I + R$, rather than S, as is the case in the predator-prey model. Note that if only susceptibles can reproduce, and if we relax the assumption that birth and death rates must balance out, then we obtain the Lotka-Volterra model, pure and simple:

$$\frac{dS}{dt} = \mu S - \beta SI$$

$$\frac{dI}{dt} = \beta SI - \gamma I$$

(4.48)

Here μ is reinterpreted as the difference between per capita birth and death rates of susceptibles, while γ is the death rate of infectives (now pathogen-induced).

This comparison shows that epidemic models belong to the general class that contains predator-prey, parasitoid-host, and herbivore-plant models. In short, they all are trophic models, and all are prone to oscillatory dynamics. Simple models for epidemics are constructed using the same general principles as the ones determining the structure of predator-prey models. In particular, the coupling between the state variables is based on the mass action principle. Other realistic features such as density-dependent population growth can be added to the epidemic models in the same way that they are added to predator-prey models. There is one important difference, however. Whereas in the predator-prey models the interaction term is based on the hyperbolic functional response, in epidemic models the coupling term is based on the equivalent of the linear functional response. Thus, we expect that epidemic models may be somewhat more stable than predation models (since the hyperbolic response is generally a destabilizing feature of trophic models). Furthermore, epidemiological theory suggests that one important factor that has a strong effect on the transmission rate, as well as the resulting dynamics, is the heterogeneity between hosts in their likelihood to become infected (often modeled with negative binomial transmission rate). This heterogeneity in risk of infection is also widely prevalent in natural populations, making it an important feature to add to models to make them more realistic.

As we shall see below, heterogeneity of risk is another factor that leads to greater stability in pathogen-host interactions.

4.5.3 Macroparasitism Models

The foundations of the theory for host-macroparasite interactions were laid by Anderson and May (Anderson and May 1978; May and Anderson 1978). These authors assumed that both hosts and parasites reproduce continuously and that parasite transmission is virtually instantaneous, allowing them to formulate their models as systems of ordinary differential equations. Let H be the population density of hosts, and P parasite density. Thus, the average number of parasites per host is P/H. Furthermore, there must be some frequency distribution describing the probability that any particular host will have i parasites.

In the basic Anderson-May model, host population is assumed to grow exponentially in the absence of parasites. Parasite loads (the number of parasites that a host bears) increase the probability of death of infested hosts. Assuming a linear relationship between the host death rate and parasite load, Anderson and May showed that the average per capita death rate will be directly proportional to the average number of parasites per host, $\alpha P/H$. The total death rate of hosts will be then $\alpha(P/H)H = \alpha P$. Thus, for the host equation we have

$$\frac{dH}{dt} = r_0 H - \alpha P \qquad (4.49)$$

If parasite fecundity is density independent, then the rate of production of transmissible stages (eggs, spores, or cysts) per parasite, λ, is constant, and the total rate at which transmissible stages are produced is λP. Let us temporarily introduce another variable, the density of free-living transmissible stages, W (by the time we get to the basic Anderson-May model, we will get rid of this variable, but employing it at intermediate stages makes it easier to understand the derivation). Transmissible stages suffer a constant per capita death rate, γ, while they are waiting to infest a host. In addition, their density is decreased when they encounter and successfully infest a host, at the rate βWH (mass action principle):

$$\frac{dW}{dt} = \lambda P - \gamma W - \beta WH \qquad (4.50)$$

The density of parasites, P, will be incremented by the rate at which infective stages encounter and parasitize hosts, βWH. It will be decreased by density-independent death rate, μ (this rate has two components: density-independent death rate of parasites within hosts, and parasitism-independent death rate of hosts, which causes death of all their parasites). Additionally, parasites will be lost as a result of parasitism-dependent host death rate. Anderson and May showed that this term has a component that depends on the variance of the frequency distribution of parasites per host. Assuming that the distribution is clumped, and can be approximately described by the negative binomial, the parasitism-dependent death rate is $\alpha + \alpha(k + 1)P/(kH)$ (this is a per capita rate). Here k is the clumping parameter of the negative binomial distribution (smaller k values indicating greater degree of clumping; recollect that as $k \to \infty$, the negative binomial distribution converges to the Poisson). The parasite density equation, therefore, is

$$\frac{dP}{dt} = \beta WH - (\mu + \alpha)P - \frac{\alpha(k+1)P^2}{kH} \qquad (4.51)$$

Putting together the three equations for H, W, and P, we have a model of host-parasite interaction. This model, however, can be simplified without a great loss of biological realism. Note that in many host-parasite systems transmissible stages are very short-lived compared with the long-lived parasitic stages. This observation suggests that W may act as a *fast* variable, quickly approaching a quasi equilibrium, $\widehat{W}$, determined by current (and very slowly changing) densities of hosts and parasites. We solve for $\widehat{W}$ by setting the derivative of W to zero:

$$\widehat{W} = \frac{\lambda P}{\gamma + \beta H}$$

Substituting this value instead of W in the equation for P, we finally obtain the basic Anderson-May model:

$$\frac{dH}{dt} = r_0 H - \alpha P$$

$$\frac{dP}{dt} = \frac{\lambda HP}{\delta + H} - (\mu + \alpha)P - \frac{\alpha(k+1)P^2}{kH} \qquad (4.52)$$

where the new parameter $\delta = \gamma/\beta$.

Note that we can rewrite the second equation of this model as

$$\frac{dP}{P\,dt} = \left(\frac{\lambda H}{\delta + H} - \mu - \alpha \right) - \frac{\alpha(k+1)}{kH} P \qquad (4.53)$$

The right-hand side of this equation is linear in P and formally is the same as the logistic equation (but with intrinsic rate of change and carrying capacity being functions of host density, H). The self-limitation term includes the clumping parameter as $(k+1)/k$. Recollect that large values of k correspond to essentially random (Poisson) distribution of parasite burden, while small k imply highly aggregated distribution. As the degree of clumping increases ($k \to 0$), the ratio $(k+1)/k \to \infty$. In other words, the self-limitation term becomes increasingly more important the more clumped parasites are. This mechanism appears to be the source of the connection between the degree of clumping and dynamical stability of the Anderson-May model.

One important application of the Anderson-May framework is the model of red grouse–nematode parasite interaction by Dobson and Hudson (1992). I defer the discussion of the parameterization and dynamics of this model until section 11.2.2.

4.6 TRITROPHIC MODELS

The final topic of this chapter is a brief discussion of modeling tritrophic systems, such as the interaction of vegetation-herbivores-predators. This discussion will be brief because the preceding material has provided us with all the elements we need to use in constructing tritrophic models. The key idea is that we take a vegetation-herbivore model, and add a predation term to the right-hand side of the herbivore equation. The vegetation and the predator equations remain unchanged. Thus, the only issues to be resolved are which components to use (vegetation growth terms, functional and numerical responses of herbivores and predators).

Probably the most common model used in describing tritrophic systems is the **Oksanen model**, which results from stacking two

Rosenzweig-MacArthur models, one for herbivores and one for predators:

$$\frac{dV}{dt} = v_0 V \left(1 - \frac{V}{m} \right) - \frac{aVN}{b+V}$$

$$\frac{dN}{dt} = \xi N \left(\frac{aV}{b+V} - \eta \right) - \frac{cNP}{d+N} \qquad (4.54)$$

$$\frac{dP}{dt} = \chi P \left(\frac{cN}{d+N} - \mu \right)$$

This model was graphically analyzed by Oksanen et al. (1981; actually, they considered a more general version, as well as extending their approach to quadritrophic systems).

The only self-limitation term in the Oksanen model is the logistic growth of vegetation. As a result, this model is easily destabilized. One particular route from stability to oscillations, considered by Oksanen and coworkers, is the increased productivity of the vegetation, or the **paradox of enrichment**. As I discussed in section 4.4.3, the Rosenzweig-MacArthur model should be in the oscillatory regime for the parameter values characterizing most herbivores. When on top of the vegetation-herbivore instability we add the potential for herbivore-predator cycles, we end up with a potentially very unstable model. Numerical investigations show that when both vegetation-herbivore and herbivore-predator links are oscillatory, it becomes difficult to keep the tritrophic system from falling apart. Usually, predators are the ones that go extinct (or effectively go extinct if their trough densities reach exceedingly low values).

This observation should not be interpreted as a theoretical statement that all tritrophic systems should be highly unstable, because in real-life systems there are other stabilizing influences, apart from the logistic vegetation growth, which is the only one assumed by the Oksanen model. To take one specific taxon, consider mammalian herbivores. Although this is a very diverse group (e.g., the spectrum of body sizes covers four orders of magnitude: from a 50 g vole to a 500 kg moose), I believe that it is possible to write a generic model for most of them (at least, a model that can serve as a useful starting point). First, I assume that vegetation is characterized by regrowth, rather than logistic dynamics (which already introduces a stabilizing

influence, compared with the Oksanen model). Second, I assume that
there is no direct density dependence in the herbivore, but there is
such a mechanism in the predator (e.g., territoriality). These modifications lead to the following *generic mammalian herbivore model*:

$$\frac{dV}{dt} = u_0\left(1 - \frac{V}{m}\right) - \frac{aVN}{b+V}$$

$$\frac{dN}{dt} = \xi N\left(\frac{aV}{b+V} - \eta\right) - \frac{cNP}{d+N} \qquad (4.55)$$

$$\frac{dP}{dt} = \chi P\left(\frac{cN}{d+N} - \mu\right) - \frac{s_0}{\kappa}P^2$$

The last term in the predator equation represents self-limitation in
the predator (in the Bazykin form). It is written in a somewhat more
mechanistic way, with s_0 and κ representing the intrinsic rate of
increase and the carrying capacity due to territoriality, respectively
(this way of writing the term makes it easier to estimate it). Although
a priori it may seem unlikely that a model such as (4.56) could serve
as a reasonable description of many mammalian systems, this appears,
nevertheless, to be the case. We shall see that this model, with minor
modifications, applies to voles, hares, and ungulate systems (but not
lemmings!).

A very useful survey of dynamics in linear chain ecosystems can
be found in Gurney and Nisbet (1998:185–200). These authors first
consider trophic systems based on primary producers growing according to the regrowth model (which they call "constant production").
If functional responses of herbivores and predators are linear, then
all the resulting mono-, di-, and tritrophic chains are characterized
by a single nontrivial stable-point equilibrium. If primary producers
grow according to the logistic equation (but we still assume that all
functional responses are linear), then again the dynamics of mono-,
di-, and tritrophic systems are characterized by stable point equilibria.
The situation changes when we replace linear functional responses
with hyperbolic ones. The ditrophic model with regrowth vegetation
is still globally stable, but the tritrophic one can exhibit stable cycles
for certain parameter values. Finally, the logistic growth combined
with the hyperbolic functional response is capable of exhibiting unstable dynamics even with only two trophic levels (this is, of course,
the Rosenzweig-MacArthur model). To summarize, consideration of

linear chain models confirms the insights from two-species ones: regrowth-based models are more stable than those based on logistic growth, and hyperbolic functional responses are destabilizing. Furthermore, greater food chain length is also destabilizing (the message that has been enunciated by May 1973b).

4.7   SYNTHESIS

Although the functional class of trophic interaction (predators, grazers, etc.) and details of specific systems affect the form of equations, at a deeper level all trophic models are based on the same logical foundations. For example, the Nicholson-Bailey reduces to the Lotka-Volterra model as the time step is made increasingly small (May 1973a). Similarly, in section 4.5.2 I show that simple microparasite-host models also reduce to the Lotka-Volterra equations. Thus, the structure of consumer-resource interaction makes trophic models inherently susceptible to oscillations. Pure trophic systems (i.e., systems lacking first-order feedbacks) exhibit unstable oscillations, such as the neutrally stable cycles in the Lotka-Volterra model, or diverging Nicholson-Bailey oscillations. Furthermore, trophic oscillations are typically **second-order** cycles.

Adding various realistic features to models may make them either more stable or more prone to violent oscillations, but again there are several unifying themes. First, self-limitation terms in both resource and consumer are inherently stabilizing features (with a caveat that in discrete systems strong intraspecific regulation may lead to *first-order cycles*). It is important to note that a self-limitation term can arise indirectly, as a result of some other feature of the system. For example, aggregated distribution of parasites within hosts indirectly introduces a first-order feedback term (see equation 4.53).

Second, trophic chains based on primary producers growing logistically are less stable compared with those based on producers obeying the regrowth model. Third, saturating functional responses (e.g., the hyperbolic one) are a destabilizing feature of models, compared with the linear response. S-shaped functional response can result in multiple equilibria. Finally, greater food chain length is a destabilizing influence.

CHAPTER 5

Connecting Mathematical Theory to Empirical Dynamics

5.1 INTRODUCTION

In this chapter, I review different kinds of dynamical behaviors that ecological models can exhibit, and interpret these mathematical predictions in terms of observable variables. The basic premise underlying the material here is that inasmuch as mathematical models reflect ecological reality, the "bestiary" of model-predicted dynamics provides us with patterns that might be matched with behaviors of real populations. Until recently, ecologists interested in nonlinear dynamics tended to focus exclusively on behaviors of deterministic models, that is, population fluctuations resulting only from endogenous factors (Schaffer 1985; Pimm 1991). Real-world populations are always affected by exogenous factors ("noise"), and the taxonomy of dynamics should reflect this fact. However, it is easier to start describing dynamical types by first focusing on models without noise. Accordingly, section 5.2 introduces basic types of dynamics that characterize deterministic population models. Next, in section 5.3, I review the kinds of dynamics that are exhibited by mixed deterministic/stochastic models. Finally, armed with the material in the first two sections, I address the question of what is population regulation, an issue that has caused a considerable degree of controversy (section 5.4).

Definitions First, however, let us make some definitions. **Exogenous**[1] refers to density-independent factors that affect population density, but are not, in turn, affected by it. Thus, there is no dynamic feedback between these factors and population density. The exogenous component of population change is often modeled as a stochastic process, but it does not have to be purely random. Nonrandom exogenous effects include trends and periodic changes in the environment, such as seasonality. By contrast, **endogenous** refers to the density-dependent component of dynamics, or population feedbacks. These feedbacks can occur with a lag, for example, the effect of specialist predators on prey. Thus, we distinguish two kinds of endogenous factors. **First-order** feedbacks act on the realized per capita rate of change without an appreciable lag (they sometimes are referred to as *direct density-dependent* mechanisms). **Second-order** feedbacks are mediated by variables that change relatively slowly, and therefore their effect on the realized rate of per capita change occurs with a time lag (they are sometimes referred to as *delayed density-dependent* mechanisms). In reality all population feedbacks involve some lag, so we operationalize the distinction between the first- and second-order feedbacks by whether the lag time involved is less or greater than the generation time (see section 2.5).

For completeness of definition, let us also add two further categories. First, **null factors** are those that do not affect the realized per capita rate of population change. Finally, we note that exogenous and endogenous factors explain *variation* in $r(t)$, and therefore in population density. Yet there are other important quantities, for example, the mean density. Nondynamical factors that set mean density are called **parameters**. Parameters do not change with time, but may vary from place to place. In addition to setting statics (e.g., mean level of fluctuations), they may also influence other structural properties of population regulation, such as process order, periodicity, and Lyapunov stability.

Fluctuations is my generic term for any kind of population dynamics, as long as there is some element of temporal change. **Oscillations** is a general term for population dynamics that have

---

[1] As usual, the definitions of concepts emphasized in bold type are given in the glossary.

some element of regularity, which makes population density some-what predictable at least a few time steps ahead. Typically, population oscillations have some sort of periodic or chaotic deterministic attractor at their heart. Oscillations can be classified by whether they are first- or second-order, whether they are periodic, and whether they are stable or chaotic.

5.2 QUALITATIVE TYPES OF DETERMINISTIC DYNAMICS

5.2.1 Attractors

Purely endogenous, or deterministic, dynamics are basically classified by the qualitative type of the attractor. An **attractor** is a geometrical object in the phase space that attracts all trajectories starting within its **domain of attraction** (recollect that **phase space** is constructed by representing each dynamical variable with its own axis in multidi-mensional euclidean space). The simplest kind of purely endogenous dynamics is stability around a point equilibrium. A *stable equilibrium* is the point in phase space to which the trajectory returns after a small perturbation (a large perturbation may take the trajectory outside the domain of attraction of the stable point, in which case the trajec-tory will not return to it). Because stable point equilibrium *attracts* nearby trajectories, it is one kind of an *attractor*. The approach to the stable point may be either **monotonic** (also known as exponen-tial) or **oscillatory** (also known as damped oscillations). In discrete systems, monotonic approach is sometimes called "undercompensa-tion," because it takes several time steps for the dynamical system to compensate for perturbation (return to the equilibrium). Oscillatory approach to equilibrium, by contrast, leads to "overcompensation," since the system overshoots the equilibrium. Finally, "perfect com-pensation" occurs when the system returns exactly to the equilibrium in one time step.

A more complex dynamical behavior is a **stable limit cycle.** In continuous models, the limit cycle attractor is a closed curve in the phase space that all trajectories approach. In order for a differential model to be able to exhibit stable cycles, it has to be of order two or higher (first-order models can exhibit only one kind of attractor—the stable point). In discrete models, a limit cycle is a finite set of points, visited in turn by the trajectory. In both cases, a limit cycle is a periodic attractor in the strict mathematical sense because the trajectory on the attractor repeats itself exactly after some time (called the period). **Quasiperiodicity** is a very similar behavior to limit cycles. A quasiperiodic attractor can occur in discrete models with two or higher dimensions and consists of an infinite set of points lying on a closed curve. It is like a stable limit cycle, but with an irrational period, so that the trajectory never exactly repeats itself. In continuous models, the equivalent of quasiperiodicity is motion on a torus, which can occur in three- or higher-dimensional systems.

Unlike zero-dimensional (e.g., a stable point) or integer-dimensional attractors (e.g., quasiperiodicity), a *chaotic attractor* is usually fractal. A chaotic trajectory of a purely deterministic system never repeats itself, and is "random-looking." A formal definition of chaos will be discussed in detail in the next section (5.2.2). In continuous systems, chaos cannot arise unless the dimensionality of the system is three or higher. In discrete systems, chaos can arise even when the dimension is one, but higher dimensionality increases the likelihood that biologically reasonable parameters will lie in the chaotic region.

Nonlinear dynamical systems may also be characterized by **multiple coexisting attractors**—fixed points, cycles, and fractal structures each having its own attraction basin within the phase space. The trajectory will go to one of the attractors, depending on initial conditions (whose attractor's basin it starts in). Such systems are sometimes called "metastable." For example, a one-dimensional continuous system can have three point equilibria: two stable ones, and one unstable one serving as the boundary between the attraction domains of the stable equilibria (Berryman et al. 1984).

5.2.2 Sensitive Dependence on Initial Conditions

Before discussing how the taxonomy of deterministic dynamics described above is affected by dynamical noise, I need to define precisely what I mean by chaos, and also discuss how chaotic dynamics can exist in the noisy world. The best definition of chaos for our purposes is *bounded fluctuations with sensitive dependence on initial conditions* (Eckmann and Ruelle 1985; Ellner and Turchin 1995). A measure of sensitive dependence is the dominant Lyapunov exponent, which is a generalization of the notion of stability from point equilibria to trajectories.

Stability of Point Equilibria   Before discussing Lyapunov exponents and stability of trajectories, let us review the stability of point equilibria. Consider a discrete deterministic system such as

$$Z_t = F(Z_{t-1}) \tag{5.1}$$

where Z_t is the state variable, which could be a vector. For example, Z_t could consist of two values: the prey density, N_t, and the predator density, P_t. The steady-state solution, or equilibrium, of (5.1) is a value of Z^* that satisfies the equation

$$Z^* = F(Z^*)$$

A steady state is stable if it attracts all nearby trajectories, and unstable if nearby trajectories diverge away from it. Let us consider two trajectories: one that starts right on the steady state, $Z_t^0 = Z^*$, and another one very close to the equilibrium, Z_t^1. Because the first trajectory starts on the equilibrium, it will remain there indefinitely. The second trajectory, however, can either approach or diverge from the equilibrium. The difference between the two trajectories, $Z_t' = Z_t^1 - Z_t^0 = Z_t^1 - Z^*$, thus, will either grow or shrink. If Z_t' is small (trajectory Z_t^1 close to the steady state), then its behavior will be largely governed by the *linearized* equation:

$$Z_t' = J(X^*) \cdot Z_{t-1}' \tag{5.2}$$

where $\cdot$ denotes matrix multiplication, and J is the *Jacobian* matrix, whose components are partial derivatives of F. For example, in a two-dimensional system

$$x_t = f(x_{t-1}, y_{t-1})$$

$$y_t = g(x_{t-1}, y_{t-1})$$

the Jacobian matrix is

$$J(x, y) = \begin{pmatrix} \partial f/\partial x & \partial f/\partial y \\ \partial g/\partial x & \partial g/\partial y \end{pmatrix}$$

Equation (5.2) is obtained by using Taylor series expansion of the function F (for a very clear explanation of the stability analysis of nonlinear discrete equations, see Edelstein-Keshet 1988:55). Equation (5.2) is linear because $J(Z^*)$ does not depend on Z' (it is a constant matrix).

The Jacobian evaluated at the equilibrium point, $J^* = J(Z^*)$, can be of two fundamental kinds: it either stretches or shrinks Z'. If the magnitude of Z' increases as we iterate equation (5.2), then the Jacobian is of the stretching kind. In other words, small deviations from the steady state will be amplified, and therefore the steady state is unstable. If the magnitude of Z' is decreased with each iteration, then the Jacobian is of the shrinking type, which implies stability, since small perturbations from the steady state will be damped. There is also a borderline state of neutral stability, when J^* will leave the magnitude of Z' unchanged (this is what happens in the Lotka-Volterra predation model). Whether the Jacobian is of the amplifying or damping kind can be determined simply on the computer, by iterating equation (5.2) and observing whether Z' grows or shrinks. Alternatively, the stability can be investigated analytically (at least for lower-dimensional models) by calculating J^*'s eigenvalues (Edelstein-Keshet 1988). To summarize, the stability of a steady state is determined by the Jacobian matrix of partial derivatives of F, evaluated at the steady state.

Stability of Trajectories: Deterministic Systems   This notion of stability for equilibrium points can be extended to the notion of stability of trajectories on other kinds of attractors. Again, consider two nearby trajectories, but this time the "reference" trajectory X_t^0 is evolving with time. For example, it could "sit" on a limit cycle. The other trajectory, X_t^1, starts very close to X_t^0, and again, let X_t' be the vector separating the two trajectories. Analogously to the case of a point equilibrium, X' will either grow or shrink depending on the nature of the Jacobian. However, because the reference trajectory moves in the phase space, the Jacobian will be evaluated at each successive point of X_t^0, and thus it will change. (In a linear system the Jacobian will be constant.) Thus, the divergence/convergence rate of trajectories will vary. In fact, in some regions of the phase space, the Jacobian may be of the shrinking kind, while in other regions it could be of the stretching kind. This means that we need to talk about what will happen to two nearby trajectories in the long term. Iterating equation (5.2), we observe that the Jacobians are multiplied:

$$Z'_t = J_{t-1} \cdot J_{t-2} \cdots J_0 \cdot Z'_0 \tag{5.3}$$

where $J_t = J(Z^0_t)$, the Jacobian evaluated at the point Z^0_t on the reference trajectory. Whether Z'_t has grown in comparison to Z'_0 depends on the product of the Jacobians. The quantity

$$\Lambda_\infty = \lim_{t \to \infty} \frac{1}{t} \log \|J_{t-1} \cdot J_{t-2} \cdots J_0\| \tag{5.4}$$

measures the average exponential rate of growth of Z', and is called the Lyapunov exponent. Here $\| \ \|$ is any matrix norm, for example, the dominant eigenvalue. *Positive Lyapunov exponent implies trajectory divergence and chaos; negative Lyapunov exponent implies trajectory convergence, and stability in the general sense that includes not only stable points but also stable limit cycles* (figure 5.1).

Some observations are in order. First, Λ_∞ measures divergence between two trajectories very close to each other. In practice, two chaotic trajectories that start a finite distance from each other will eventually stop diverging (see figure 5.1b). This is a consequence of the second part of the definition of chaos—boundedness. Second, the convergence rate is averaged over all the points of the attractor, in proportion to how often they are visited. This is an automatic consequence of following the reference trajectory. A third, and most important, point is that Λ_∞ is the long-term growth rate, thus the infinity in equation (5.4). The rate of convergence/divergence varies as the trajectory visits different regions of the attractor; it even may switch sign (this is a consequence of nonlinearity of dynamics). Therefore, the long-term average Λ_∞, being only a single number, retains only the information about the mean convergence rate, not about short-term fluctuations in sensitivity to initial conditions. However, short-term variation in **local Lyapunov exponents** can be quite significant, and another important way to characterize dynamics of the system, in addition to the **global Lyapunov exponent** (Ellner and Turchin 1995; Turchin and Ellner 2000a). A local Lyapunov exponent is defined as

$$\Lambda_t = \frac{1}{t} \log \|J_{t-1} \cdot J_{t-2} \cdots J_0\|$$

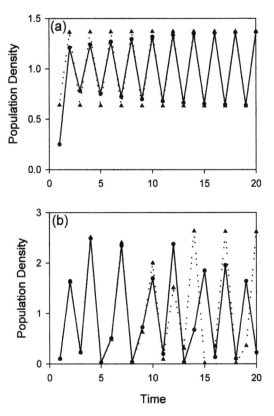

FIGURE 5.1. Sensitive dependence on initial conditions in purely endogenous dynamical systems. In (a) trajectories converge, implying lack of sensitive dependence ($\Lambda_\infty < 0$). In (b) trajectories diverge, implying sensitive dependence on initial conditions ($\Lambda_\infty > 0$) and chaos.

which is the same as the definition in equation (5.4), but without taking the limit $t \to \infty$ (as a notational convention, I emphasize the difference between the global and local Lyapunov exponents by using subscripts ∞ and t, respectively. Thus, Λ_∞ is the *long-term* average trajectory divergence rate, while Λ_t is the *short-term* rate (over the period of 1 year, for example).

Stability of Trajectories: Systems Affected by Noise In mixed stochastic/deterministic systems, trajectories will diverge because of exogenous noise, without respect to whether the endogenous

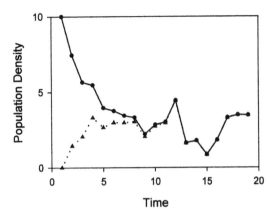

FIGURE 5.2. Trajectory convergence in a stable mixed endogenous/exogenous dynamical system. This is a linear (and thus globally stable) system with large amounts of noise. Two trajectories, affected by the same sequence of random perturbations, start far from each other, but rapidly converge, thus implying lack of sensitive dependence ($\Lambda_\infty < 0$).

component is chaotic or stable. However, the Lyapunov exponent of a mixed system should measure sensitive dependence on initial conditions that is due only to the endogenous part. One way of doing this is to use the same definition as the one used for the deterministic systems (5.4). The difference between systems with and without noise, then, would only be that the sequence of points visited by the trajectory, $\{Z_t\}$, will be affected by process noise in the latter case. One way to think about Λ_∞ in a stochastic system is the rate of divergence or convergence of two nearby trajectories that are influenced by random, but *exactly the same* sequence of, external perturbations (figure 5.2). Note that, in general, adding process noise to a purely deterministic system will change the numerical value of Λ_∞ (indirectly, by affecting which parts of the phase space are visited more often by trajectories). As will be discussed below (section 5.3.3), adding stochasticity can even flip the sign of Λ_∞.

Chaos and Periodicity The defining feature of chaotic oscillations is sensitive dependence on initial conditions. This definition does not preclude the possibility of imperfect statistical periodicity in chaotic dynamics. This is in contrast to the common characterization of chaos as "deterministic nonperiodic flow" (Lorenz 1963), but the "nonperiodic" is used in its mathematical sense—the trajectory never repeats itself. In fact, many examples of chaotic dynamics arising in ecological models do exhibit strong statistical periodicities (Kendall et al. 1993). After a bifurcation from a limit cycle to chaos, for example, the new strange attractor usually consists of several distinct pieces, each visited by the trajectory in turn. Such dynamical behavior will look somewhat like a noisy limit cycle. In addition, a trajectory may temporarily exhibit bouts of near periodicity, by "shadowing" an unstable periodic orbit within the chaotic attractor, or by getting trapped in the neighborhood of a "semi-periodic semi-attractor" (Kendall et al. 1993). In short, chaos and statistical periodicity are not mutually exclusive concepts.

5.3   POPULATION DYNAMICS IN THE PRESENCE OF NOISE

5.3.1   Simple Population Dynamics

The simplest stochastic model that has only an exogenous but no endogenous component is **stochastic exponential growth** (or decline), in which the per capita rate of population change is completely unaffected by any population feedbacks, but only by exogenous factors. Because the stochastic exponential growth/decline implies the absence of population regulation, it is the starting point, or null hypothesis, for any investigation of population regulation. Note that exponential growth (with or without noise) is characterized by sensitive dependence on initial conditions. However, we do not consider such dynamics chaotic, because they fail the condition of boundedness.

The next simplest model to consider is the one whose endogenous component is characterized by a monotonic convergence to a point equilibrium (e.g. the logistic model). When adding a purely stochastic exogenous component to monotonically stable dynamics, we do not get a qualitative change in dynamical behavior of the system. Depending on relative strengths of each component, such a **stability with noise** system will either hover very near the equilibrium (tight regulation), or go on long excursions away from the equilibrium, but always eventually returning to it (weak regulation).

Stochastic exponential dynamics and monotonic stability with noise together define **simple population dynamics**. Simple dynamics combine an essentially linear endogenous component with an additive noise; they have no statistical periodicities, nor are they chaotic. These dynamics, zero-order (exponential growth) and first-order (monotonic stability) simple dynamics, provide two null models, or starting points, for an investigation into complex population dynamics.

5.3.2   Stable Periodic Oscillations

When the endogenous component of a mixed deterministic-stochastic model is characterized by an oscillatory approach to the equilibrium, the exogenous component will prevent the system from settling to the equilibrium. The trajectories of such a stable periodic system will look like noisy cycles. This is the first example of how an interaction between the endogenous and exogenous components can create something novel. Without noise, the system would simply sit at the equilibrium forever. Without the endogenous component, there would be no periodicity. Put them together, and you have sustained oscillations.

When we add noise to stable limit cycles or to a quasiperiodic system, we obtain the same qualitative type as described above—noisy-looking fluctuations characterized by statistical periodicities. Quantitatively, the strength of periodicity is likely to be more prominent when the underlying deterministic dynamics are limit cycles, as opposed to a stable oscillatory point, but qualitatively it is the same type of dynamics. Assuming that the addition of noise did not change the sign of the Lyapunov exponent (i.e., it is still negative), I refer to all these dynamics collectively as **stable periodic oscillations**. The defining features are $\Lambda_\infty < 0$ and a statistically significant periodicity, as measured, for example, by ACF.

5.3.3   Chaotic Oscillations

Chaotic oscillations are defined as bounded population dynamics that are characterized by sensitive dependence on initial conditions ($\Lambda_\infty > 0$), as defined in section 5.2.2. As a result of sensitive dependence on initial conditions, a chaotic system is a "noise amplifier" (Ellner and Turchin 1995). The effects of random perturbations are amplified by the nonlinearities of the endogenous part, and at least some of the system unpredictability is due to the endogenous factors. A stable system, by contrast, damps the effects of exogenous perturbations. It is a "noise muffler." In a stable, nonchaotic system unpredictability is entirely due to the action of exogenous factors.

The deterministic attractor underlying noisy chaotic oscillations will often be chaotic itself. However, it is also possible for dynamical noise to transform a deterministically stable attractor into a chaotic one ("noise-induced chaos"). Vice versa, noise can also transform a deterministically chaotic attractor into a stable one.

Noise-induced chaos is another striking example of how nonlinear endogenous dynamics can create something novel when interacting with exogenous noise (Rand and Wilson 1991). One possible route to this behavior is when a dynamical system is characterized by chaotic transients that in the absence of stochasticity would eventually die out, allowing the trajectory to settle on a stable point equilibrium or a limit cycle. In the presence of even a little noise, however, the chaotic transients never die out, so that the system behavior is chaotic.

Here is an informal example illustrating this idea. Consider a simple discrete one-dimensional model $N_{t+1} = f(N_t)$, illustrated in figure 5.3a. The function f was constructed by taking the Ricker equation in the chaotic regime ($r_0 = 3.5$), and adding a little "hump" in the vicinity of the equilibrium (where f intersects the $N_{t+1} = N_t$ line), to make the slope of f there less in magnitude than 1. Thus, the equilibrium is stable, but elsewhere f has steep slopes that are conducive to chaos. If we provide an initial condition, N_0, and iterate this dynamical model without noise, then the trajectory will jump around chaotically, until it hits the vicinity of the equilibrium, and will rapidly converge to it (figure 5.3b). Thus, the deterministic system is

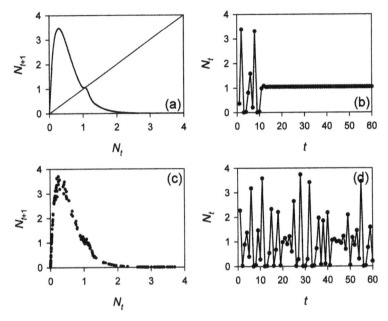

FIGURE 5.3. A chaotic system with a stable deterministic "skeleton." Noise-free dynamics: the shape of the $N_{t+1} = f(N_t)$ relationship (a), and a typical trajectory (b). Dynamics with noise: in the phase space (c) and a typical trajectory (d).

stable—no matter what initial conditions, the trajectory will eventually reach the vicinity of the equilibrium, and will be trapped there. Adding just a little noise, however, is enough to knock the trajectory away from the equilibrium whenever it gets into its vicinity, leading to persistent oscillations (figure 5.3d). In other words, noise "stabilizes" the chaotic transient, and we have a paradoxical situation, in which the system is stable without noise, but chaotic in its presence. But how can we be sure that the fluctuations in figure 5.3d are really chaotic, and not simply noisy? This is made clear by plotting the trajectory generated by the model with noise in the $N_{t+1} - N_t$ phase plot (figure 5.3c). The data points clearly trace out the function f, with very little "fuzz" resulting from the exogenous noise. The inescapable conclusion is that the dynamics of this mixed deterministic/stochastic system are largely driven by the endogenous chaos. In sum, noise is necessary for pushing the system from stability into chaos, but the system is chaotic in the usual sense, since practically all irregularity in the trajectory is due to the endogenous part.

This example shows that endogenous chaos and exogenous noise are not mutually exclusive and opposed explanations of patterns in population dynamics. "Noise" is always present in real population systems. In fact, exogenous ecological factors are an integral part of population dynamics, and are of no less intrinsic interest than the endogenous ones. Noise can creatively interact with nonlinearities (e.g., transforming a stable system into a chaotic one). The erratic component in population dynamics can be entirely due to noise, or it can be a sum of exogenous stochasticity and endogenous irregular motion—chaos.

Certain authorities in modern nonlinear time-series analysis see the goal of analysis as the characterization of the purely endogenous part underlying fluctuations, also known as the "deterministic skeleton" (Tong 1990; Dennis et al. 1995). I believe that such a focus is not productive, and can be seriously misleading, as the thought experiment illustrated in figure 5.3 shows. Although the specific example was carefully contrived to produce the desired result, the general phenomenon, noise-induced stabilization of chaotic transients, is ubiquitous in ecological models. When a complex chaotic attractor loses stability to a limit cycle, or a fixed point, the attractor may persist as a weakly repelling invariant set. Trajectories starting near this set exhibit a long chaotic transient, before eventually collapsing on the simple attractor and staying there. With noise, trajectories are soon kicked back to the complex, weakly repelling invariant set, initiating another long chaotic transient. Thus, in the presence of noise the trajectories act as if the invariant set was still attractive, and "ignore" the deterministically stable attractor. Ecological examples include the epidemiological model parameterized for measles (Rand and Wilson 1991), a spatial model of Dungeness crab dynamics (Hastings and Higgins 1994), and a predator-prey model parameterized for Fennoscandian voles (Turchin and Ellner 2000a). Note that these are not contrived examples, but a behavior exhibited by the best model with empirically estimated parameter values. As Steve Ellner and I argued in a paper titled "Chaos in a Noisy World" (Ellner and Turchin 1995), both our conceptual approaches and our specific methodologies must explicitly take into account the fact that ecological systems are governed by mixed endogenous and exogenous mechanisms. This means, in particular, that we should abandon deterministic approaches that do not generalize well to systems affected by noise, including those that rely on characterization of deterministic skeletons.

5.3.4 Quasi-Chaotic Oscillations

Fully embracing "noise" (exogenous influences) as an integral component of population dynamics has an important consequence: disappearance of sharp boundaries between different classes of dynamics. Above I have already referred to the observation that, in the presence of noise, there is no sharp boundary between stable oscillatory dynamics and limit cycles. Consider, for example, the Ricker model as it goes through the first bifurcation (at $r_0 = 2$), in which the stable point equilibrium becomes a two-point limit cycle. In the absence of noise, there is indeed a very visible difference between the attractor for $r_0 = 1.99$ and $r_0 = 2.01$. In the first case, the dynamics eventually approach the equilibrium point $N^* = 1$, while in the second case, the trajectory jumps between $N_1 = 0.88$ and $N_2 = 1.12$. But if we add even a tiny amount of additive noise to r_t (e.g., $\sigma = 0.01$), we find that dynamics for both r_0 values are essentially the same: both show a pronounced 2-point periodicity with somewhat variable amplitude. The variation between different realizations far outweighs the small difference in the average amplitude. Adding even a little noise destroys, for all intents and purposes, the sharp boundary between dynamics observed in the deterministic case. The quantitative difference remains, so that dynamics of the Ricker model with the dominant eigenvalue near zero differ greatly from those for which the eigenvalue greatly exceeds 1, even in the presence of large amounts of noise, but the difference between eigenvalues slightly greater or less than 1 is not noticeable.

The same insight applies to the boundary between chaotic and stable oscillations. Formally, we distinguish between nonchaotic dynamics with the dominant Lyapunov exponent $\Lambda_\infty < 0$, and chaotic dynamics with $\Lambda_\infty > 0$. In practice, however, the difference is not qualitative but quantitative, and the mechanism is analogous to the one discussed in the previous paragraph, with one important difference: whereas stability of a point-equilibrium can be described with a single number (the dominant eigenvalue evaluated at the steady state), trajectory stability is *not* fully described by the global Lyapunov exponent.

As I discussed in section 5.2.2, the global **Lyapunov exponent**, Λ_∞, is the long-term exponential rate of trajectory divergence (in the limit $t \rightarrow \infty$). Negative Λ_∞ indicates that trajectories are converging rather than diverging. Another, and arguably better, measure of sensitive dependence on initial conditions (Ellner and Turchin 1995; Turchin and Ellner 2000a) is the distribution of local **Lyapunov exponents**, $\{\Lambda_t\}$ (recollect that I am using subscripts ∞ and t to emphasize the difference between the global and local Lyapunov exponents, respectively). Local exponents measure the divergence over a finite interval of time (for example, over one year, or one generation), and they depend on the current state of the system and on the trajectory that it follows over the time interval (Bailey et al. 1997). Local Lyapunov exponents, thus, characterize the variation in sensitivity to perturbations in different parts of state space, which provides more information about the underlying dynamics than a single Λ_∞ (Bailey et al. 1997).

A dynamical system characterized by a value of Λ_∞ near zero will typically have a spectrum of local Lyapunov exponents extending from negative to positive values. Thus, the system goes through recurrent stages of trajectory divergence interspersed with instances of trajectory convergence (Ellner and Turchin 1995). The short bouts of "local chaos" can be quite intense. For example, the best model for the vole *Clethrionomys rufocanus* in northern Finland suggests that one year in four (i.e., a typical cycle) the amplification factor is seven or larger, and one year in ten, the amplification is twenty or larger (Turchin and Ellner 2000a). Thus, small exogenous perturbations (if they occur during the period of local chaos) have large effect. Even relatively small demographic stochasticity can be amplified into periods of high unpredictability (Ellner et al. 1998).

In sum, the boundary between chaotic and nonchaotic dynamics is not abrupt. Population dynamics characterized by Λ_∞ near zero (both positive and negative values) are not qualitatively different from "strong chaos": in both cases systems go through recurrent periods of trajectory divergence interspersed with periods of convergence. The difference is quantitative: what is the relative strength of the convergence/divergence tendencies, and does one or the other dominate in the long term, or are they evenly balanced? Because there is no clear-cut boundary between the chaotic and nonchaotic dynamics, we recently proposed that population dynamics characterized by a global Λ_∞ near zero, let us say in the range $-0.1 < \Lambda_\infty < 0.1$, should be called **quasi-chaotic** (Turchin and Ellner 2000a). As we shall see later, quasi chaos is an important category that is frequently encountered in the analysis of empirical ecological systems.

5.3.5 Regular Exogenous Forcing

The final point to make in the discussion of mixed exogenous/endogenous dynamics is that the exogenous part does not have to be purely stochastic, "white" noise. The exogenous component can have a strong periodic component (section 3.2.2), in which case it is often called **periodic forcing**. Periodic forcing can also interact with the nonlinear endogenous dynamics, and produce unexpected patterns. For example, a purely deterministic periodic exogenous component combined with an endogenously driven limit cycle, under certain conditions, can produce chaos (Schaffer and Kot 1985; Kot et al. 1992). This is yet another example of how nonlinear population dynamics can create something novel when interacting with an exogenous driver. In this case, two regular, periodic kinds of motion in the absence of noise produce a seemingly noisy, irregular trajectory.

5.3.6 Synthesis

Nonlinear endogenous dynamics combined with stochastic exogenous factors and periodic forcing can produce a variety of complicated dynamical behaviors. Furthermore, ecological theory is rich in multidimensional models (e.g., species interactions; age, stage, or size structure; and spatial dynamics) that are much more likely to exhibit complex dynamics than simpler one-dimensional models. Trying to impose a discrete classification system on dynamics is not productive.

However, a more quantitative scheme is possible. Thus, dynamical systems may be periodic or not, and if periodic, the degree of periodicity may vary. Second, systems can be characterized as stable, quasi-chaotic, or chaotic, but even better simply by the estimated value of the global, and the spectrum of local, Lyapunov exponents. Third, different systems are characterized by different strengths of endogenous versus exogenous components of dynamics. In sum, instead of a discrete classification, such as the one based on classifying the deterministic "skeletons," we should estimate such quantities as the strength of periodicity, trajectory stability, and noise/signal ratio. Practical issues concerned with the estimation of these and other "probes" will be taken up in chapter 7.

5.4   POPULATION REGULATION

Although population regulation is one of the central themes in ecology (Royama 1981; Berryman et al. 1987; Lawton 1991; Murdoch 1994), it has also been a most contentious issue, and has been the subject of a highly acrimonious debate since the very beginning of population ecology (history of the early debate in Kingsland 1995; some amusing quotes in Turchin 1995b). Much of the controversy was a result of ecologists speaking past one another (this particularly affected the dialogue between theorists and empiricists). Furthermore, until recently many ecologists were unfamiliar with the basic concepts of dynamical theory. Now that dynamical thinking has become much more prevalent, the debate seems to be finally dying out, or at least shifting to secondary issues (Turchin 1999).

My goal in this section is to discuss the notion of population regulation, and proffer a perhaps not ironclad but workable definition of the concept. I begin by defining a subsidiary notion, density dependence. Then I discuss some early attempts at defining regulation, before moving to the one I prefer, based on the notion of stationarity. Finally, I discuss some of the limitations of the stationarity definition.

5.4.1 Definition of Density Dependence

Different ecologists sometimes mean different things when they talk about density dependence, a tendency that made an important contribution to the controversy surrounding the related concept of regulation. As a result, many ecologists prefer to use other terms that do not carry as much conceptual baggage, for example, stabilization (Den Boer and Reddingius 1996) or resiliency (Pimm 1991). Some even propose to do away with this concept (Berryman et al. subm.). The situation is made worse by the tendency not to specify the context, so that density dependence may refer to the general property of a system (e.g., detecting density-dependent regulation), to a property of the vital rates (e.g., birth rate is, or is not, density dependent), or to a specific mechanism that is responsible for bringing about this property (e.g., territoriality results in density dependence). Some authors include both the state and a mechanism in their definition: "Ecological density dependence . . . is a return tendency in population abundance coupled with a scientifically defensible identification of a regulatory mechanism" (Wolda and Dennis 1993:589).

Here is the definition that I prefer: **density dependence** is some (nonconstant) functional relationship between the per capita rate of population change and population density, possibly involving lags (Murdoch 1994). Thus, density dependence is a general property of the dynamical system, and does not imply that any particular vital rate must be density dependent. Neither does this definition require a specification of the ecological mechanism that is responsible for density dependence (pace Wolda and Dennis 1993): the *property* of a dynamical system and the *mechanism* explaining it are two logically distinct issues.

Translating this definition into more formalized language, we have (using the discrete-time framework):

$$r_t = f(N_{t-1}, N_{t-2}, \ldots, \epsilon_t) \tag{5.5}$$

where $r_t \equiv \ln(N_t/N_{t-1})$ is the per capita rate of population change, N_t is the population density at time t, and ϵ_t represents the action of exogenous factors.

5.4.2   Regulation: Evolution of the Concept

The first clearly formulated definition of population regulation employed the concept of a stable point equilibrium. For example, Varley et al. (1973:19) gave the following definition: "a regulated population (is) ... one which tends to return to an equilibrium density following any departure from this level." The applicability of the notion of a stable point equilibrium to natural populations was challenged by Andrewartha and Birch (1954), as well as more recently (e.g., see Wolda 1989). There are, indeed, two serious limitations of this definition. First, nonlinear dynamical systems, such as natural populations, can be characterized by dynamical behaviors that are more complex than a stable point equilibrium, but do nevertheless fit our intuitive notion of being regulated. Second, the definition relying on a stable point equilibrium does not explicitly acknowledge that population density is affected by both density-dependent factors (that may be responsible for a stable point equilibrium) and density-independent factors that provide a continuous source of stochastic fluctuations in population density.

A generalization of the stable point equilibrium that addresses its first shortcoming is the notion of a deterministic **attractor** (section 5.2.1). The notion of a deterministic attractor, however, is still not general enough to be useful in ecological applications, because it does not explicitly incorporate the stochastic sources of population change.

In the presence of noise, a deterministic attractor becomes a stationary probability distribution of population density (May 1973b). In intuitive terms, the "equilibrium" is no longer a point, but a cloud of points (Wolda 1989; Dennis and Taper 1994). Similarly, periodic attractors (a finite number of points) and chaotic attractors (fractals with fine complex structure) are "smeared" into probabilistic clouds of points (for examples, see Schaffer et al. 1986).

5.4.3   The Stationarity Definition of Regulation

Thus, one defensible notion of population regulation is to identify it with the presence of a long-term *stationary probability distribution* of population densities (Dennis and Taper 1994; Turchin 1995b). Other names for the same thing are May's (1973b) stochastic equilibrium probability distribution, Chesson's (1982) convergence in distribution to a positive random variable, and Wolda's (1989) probabilistic cloud of points. In addition to noisy stable points, periodic oscillations, and chaos, this notion also includes multiple stable states, or "metastable dynamics." The latter kind of a behavior results when there are two or more deterministic attractors, each with its own basin of attraction, which in the presence of noise become "connected" into one stochastic attractor. The stationary probability distribution in such populations may have a bimodal shape.

A generalization of the "stationarity" definition of population regulation is "stationarity in the wide sense" of Royama (1977). Royama's definition does not require a stationary probability distribution of population density, but only that the mean and the variance of the distribution do not change with time. The distinction between narrow-sense and wide-sense stationarities is a rather minor one, and unlikely to be important in practice.

5.4.4 Beyond Stationarity: Stochastic Boundedness

One shortcoming of the stationarity distribution of population regulation is that it excludes certain dynamical behaviors that we intuitively feel should be regarded as examples of regulation. In particular, a population fluctuating around a trend is not stationary, since its mean is changing with time. Thus, by the stationarity definition, the population is not regulated. Nevertheless, if the trend does not lead to zero or infinite density, many ecologists would prefer to include this kind of behavior in our notion of regulation. One answer to this quandary may be provided by the notion of stochastic boundedness (Chesson 1978, 1982). Stochastic boundedness is a characteristic of persistent populations. Mathematical details are given in Chesson (1982), but the main intuitive idea behind stochastic boundedness is that "small populations are seen infrequently. Stochastic boundedness implies a steadiness to population fluctuations. No trends to ever lower population densities are possible and the average frequency of fluctuations to low density does not increase with time" (Chesson 1982).

In general, the presence of a lower bound on population density seems to be much less controversial than stronger notions of population regulation. In favorable environments and when population density is low so that there is plenty of food and few natural enemies, the average rate of population change has to be positive, and the population should increase; otherwise it would go extinct and we would not have to worry about it. Unfortunately, further consideration shows that the concept of a lower bound is much more complex than one would think (see section 2.3.1). Additionally, in order for the concept to be useful, we need to know how to detect stochastic boundedness in practice. While we have well-developed statistical tests for stationarity, the same can hardly be said for boundedness. It is also clear that a test for boundedness would require massive amounts of data, simply because typical populations spend a small proportion of time near the lower bound on density fluctuations.

Practical considerations also limit the usefulness of tests that would attempt to detect "regulation around a trend." Our typical time series are, at best, several decades long. A very high proportion of time series of this length generated by unregulated dynamical systems (e.g., random walk) look just like "regulation around a trend." The Statistical power of tests that distinguish stationary from unregulated processes is not very high, given the typical length of data sets (Dennis and Taper 1994). If we extend the notion of *regulated* to include *regulated around a trend*, then we will need such massive amounts of data that we will not be able to make much headway in practical applications. For these reasons, I will continue to stick to the stationarity definition of regulation, even while acknowledging its limitations.

5.4.5   Synthesis

Although population ecologists continue to debate regulation, the broad outlines of a consensus are emerging (Royama 1981, 1992; Berryman et al. 1987; Berryman 1999; Hanski 1990; Murdoch 1994; Dennis and Taper 1994; Den Boer and Reddingius 1996; Hassell et al. 1998). I believe that most population ecologists are in agreement on the major, strategic issues in population regulation, while the ongoing debate increasingly focuses on narrow tactical questions

(Turchin 1995b). Here are some areas of agreement (Turchin 1999):

1. The central quantity of interest in the analyses of population regulation is the realized per capita rate of population change, defined as $r_t = \ln(N_t/N_{t-1})$, where $\ln N_t$ is the natural logarithm of population density at time t. As an aside, this point follows directly from the first law of population dynamics (chapter 2).

2. The realized per capita rate of change, r_t, is affected by both exogenous and endogenous factors. Exogenous factors are not "noise" to be tuned out. They represent important biological processes affecting population change, and are a legitimate and an interesting subject for study in their own right.

3. Some negative feedback between r_t and population density is a necessary (but not sufficient) condition for population regulation.

4. Population dynamics are inherently nonlinear. A wide variety of functional relationships between the expected r_t and population density are possible, including monotonic—either convex or concave—but also more complex relationships such as the Allee effect or metastable dynamics. For some ranges of density the expected per capita rate of change may be flat, with population dynamics dominated by exogenous factors (the so-called density-vagueness).

5. The rate of population change may be affected not only by the current population density but also by lagged density. Specific mechanisms may involve either intrinsic or extrinsic factors (reviewed in chapters 3 and 4, respectively). The lag structure of population regulation may be quite complex, with more than one time delay affecting r_t.

6. Finally, a focus on testing null hypotheses against unspecified alternatives has proved to be unproductive in investigations of population regulation. In other words, the interesting question is not whether we can reject the hypothesis that a population is "unregulated." A much more fruitful approach is to investigate the *structure of population regulation* (see section 6.2.1). Ultimately, we need to determine what ecological mechanisms explain population regulation in any particular case study.

PART II
DATA

CHAPTER 6

Empirical Approaches: An Overview

6.1 INTRODUCTION

There are three general approaches to studying population fluctuations: statistical analysis of observational (e.g., time-series) data, mathematical modeling of mechanisms, and experiments. Until recently, ecologists (at least, in North America) have tended to emphasize manipulative experiments as *the* way to address ecological questions. For example, the eminent empirical ecologist C. J. Krebs argues against both time-series analysis and mathematical models. He suggests that we should avoid collecting long-term data simply for the purpose of analyzing it for density dependence, because "this approach has been a bankrupt paradigm" (Krebs 1991:6). Furthermore, "mathematical models are more seductive than useful at this stage of the subject" (Krebs 1988). Krebs argues that progress in elucidating the mechanisms of population regulation can be achieved only by careful experimentation (Krebs 1995). I strongly disagree with this philosophical stance. In fact, I believe that the reason why it took us so long to begin understanding the mechanisms of population dynamics is precisely because of the narrow focus on experimental approaches by empirical ecologists (especially in North America; Europeans have adopted a much more synthetic framework), and lack of integration between experimental, theoretical, and data-analytic approaches. In contrast to Krebs, who argues that we should start with experiments, I propose that we should *end* with them. The experimental approach is most powerful during the later stages of an investigation into dynamics of any particular population, after time-series analyses have reduced the number of viable hypotheses, and the potential mechanisms have been modeled mathematically to obtain quantitative predictions that can now be tested experimentally.

I begin this chapter by discussing the role of time-series analysis in the investigation of population fluctuations (section 6.2). Then I discuss how experiments can be integrated into the synthetic framework (section 6.3). The next two chapters are devoted to two general, but also rather technical, subjects of population data analysis: phenomenological time-series analysis (chapter 7) and fitting mechanistic models to data (chapter 8). Because many issues of experimental approach are system- or question-specific, I chose not to devote a specific chapter to a general treatment of experiments, but instead review experimental results for specific empirical systems in appropriate chapters.

6.2   ANALYSIS OF POPULATION FLUCTUATIONS

6.2.1   The Structure of Density Dependence

Time-series analysis is particularly useful during the initial stages of investigation, when little is known about mechanisms underlying the pattern of population change. One goal of time-series analysis is a quantitative description of the fluctuation pattern (mean, variance, autocorrelations, etc). Another is to characterize the **structure of density dependence**, by which I mean the following aspects of dynamics: process order, the shape of the functional relationship between r_t and lagged population densities, trajectory stability, and signal/noise ratio.

Time-series analysis has the potential of reducing the list of hypotheses for potential ecological mechanisms underlying the observed dynamics. For example, during the 1980s the most popular explanation for southern pine beetle dynamics was that its fluctuations were driven by variable climate (see chapter 10). Yet our analysis of time-series data showed that the greatest proportion of variance in the realized per capita rate of change was explained by second-order endogenous factors. Because the predictions of the hypothesis and data patterns did not match, we concluded that we need to investigate other mechanisms to understand this system.

Because any given pattern of density dependence may be produced by many different ecological mechanisms, time-series analysis alone cannot prove that a specific mechanism is the one underlying oscillations. For example, an apparent Allee effect, a positive relationship between r_t and N_{t-1} at low N_{t-1}, could be either a result of direct cooperation between organisms, or a side effect of Type II functional response by natural enemies, to name just two mechanisms. Similarly, delayed density dependence can arise as a result of a variety of mechanisms, as was extensively discussed in chapters 3 and 4. It is important not to oversell the value of time-series approaches, since characterizing the structure of density dependence can, at best, get us only a part of the way to an understanding of mechanisms. Yet, it is equally important not to neglect the usefulness of this approach. And we should not forget that other scientific approaches (including manipulative experiments) also cannot *prove* a hypothesis.

In section 1.2.2 I suggested that there is a sequence of model types used in population ecology, ranging from most mechanistic (individual-based models) to less mechanistic (models employing functions of summarizing features of individuals). Models of density dependence are the next step toward the phenomenological end of the scale. Thus, I view "density dependence" not as a hypothesis to be falsified or corroborated but as a research program: a theoretical framework with which to investigate the causal factors of population fluctuations. Although quantifying density dependence does not uniquely identify the biological mechanisms responsible for various aspects of population dynamics, it allows us to make the first step toward this goal.

6.2.2   *Probes: Quantitative Measures of Time-Series Patterns*

To make progress toward identifying the mechanism(s) of population fluctuations, we must define a set of alternative hypotheses, translate them into specific models, and use the models to generate predictions about time-series patterns (section 1.2). We can then attempt to distinguish between the competing models by doing parallel time-series analyses on the data and on model outputs, to obtain quantitative measures of their relative success at matching the patterns in the data.

A basic premise of this book is that there is a correspondence between patterns in time-series data and ecological mechanisms that can explain population fluctuations (Kendall et al. 1999; Turchin and Ellner 2000b). Clearly the correspondence between mechanisms and patterns is not one-to-one (see the previous section). Nevertheless, there is a sufficient mechanism/pattern congruence to allow us to narrow down the field of competing hypotheses. Thus, even in circumstances where experimental studies are not possible, characterization of the fluctuations by time-series analysis may provide useful information for testing hypotheses. In summary, time-series analysis can point out promising avenues to explore, and when combined with mathematical modeling and short-term experiments it can also distinguish between rival hypotheses.

A critical ingredient in the program of matching ecological mechanisms with data on population fluctuations is some systematic quantification of time-series patterns. Some quantitative measures—let us call them time-series probes—are straightforward, and require conceptually simple analytical methods. The sample mean and variance of population data are the most obvious simple probes. Another set of simple probes is the average period and some measure of periodicity strength.

Other probes are less straightforward. One very useful probe would be an estimate of the order of the dynamical process. From the statistical point of view, the order of a process is the number of values that are useful for predicting future changes, for example, the number that gives predictions with the lowest mean square error (Cheng and Tong 1992). Another way to characterize dynamics is by their stability (section 5.3) with, for example, Lyapunov exponents (section 5.2.2). An even more basic distinction is between regular oscillations and random fluctuations. These two types of dynamics are distinguished by whether future values of population density are at all predictable based on past values. Thus, a useful probe would be a measure of the relative strengths of endogenous versus exogenous contributions to population fluctuations, because many ecological mechanisms fall into one or the other category.

It is important to choose probes that are "physically" meaningful and relate directly to properties of the real-world dynamics. Any good model of the system should be able to approximate probes of this sort in a robust fashion and not yield grossly biased estimates due to inevitable inaccuracies of any fitted model. An example of a bad probe, in this sense, is the fractal dimension of the attractor (Ellner 1989). The presence of dynamic noise means that the attractor dimension is really infinite, so any finite dimension that we compute is simply a reflection of how we choose to view the attractor in finite dimensional space.

Amplitude and periodicity can be directly calculated from an observed time series of population numbers, but process order, trajectory stability, and predictability cannot. To estimate these probes, we need to fit models to data, and use the models to calculate the probes. This brings us to a central topic: choosing and fitting time-series models.

6.2.3 *Phenomenological versus Mechanistic Approaches*

Analysis of time-series data can be approached with a spectrum of models, ranging from mechanistic to purely phenomenological. One extreme is a **fully specified model**: it is assumed that all functional relationships between variables are completely known, and there are independent estimates for all parameters. Such a model can be used to predict patterns in time-series data. A quantitative comparison between model predictions and data patterns constitutes a test of the hypothesis that mechanisms underlying population oscillations are understood and correctly modeled.

At the opposite end of the spectrum are completely phenomenological models, in which some flexible functional form is used to describe the pattern of temporal relationships in the data. The extreme of flexibility is to use nonparametric regression, such as smoothing spline, neural net, or kernel regression (Nychka et al. 1996), which are now generally available in commercial statistical software. The functional form and the order of the process may be left unspecified, and are estimated from the data. In such a model, the only assumption being made is that the population dynamics are driven by a combination of density-dependent feedbacks and random exogenous perturbations.

Both approaches have their strengths and weaknesses. Phenomenological models make fewer assumptions, but require more data, especially for fitting high-dimensional models. More mechanistic models, on the other hand, are much more parsimonious with data, but may completely miss the mark by including a grossly wrong assumption.

As a result, most practical approaches to modeling time-series data fall in between the two extremes. For example, we might develop a model in which all variables and parameters have a clear biological interpretation, and in which specific functional relationships are assumed, but values of parameters are not known and need to be estimated. Such a model would fall near the mechanistic end of the spectrum. Moving further to the phenomenological end, we may partially specify the mechanistic structure of the model, but leave some functional relationships unspecified and fit them empirically using flexible general forms (Ellner et al. 1998; Wood 2001).

Thus, it can be argued that the art of time-series modeling consists of "hardwiring" in the model what is definitely known, and using the time-series data to estimate what is not known. Clearly, the optimum mix of mechanism and phenomenology will depend on how much is already known about the system and how much data are available. As we learn more about the system, we should expect to employ increasingly more mechanistic models.

6.3   EXPERIMENTAL APPROACHES

I define **experiment** as a planned comparison between data and a novel, nontrivial prediction derived from a hypothesis. The purpose of the scientific experiment is to test a theory, that is, to increase or decrease its empirical support. A *novel* prediction is the one whose truth is not known to the investigator at the time when he/she derives it from the hypothesis. Often predictions are made about some future (as yet unobserved) event or measurement, but logically it is not necessary that the event take place in the future. For example, if one investigator makes a prediction about some data already collected by another scientist, then I consider it a strong prediction if the first investigator had no knowledge of the data when making the prediction.[1] The key to novelty is not predicting the future, but the potential

---

[1]Logically, in fact, it should not matter whether the investigator knows the outcome while making a prediction, but from the psychological point of view predictions about a currently unknown outcome carry more weight.

for falsification (understood broadly as the potential for increasing or decreasing empirical support of a hypothesis).

The second aspect of prediction, its lack of triviality, is also related to the falsifiability potential. Intuitively, a trivial prediction is one that we could easily make without the theory we are aiming to test. Trivial predictions could be simply vague and all-encompassing ("population density will fluctuate in the future"), or something that we would expect to happen anyway. For example, practically all ecological models predict that if population density increases much above the normal level of fluctuations, then we should expect a population decline. Deriving this "prediction" from, say, a predation-based hypothesis and determining that, indeed, a population decline occurred provides very little empirical support for the predation hypothesis. Thus, potential for falsifiability is an important feature of an empirical test of a theory. Clearly a theory that makes a strikingly unexpected prediction that eventually turns out to be true gains more empirical support than a theory that makes a rather trivial prediction that also is found to be true. But what constitutes a striking, unexpected prediction is difficult to define, especially because people's expectations tend to change. What was striking and unexpected ten years ago may now be commonplace. The only way I know that allows us to measure this property of prediction (and therefore the empirical value of the experiment) is by contrasting it with a prediction derived from some alternative theory. In other words, I do not believe that a hypothesis can be tested in isolation; we can only test predictions from two or more hypotheses against each other (see also section 1.2). In the context of multiple hypotheses, the idea of nontriviality is much easier to define. Essentially, a nontrivial prediction is the kind of outcome that is predicted by hypothesis 1 with high probability while all other hypotheses assign this outcome a low probability. This definition directly leads to the quantification of empirical support for one hypotheses relative to others in terms of likelihood ratios or Bayesian posteriors.

An important aspect of nontriviality is making predictions quantitative. A qualitative prediction of the kind "upon increasing factor X the theory predicts that factor Y will decrease" is of little use. If under the null hypothesis, the factor Y can increase with probability 0.5 or decrease with the same probability, then a decrease in Y

could have happened by chance alone. Thus, an experiment showing that Y indeed decreased provides little support to the tested hypothesis. In somewhat more technical terms, this experiment does not allow us to distinguish between the tested and the null hypotheses, because the probability of the outcome "Y decreases" is rather high under both hypotheses. Making the prediction quantitative ("Y will be in the region between Y_{min} and Y_{max}") decreases the probability of this particular outcome under the null hypothesis, and thus makes the experiment more meaningful.

Obtaining predictions to be tested experimentally is a serious business, requiring much thought and effort. It is amazing to me that many ecologists will lavish an enormous amount of energy on the construction of the experimental apparatus and data collection while paying very little attention to predictions that they are trying to test. There seems to be a stock of "the usual suspects" that are supposed to be tested. One of the worst examples of this channelized way of thinking is empirical tests of the role of predators in population cycles. Somehow a notion became established that the only way to "prove" that predators are driving cycles is to remove them and observe whether the cycle stops. One problem with this idea is that it assumes that the only definitive experiment has to be the *manipulative* kind. However, from the logical point of view, a *mensurative* experiment has equal validity for testing hypotheses. If two alternative hypotheses make very different predictions about some feature of population dynamics, then we can distinguish between the two hypotheses simply by measuring the appropriate variables. For example, food-based and predator-based explanations will make different predictions about predation mortality patterns and about some food-related patterns (e.g., how body weights should change throughout the cycle). If we can figure out a way to measure these variables during the complete cycle, then we should be able to determine which of the explanations comes closer to the patterns observed in the data. The ability to manipulate an ecological system is not an essential feature of the experimental approach (Krebs 1991; see also Chitty 1996:55). However, although ecologists are well familiar with this idea in theory, in practice they still believe that a manipulative experiment somehow provides a stronger test of a hypothesis.

Another problem with "stopping the cycle" is that it is very difficult to do in practice. In fact, as far as I know nobody has succeeded in this yet (this is true at least with respect to the main case studies I review in part III). For example, in the remarkable experiment at Lake Kluane, Krebs et al. (1995) carved the boreal forest into giant 1 km × 1 km chunks. They put up an electrified fence to exclude ground predators and an aerial net to exclude avian predators from some experimental areas, and did other manipulations in others (more on this in chapter 13). Their intent was to "stop the cycle" of the snowshoe hare population. Instead what happened was that removing predators increased the peak density and delayed the population collapse, *but it did not stop the cycle*. Does this experimental result falsify the predation hypothesis? Dennis Chitty (1996:45) thinks so, while Krebs and coworkers disagree. We shall return to the question of what this experiment tells us in chapter 13, but what is important for the point in hand is that even a high-powered group of accomplished field ecologists, after decades of studying snowshoe hares, designed a multimillion dollar experiment and still were unable to stop the hare cycle. What's more, even after failing to stop the cycle, these authors nevertheless concluded that predators were one of the two factors driving the hare cycle. This seems clear evidence to me that stopping the cycle is not necessary to establish empirical support for a hypothesis! Another experimental study (on the red grouse) came even closer to stopping the cycle (Hudson et al. 1998), but still failed (Lambin et al. 1999; response in Hudson et al. 1999). The problem is that if a predator (or parasite) removal manipulation fails to stop a cycle, then several explanations are possible:

1. Predation is not the mechanism that drives the cycle.
2. Experimenters did not manage to remove enough predators to obtain a detectable effect.
3. Prey diffused out of experimental areas into low-density areas outside.
4. Removal of predators caused prey density to increase to the point where another factor came into play (exhaustion of food, an epidemic, etc.) that caused the decline of prey density.

Thus, a negative result by no means amounts to a "refutation" of the predation hypothesis (item 1 in the preceding list), since one also needs to exclude all the other logical possibilities (items 2–4).

I think it is time to admit that the "stop the cycle" attitude is actually counterproductive. By raising the level of proof that we require of a hypothesis very high, and then failing to achieve that level for any current theory, we ignore equally valid, although less spectacular, advances in testing alternative hypotheses, and create an impression that no progress has been accomplished since Elton's 1924 paper. Furthermore, those who criticize trophic hypotheses for various empirical systems on the grounds that no experiment has managed to stop the cycle by removing consumers are typically adherents of some intrinsic mechanism. Yet, I know of no instance of where an experimental manipulation of an intrinsic factor resulted in stabilizing a cycling population.

"Stopping the cycle" is a prediction that is intellectually easy to make—it requires no hard thinking and diligent modeling—but it turns out to be very difficult to do in the field. Perhaps it would be better to turn it around, and spend some intellectual effort on deriving equally informative predictions that are *easy* to test in the field. Perhaps we should also take some lessons from physicists, who never tried to do a fool thing like stopping a planet from revolving around the Sun to prove that Newton was right. In fact, most of the experiments that changed the course of physics were of the mensurative kind. Perhaps the most influential physical experiment of the 20th century was the observation that the gravitational field of the Sun bends Mercury's light. This mensurative experiment in one stroke changed Einstein's theory of relativity from an esoteric hypothesis to the established theory (Oksanen and Oksanen 2001).

CHAPTER 7

Phenomenological
Time-Series Analysis

At the start of an investigation into population dynamics of some specific system we typically do not know enough about it to begin formulating intelligent hypotheses about its behavior. Thus, the first phase
of the investigation should be exploratory, and we need to answer the
following questions (see chapter 5): Are dynamics periodic? What
kind of stability does the system possess? What is the process order
of fluctuations? What are relative contributions of endogenous and
exogenous factors? To answer these questions, we need to fit to data
some generic, or **phenomenological**, models. The goal of this chapter,
therefore, is to review phenomenological approaches to time-series
analysis of population fluctuations. In addition to discussing general
approaches, I describe one particular implementation, the *nonlinear
time-series modeling* (NLTSM) approach, which evolved from the
original paper by Turchin and Taylor (1992; see also Turchin 1993,
1996).

7.1 BASICS

7.1.1 Variance Decomposition

The traditional approach to time-series analysis is based on decom-
posing the variance of a series into trend, seasonal variation, other
cyclic oscillations, and the remaining irregular fluctuations (Chatfield
1989). This framework can be adapted to ecological problems with a
few modifications, as follows.

Trend is a long-term, exogenously driven, systematic change in the environment. The most frequent consequence for population dynamics is a change in the mean level of fluctuations. The change can be either gradual or abrupt (a step trend). Trend is one example of **nonstationarity**. Other kinds of nonstationarity include systematic changes in the variance, and in the dynamical type of fluctuations.

Periodic changes in the environment are another kind of nonstationarity. The most common example of periodic environmental "forcing" is seasonality. Although seasonality affects most natural populations, I largely sidestep the issue of how to model it, because the main goal of methods reviewed in this chapter is the understanding and quantification of *multiannual* population fluctuations. If data are collected several times a year, they can be aggregated into yearly indices of population density either by averaging within a year or by subsampling once per year. Some approaches to the explicit modeling of seasonal oscillations can be found in Ellner and Turchin (1995), and an example of fitting a mechanistic seasonal model will be discussed in chapter 12.

The sources of variation discussed above represent systematic **exogenous** influences (changes in the environment that affect population change, but are not themselves influenced by population numbers). The **endogenous factors** (dynamical feedbacks affecting population numbers, possibly involving time lags) are another important source of variation, particularly if the underlying endogenous dynamics belong to the class of complex dynamics (limit cycles, quasiperiodicity, and chaos).

Irregular **fluctuations**, finally, are the residuals left after systematic environmental changes and endogenous oscillations have been extracted from the series. These residuals are usually interpreted as exogenous "environmental stochasticity," but they will actually contain at least two fundamentally different sources of variation. **Process noise** is the technical term I use in the book for exogenous environmental stochasticity. **Measurement noise** arises from observation errors. Process noise and measurement errors are treated very differently in time-series analysis. A perturbation due to process noise affects not only N_t but also the following densities, since there is a functional dependence between N_{t+1} and N_t, due to the endogenous dynamics of the system. By contrast, a perturbation due to measurement error affects only the data point N_t that we use in the analysis, not the actual density. As a consequence, there is no influence (from this particular perturbation) on subsequent density values.

7.1.2 Data Manipulations Prior to Analysis

An ideal time series to analyze is long, has been collected at completely regular time intervals, has no zero measurements or missing values, and comes from a stationary dynamical system. Unfortunately, in the real world such ideal data sets are rare. Below I discuss some practical approaches for dealing with two particularly problematic issues, zero values and nonstationarity.

Handling Zero Values Presence of zero values in the data presents a serious difficulty, because data will usually be subjected to various transformations, most notably logarithmic. The standard approach to dealing with this situation, which I do not recommend, is to add 1 to the whole series. The problem with this approach is that it can seriously distort the data patterns, because it ignores the natural scale of variation in the data. For example, suppose that the minimum (apart from zeros) and the maximum of a data series is 0.0001 and 0.01, respectively. Adding 1 to such a series will cause it to fluctuate between 1 and 1.01, completely hiding the true variation. Similarly, if nonzero numbers vary between 100 and 200, then adding 1 to zeros will result in exaggerating the actual amplitude.

A good practical approach is to replace zeros with the value that could be the smallest potentially observable number in a given study, as determined by the apparatus used to measure density. For example, if in a mammal-trapping program we have used 100 traps per hectare, then the smallest nonzero measure of population density that we could observe would be 1/100 individuals per ha (if we capture just one individual). Thus, we can substitute 0.01 for zeros in this data set. Another approach, which can be used when no information about measuring apparatus is available, is to substitute the smallest observed value for zeros.

Detrending Presence of exogenous trends in the data, for example, a gradual change in the mean level of fluctuations, leads to a lack of stationarity, and may seriously degrade our ability to extract endogenous dynamics (Nychka et al. 1992). In nontechnical terms, a dynamical system is **stationary** when the mechanisms generating its fluctuations do not change with time. In other words, a stationary system is dynamical and its state constantly changes with time, but the rules underlying change do not. If we attempt to fit a statistical model assuming stationarity to data generated by a nonstationary system, then the model will not fit data well, and its estimate of the random element in dynamics ("irregular fluctuations") will be inflated as a result of failing to take the systematic trend into account.

A commonly used approach in time-series analysis is first to remove the suspected trend (this is called *detrending*), and only then attempt to characterize endogenous dynamics. Unfortunately, detecting and properly characterizing trends is not particularly easy (Berryman et al. 1988). Furthermore, when detrending is done in a completely phenomenological manner, without any biological information, it can lead to highly spurious results. For example, detrending a random walk (an unregulated population) will introduce spurious autocorrelations at small lags (Jassby and Powell 1990), thus possibly giving an illusion of regulation. In other words, a time series of an unregulated or weakly regulated population can be "decomposed" into a spurious trend and equally spurious endogenous regulation component. Another potential problem is that not all systematic changes in population density are a result of environmental trends. For example, a population may appear to exhibit a gradual trend of increasing numbers, suggesting that the carrying capacity of the environment is increasing while in reality the population is recovering from a catastrophic event just prior to the beginning of observations. Thus, it is best to reserve detrending for situations where there is some external evidence of environmental change. In case such evidence is lacking, but there are strong indications of nonstationarity, data should be analyzed both with and without detrending, to document the effect of detrending on results. Finally, if data are long enough, then they can be split into shorter pieces and analyzed separately (see next page).

As to specific methods to detrend a time series, the classical ARIMA approach (Box and Jenkins 1976) employs differencing, which would be equivalent to constructing a new series based on realized per capita rates of change, $\{r_t\}$. In my opinion, differencing is usually not a useful approach in population dynamics, because results of fitting models to the r_t series are much more difficult to interpret in ecological terms than doing the same for the nondifferenced data. While we have some intuition and even formal results allowing us to interpret results of fitting r_t to $N_{t-1}, N_{t-2}, \ldots$, there is no comparable framework for interpreting models like $r_t = f(r_{t-1}, r_{t-2}, \ldots)$.

The most straightforward and useful approach to detrending in ecological applications is to fit the temporal trend with a curve, and then subtract it from the data. For example, removing a linear trend involves first fitting a line to the data by regressing $Y_t = \log N_t$ on t. This gives us the intercept and slope parameters, a and b. The detrended series then is

$$Y_t' = Y_t - (a + bt)$$

If a trend appears to exhibit curvilinear characteristics, than a second-degree (or higher) polynomial could be used in the same way.

Data Splitting   Data splitting involves cutting the series into pieces and analyzing each segment separately. This approach is particularly useful when we deal with a step trend, because we may suspect that the nature of the dynamical process has somehow changed in the middle of the observation period. Note that a step trend may affect the mean, the variance, and/or the autocorrelation structure (e.g., see the red grouse data in figure 11.1).

Data splitting should be practiced only with relatively long data series. Extensive experience with time-series analysis of ecological data suggests that data series shorter than about 20 years rarely yield satisfactory results, especially when we are investigating second-order dynamics. Of course, if we are interested in quantifying first-order oscillations, we may get away with just ten data points, which could allow us to fit a simple three-parameter model, such as the theta-Ricker. For second-order dynamics, for which period lengths are substantially longer (typically 6–12 years), we need longer series. The general rule of thumb is that we need at least three "data points"—three complete oscillations. Depending on the average period of each cycle, then, we may need from 18 to 36 years or, rounding, a series of 20–30 years. In fact, it may be a good idea to check on one's results by routinely splitting long series, and analyzing shorter pieces separately, even when no obvious nonstationarity is detected by preliminary examination of the data.

In sum, the rules of thumb that I suggest for data splitting are as follows (this assumes second-order dynamics; for shorter cycles one can correspondingly multiply the number of pieces beyond the ones I suggest here). Aim at 20–30 years per piece. Thus, a series that is 40–60 years long should be split in half, 60–90 years long should be split in three pieces, and so on. However, if cycle period is rather long, than increase the length of pieces. Attempt to have no fewer than three oscillations per piece.

7.1.3   Diagnostic Tools

The Time Plot   The very first step in time-series analysis should always be to plot observed values of population numbers at time t, N_t, against time. The graph may reveal such features of data as a trend (gradual or step), presence of periodic oscillations, and outliers. The data should be log-transformed by calculating $Y_t = \log_{10} N_t$. The base of 10 is useful because we can immediately see "orders of magnitude" in population fluctuations, by comparing the degree of population variability against a single tick (corresponding to a tenfold change) on the y-axis.

The Phase Plot   Phase portraits plot current population density, N_t, against lagged density, N_{t-1} (again, it is customary to plot these variables on a log scale). Points corresponding to successive t are connected by lines to indicate population trajectory through the phase space. A useful variation of this approach is to plot $r_t = \ln(N_t/N_{t-1})$ against N_{t-1}. The phase plot often provides useful clues. Thus, populations that are regulated by direct density-dependent mechanisms show back-and-forth fluctuations around the mean value, while populations regulated by delayed density-dependent feedbacks show circular clockwise orbits.

A Measure of Amplitude   An intuitive measure of amplitude is the ratio of peak to trough density. Unfortunately, this measure has an undesirable property of being affected by the length of data, because in longer series unusually low or high values are more likely than in short data. An alternative measure that has better statistical properties is the standard deviation of log-transformed data:

$$S = \frac{1}{n-1} \sum_{i=1}^{n} (Y_t - \overline{Y})^2 \tag{7.1}$$

where $Y_t = \log_{10} N_t$, $\overline{Y}$ is the mean of Y_t, and n is the number of observations. S was introduced as a measure of relative variability by Lewontin (1966), and used in the context of population dynamics by Stenseth and Framstad (1980). Because S is calculated based on log-transformed densities, it is linearly related to the peak/trough ratio. For example, a sine wave with peak/trough ratio of 10 and 100 has $S = 0.35$ and 0.7, respectively. In other words, $S = 0.35$ can be interpreted as one order of magnitude between peak and trough density.

Frequency Distribution of r_t   Population oscillations are often asymmetric, because the rise phase is longer than the decline phase (Moran 1953; Ginzburg and Inchausti 1997). The reason is that there is no biological limit on how fast organisms can die, while there is a limit on how fast they can reproduce. Asymmetry can be noticeable by the naked eye, but it would also be useful to be able to quantify it. A simple approach is to divide a population series into rise and decline intervals, and calculate the mean duration of each phase. This approach works well with very regular data such as the lynx fur returns (see Royama 1992:187–188; also Ginzburg and Inchausti 1997), in which peak and trough years can be assigned unambiguously. For messier data sets, a more robust procedure would be to examine the frequency distribution of r_t. Cycle asymmetry is revealed by less numerous but more extreme negative values combined with more numerous, but smaller in magnitude, positive r_t. This pattern can be quantified (and statistically tested for) by calculating the skewness of the r_t distribution.

Frequency Distribution of Y_t   Another topological characteristic of cycles, in addition to asymmetry, is the peak shape. Two general scenarios can be distinguished. In one, population density grows exponentially (linearly on the log-transformed plot) until it reaches the peak, when there is an abrupt switch and density begins declining. Such population cycles have a saw-toothed pattern characterized by "sharp" peaks. The second possibility is for density to approach the peak value and remain at high density for a more or less extended period of time. These dynamics yield a series of "blunt" peaks. As I discussed in section 4.2.3, predator-prey models in an oscillatory regime tend to produce saw-toothed predator dynamics and blunt-peak prey dynamics (see figure 4.3 in chapter 4). We can quantify peak shape by examining the frequency distribution of Y_t values, and testing whether its skewness is significantly different from zero.

Autocorrelation Function (ACF)   The most ubiquitous and a very useful diagnostic tool in time-series analysis is the autocorrelation function, or ACF. ACF is estimated by calculating the correlation coefficient between pairs of log-transformed population densities $Y_{t-\tau}$ and Y_t separated by lag τ, with $\tau = 1, 2, \ldots$. These correlation coefficients, plotted together as a function of lag τ, are known as a correlogram. By averaging over a noisy series, ACF reveals the trend and periodic patterns more clearly than the time plot (Box and Jenkins 1976; Chatfield 1989; for ecological interpretation of ACF, see Finerty 1980; Nisbet and Gurney 1982; and Turchin and Taylor 1992).

The simplest ACF pattern is exhibited by stationary monotonically damped systems (simple dynamics). ACF of such systems is positive at small lags, and then decays exponentially to zero at high lags. If ACF decreases slowly with lag, in a somewhat linear fashion, and at high lags becomes increasingly negative, the data are probably affected by a trend. Oscillatory ACF can arise either as a result of exogenous periodic forcing, or as a result of periodic endogenous dynamics. The former case is also known as "phase-remembering quasi-cycles" (Nisbet and Gurney 1982), and its ACF will not decay to zero at high lags, but continue to cycle indefinitely. The latter case is known as "phase-forgetting quasi-cycles" (Nisbet and Gurney 1982); its ACF eventually decays to zero at high lags. Thus, in theory ACF patterns can help distinguish between a series that is oscillating in response to an exogenous oscillator with a set period from a series that undergoes endogenous cycles. In practice, however, ecological series are rarely long enough to resolve this issue.

A crude (but sufficient for most applications) procedure of determining the statistical significance of ACF patterns is as follows. First, we determine the lag T at which ACF reaches its first maximum: this is the estimated **dominant period**. Next, we check whether ACF at lag T, ACF[T], is greater than $2/\sqrt{n}$, where n is the number of data points in the series. If yes, then we have *strong evidence of statistical periodicity*. If not, then we check the ACF at the lag nearest half the estimated period. If ACF[$T/2$] $< -2/\sqrt{n}$, then we have *weak evidence of statistical periodicity*. If ACF is not significantly different from zero at either period or half-period, then we conclude that there is no evidence of periodicity.

Spectrum The spectral density function, or spectrum, shows the decomposition of the variance of the process into contributions due to each frequency. The estimated spectrum (usually referred to as periodogram) exhibits peaks at those frequencies that contribute most to population oscillations. Chatfield (1989) provides a readable exposition of calculating and interpreting spectra; for an application in ecology see Finerty (1980). The spectrum provides much the same insights as ACF (in fact, the spectrum is the Fourier transform of the autocovariance function, a close relative of ACF). Spectra are widely used in engineering and physics. Their utility in ecology, however, is probably limited. First, interpretation of spectra is much less intuitive than interpretation of ACF, and a steep "learning curve" is involved in learning to use spectral analysis properly. Second, estimating spectra from data is a highly technical field, because a periodogram has to be smoothed to provide a consistent estimate of the spectrum. Finally, spectral analysis requires large amounts of data. For short series typically found in ecological applications one can often obtain more reliable results with ACF (Jassby and Powell 1990), so I chose not to include spectrum in the current implementation of NLTSM.

Partial Autocorrelation Function (PACF) While ACF is a very useful diagnostic for presence of trends and periodicities, it cannot tell us much about the lag structure of density dependence. A statistically significant ACF spike at lag 10 does not necessarily mean that there is a direct effect of density 10 years ago on present density, since it is much more likely that N_{t-10} affects N_t indirectly through the intermediate N values. For example, a model with only two lags, N_{t-1} and N_{t-2}, can generate limit cycles with period 10 yr. The question is, how many lags do we need to model population fluctuations adequately, or what is the *order* of the process? This problem is analogous to deciding on the number of terms to include in a multiple regression (Box and Jenkins 1976). Box and Jenkins recommend using the partial autocorrelation function (PACF) as an indicator of the order of the process. If dynamics of the system can be at least approximately represented as a linear (in logs) autoregressive process of order p (ϵ_t represents the exogenous dynamical component, a_i are constants)

$$Y_t = a_0 + a_1 Y_{t-1} + \cdots + a_d Y_{t-p} + \epsilon_t \qquad (7.2)$$

then the theoretical PACF will have nonzero values at lags that are less than or equal to p, and will be equal to zero at lags greater than p. This observation suggests the following practical rule (Box and Jenkins 1976): if PACF has spikes significantly different from zero at p lowest lags, and at higher lags PACF abruptly drops off, then one candidate model for the system is an autoregressive process of order p.

Partial Rate Correlation (PRCF) Functions   The major problem with using PACF as a diagnostic tool in ecology is that model (7.2) is not well suited to the analysis of population fluctuations. This is because most factors affecting population change—birth, death, and emigration—are per capita rates. Thus, in populations for which immigration can be neglected, a better model will have the general form

$$N_t = N_{t-1} F(N_{t-1}, N_{t-2}, \ldots N_{t-p}, \epsilon_t) \qquad (7.3)$$

The consequence of this functional form is that PACF at first lag (PACF[1]) does not contain much useful information about regulation at first lag. For example, if there is no regulation, PACF is still going to have a positive spike at first lag. Even in the presence of a negative effect of N_{t-1} on population change, PACF[1] will be positive if regulation is of the undercompensating kind. Only in the presence of overcompensatory regulation (tendency to overshoot equilibrium) will PACF[1] be expected to be negative.

The nature of the problem suggests the remedy. Instead of calculating partial autocorrelations between Ys separated by various lags, Berryman and Turchin (2001) proposed that partial correlation coefficients are to be calculated between $r_t = \ln N_t/N_{t-1}$ and $Y_{t-1}, Y_{t-2}, \ldots$. Thus we have an analog of PACF, which we will call the partial rate correlation function (PRCF).

7.2 FITTING MODELS TO DATA

The diagnostic approaches discussed in section 7.1.3 are essential tools in exploratory analysis of data, but in order to quantitatively characterize the **structure of density dependence**—process order, functional shapes, trajectory stability, and signal/noise ratio—we need to fit models to data. In this section, I discuss general approaches to fitting phenomenological models, and describe one particular implementation, the nonlinear time-series modeling (NLTSM) approach.

7.2.1 General Framework

The most general model underlying phenomenological approaches to ecological time-series analysis is

$$N(t + \tau_{\text{pred}}) = F[N(t), N(t - \tau), \ldots, N(t - (p - 1)\tau),$$

$$U^1(t), U^2(t), \ldots, \epsilon(t)] \tag{7.4}$$

As usual, $N(t)$ represents population numbers or density measured at time t. Reconstruction parameters τ_{pred}, τ, and p are, respectively, the *prediction time*, the *base lag*, and the **process order**. $U^i(t)$ are the exogenous factors that have been measured, and whose effects can be quantified. Finally, ϵ_t represents the **process noise**—action of exogenous factors that are not explicitly modeled (either because we have not measured them, or because we have a multitude of small-effect factors that we do not care to explicitly include in the model). Lacking data on exogenous influences, we usually have to lump all exogenous factors together into ϵ_t, but this need not be so. We shall discuss approaches to estimating effects of measured exogenous variables in section 8.2. As an important note, the formalism on which model (7.4) is based applies equally well to discrete and continuous dynamics. We can translate back and forth between a phenomenological model like (7.4) and a model formulated as a system of ODEs (an example in chapter 12).

In equation (7.4) the prediction time, τ_{pred}, is allowed to be different from the base lag, τ. Actually, we do not even have to reconstruct dynamics with densities lagged at regular intervals, as is assumed by (7.4). In theory, given unlimited data and computer power, we could use the data to help us select the best values of reconstruction parameters, using approaches similar to cross-validation discussed below. In practice, however, allowing too much flexibility in the general model is self-defeating—there are too many options that can be adjusted, while data are always limited. Experience suggests that we should limit the number of parameters that we estimate, and thus the customary approach is to set $\tau_{\text{pred}} = \tau$ (Ellner and Turchin 1995), leading to the following simplified model

$$N_t = F(N_{t-\tau}, N_{t-2\tau}, \ldots, N_{t-p\tau}, \epsilon_t) \tag{7.5}$$

Here I also dropped exogenous variables U^i, and switched to subscripts for improved readability. Furthermore, note that I switched to N_t instead of $N_{t+\tau}$ as the dependent variable in (7.5). This lag shift reflects the convention that I employ throughout the book: models for data analysis have N_t on the left-hand side (because we are trying to explain the present state of the system from knowledge of its past state), while theoretical models use N_{t+1} (because we are trying to predict the future based on the present state of the system).

A further modification of model (7.5) is suggested by the first law of population dynamics (section 2.2):

$$r_t = f(N_{t-\tau}, N_{t-2\tau}, \ldots, N_{t-p\tau}) + \epsilon_t \tag{7.6}$$

where I assume that random environmental influences affect population dynamics approximately multiplicatively, via vital rates. Log-transforming the dependent variable ($r_t \equiv \log N_t / N_{t-1}$) translates these multiplicative influences into additive ones.

To illustrate the advantages of using r_t as the response variable, consider the following example: fluctuations of a laboratory *Drosophila* population (data collected by Rodriguez 1989; analysis in Turchin 1991). When we plot these data in the $N_t - N_{t-1}$ phase plot, we observe a rather noisy relationship (figure 7.1a). We know that the

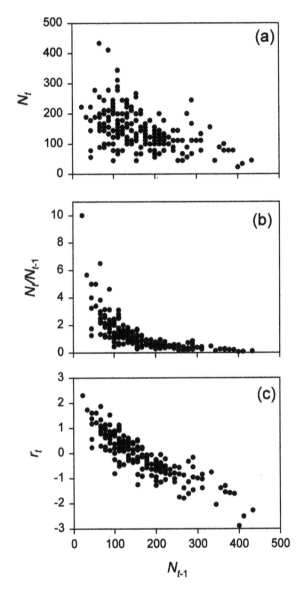

FIGURE 7.1. Three views of structure of density dependence in *Drosophila*.
(a) $N_t = f(N_{t-1}) + \epsilon_t$. (b) $N_t/N_{t-1} = f(N_{t-1}) + \epsilon_t$. (c) $r_t = f(N_{t-1}) + \epsilon_t$.

curve must go through the origin (no immigration was allowed in this population), but the data scatter obscures any underlying relationship. Plotting the per capita replacement rate N_t/N_{t-1} against N_{t-1} results in a much better definition of the functional relationship (figure 7.1b). However, the distribution of residuals is clearly heteroscedastic: the variance decreases from left to right. A logarithmic transformation takes care of this problem (figure 7.1c). We can now fit a simple linear relationship to the data (although a slight curvature suggests that employing a theta-Ricker model with $\theta < 1$ could improve the fit). The data are still noisy, but the r_t versus N_{t-1} plot reveals a clear density-dependent relationship in the data, something that is not very apparent on the N_t versus N_{t-1} plot.

Of course, we could fit the theta-Ricker model directly to the N_t versus N_{t-1} data, using nonlinear methods and an appropriate statistical model for the residuals. So the primary advantage of using r_t instead of N_t as the response variable is heuristic. Nevertheless, it is an important advantage, because it is much easier to see density-dependence relationships in the r_t-based phase plot. As a result, we are more likely to choose an intelligent model for fitting data. For example, until we plotted the data in an appropriate way, it was not apparent that we should use the theta-Ricker model to capture the downward curvature.

Model (7.6) provides the foundation of time-series analyses for the rest of this chapter. Although we have already made a number of simplifying assumptions, more work is needed before we can use it to investigate dynamics of a specific population system. We need to choose an appropriate base lag (section 7.2.2), select functional forms for approximating the relationship between r_t and lagged densities (section 7.2.3), and estimate model complexity (section 7.2.4).

7.2.2   Choosing the Base Lag

The choice of the base lag τ (which also serves as the prediction interval, since I advocate setting $\tau_{\text{pred}} = \tau$) can have a strong effect on the results of a dynamical analysis of the data. There are no universally accepted guidelines for making this choice, although we now understand that we need to avoid two pitfalls—redundancy and irrelevance

(Casdagli et al. 1991). *Redundance* occurs when τ is very small, so that the expectations of N_t, $N_{t-\tau}$, $N_{t-2\tau}$, and so on are practically the same. In such a case, knowing $N_{t-\tau}$ and $N_{t-2\tau}$ in addition to N_t does not help to characterize the state of the system any better than knowledge of N_t only. Attempting to fit a relationship between N_t and lagged densities will only model measurement noise. The opposite problem of *irrelevance* occurs when τ is so large that N_t and $N_{t-\tau}$ are not functionally related anymore, because of cumulative effects of noise (and trajectory divergence, if the system is chaotic).

An optimal choice of τ, therefore, falls somewhere between these extremes of redundance and irrelevance. If the population trajectory has been *undersampled* (i.e., sampled at too long intervals), then we will have a problem of irrevelance. The only remedy in this case, unfortunately, is to collect the data again using a smaller time interval. If the data are oversampled, by contrast, we can avoid the problem of redundance by using τ equal to multiple intervals. The general rule of thumb is to avoid using a base lag for which the autocorrelation between N_t and $N_{t-\tau}$ is too high (Casdagli et al. 1991). Thus, Ellner and Turchin (1995) recommend using the first τ at which ACF decreases below $\frac{1}{2}$. This rule typically leads to τ equal to between $\frac{1}{6}$th and $\frac{1}{8}$th of the dominant cycle.

It is also important for the choice of τ to make biological sense; that is, τ should correspond to some notion of generation time. If time delay used in the analysis approximately matches the generation time, then it becomes much easier to connect estimated process order to biological mechanisms that may be driving population dynamics. The simplest case is when we are dealing with discrete-generation systems such as many forest insects. The obvious choice is $\tau = 1$ yr, which matches both the generation times of the focal population (insect herbivore) as well as many of potential interacting species (e.g., parasitoids). As long as the dominant cycle period is less than 8–10 yr, the ACF$[\tau]$ will not exceed 0.5 too much, and so we shall not run into the problem of redundance. For longer cycles, we should repeat the analysis with longer τ (2–3 yr, depending on which choice is suggested by the ACF$[\tau] < 0.5$ rule), and check how the analysis results are affected. An annual system with long-period cycles is also one kind of system for which we should consider decoupling τ from τ_{pred} (e.g., leave $\tau = 1$ yr, but use longer τ_{pred}).

Whenever a biological system is better described by a continuous rather than a discrete model, we do not have a natural choice of the base lag. In systems with continuous reproduction, and generation times (T_c) of about a year or somewhat less (e.g., many rodents), I recommend using the same delay, $\tau = 1$ yr. Using a delay smaller than a year would force us to model seasonal dynamics explicitly, significantly increasing model complexity. Additionally, one might take into consideration the possibility that generation times of significant interacting species (e.g., predators) might be somewhat longer (or shorter). However, no specific choice can possibly satisfy all the conflicting demands, so in a phenomenological analysis setting, opting for an annual clock may not be a bad overall decision. (Different generation times for the focal species versus its interactants can be built in, but this is best done with mechanistic models; see chapter 8.)

When the generation time is much shorter than a year, than we should set τ approximately equal to it. In such cases, we may have to model seasonality explicitly, unless the whole time series fits within a single season during which we can assume that the environment does not change substantially (i.e., we can assume **quasi-stationarity**).

For systems where generation time is much greater than a year (e.g., large mammals), again the best procedure is to set τ equal to T_c. It is also a good idea to vary τ somewhat to determine how much analysis results depend on any specific choice of the delay.

In summary, I recommend that τ should generally be set equal to 1 yr, unless the generation time, T_c, is much different from 1 yr. In the latter case the base time delay should be chosen to approximate T_c. In any case, the analyst should check the ACF at τ. If $\mathrm{ACF}[\tau]$ is too close to 1, (e.g., above 0.5), then the analysis should be repeated with longer delays. Finally, in some situations the values of base lag and prediction period may have to be uncoupled.

7.2.3 *Functional Forms*

Model (7.6) does not specify the functional form of the dependence between r_t and lagged densities, $f(\cdot)$. This relationship is unknown and must be estimated from data (after all, at this—phenomenological—stage of analysis we do not yet know which mechanisms may be driving dynamics). Many approaches for approximating functional relationships exist, but it is essential to recognize that a good fit for some purposes may be a bad fit for others (Ellner and Turchin 1995). For example, if our purpose is to characterize the qualitative dynamics of the system with Lyapunov exponents, then it is important to keep in mind that the estimates of Lyapunov exponents are calculated from partial derivatives of $f(\cdot)$. A method that gives accurate estimates of f may give poor estimates of derivatives of f. For example, a piecewise constant approximation can come arbitrarily close to a smooth function, just as a staircase with many tiny steps can approximate a straight line with constant slope. But the derivative of this staircase approximation is zero at all flat portions, and undefined at vertical portions, so it cannot be used to generate estimates of the Lyapunov exponent (Ellner and Turchin 1995).

Another consideration is the amount of data one has. Physical applications often deal with tens of thousands of very accurately measured data points, and thus can employ very sophisticated but data-hungry methods for approximating functional relationships. For shorter data sets, in the range of 50–500 data points, coming from systems affected by substantial amounts of dynamical noise, two approaches seem to work well: feedforward neural networks (McCaffrey et al. 1992) and thin-plate splines (Wahba 1990). The majority of ecological data sets, however, fall in the range of 10–100 data points. In my experience, employing sophisticated and flexible approaches like neural nets and splines for such short data is an overkill. I found that a modified polynomial scheme, response surface method (RSM) of Box and Draper (1987), works well for typical ecological series.

Ellner and Turchin (1995) applied all three approaches (neural nets, thin-plate splines, and RSM) to a set of ecological time-series data, and found that all three approaches gave similar results. The great advantage of RSM over the more sophisticated approaches, however, is that it is very easy to fit a response surface model to data. Fits can be obtained with standard statistical software, such as *STATIS-TICA*, are almost instantaneous on modern personal computers, and do not suffer from the problems of being trapped in local minima, unlike more sophisticated approaches such as neural nets. The disadvantage of RSM is that it is not as flexible as the more sophisticated alternatives, and therefore it can approximate only rather simple relationships in the data. Furthermore, RSM is best at fitting data coming from low-dimensional dynamical systems (process order of no more than three to four). However, given the typical length of ecological time series, we simply do not have enough data points to characterize high-dimensional complicated attractors. Accordingly, RSM appears to provide the optimal mixture of flexibility and simplicity, and I chose to use it in the NLTSM approach.

NLTSM Implementation   Response surface method is based on generalizing polynomial regression by allowing Box-Cox (1964) transformation of the independent variables, as suggested by Box and Draper (1987). The Box-Cox transformation is basically a power transformation that includes many commonly used statistical transformations: square root ($\theta = 0.5$), inverse ($\theta = -1$), and no transformation ($\theta = 1$). Additionally, the logarithmic transformation is the same as the Box-Cox transformation with $\theta = 0$. The form of the RSM model is

$$r_t = P_q\left[N_{t-1}^{\theta_1}, N_{t-2}^{\theta_2}, \ldots, N_{t-p}^{\theta_p}\right] + \epsilon_t \tag{7.7}$$

where P_q is a polynomial of degree q, p is the process order as before, and θ_is are transformation parameters. For example, for second-order process ($p = 2$) and assuming second-degree polynomials ($q = 2$), the general RSM model becomes

$$r_t = a_0 + a_1 X + a_2 Y + a_{11} X^2 + a_{22} Y^2 + a_{12} XY + \epsilon_t \tag{7.8}$$

where $X = N_{t-1}^{\theta_1}$ and $Y = N_{t-2}^{\theta_2}$ are Box-Cox transformed lagged densities (remember that by convention N_t^0 is interpreted as $\log N_t$).

Once the structural RSM parameters—p, q, and θ_i—are selected by cross-validation (see section 7.2.4), fitting model (7.7) is done simply by least squares (the model is linear in parameters). The simplest approach, which appears to work well, is to treat transformation parameters θ_i as discrete quantities, to be selected together with p and q (see p. 191). Experience shows that the set $\{-1, -0.5, 0, 0.5, 1\}$, corresponding to most commonly used transformations, is sufficient in most applications. An alternative approach is to estimate transformation parameters together with coefficients a_i using nonlinear fitting methods (this approach was advocated, e.g., by Perry et al. 1993). In practice, allowing θ_i to vary continuously usually has only a slight effect on the estimates of such quantities as R^2_{pred} and Λ_∞. However, it is a good practice to check on this by fine-tuning parameter estimates with nonlinear least squares that treat θ_i as continuous parameters, as a final step in the analysis.

7.2.4 Model Selection by Cross-Validation

The most critical step in nonlinear time-series analysis (and this applies to all methods, not just RSM) is choosing the complexity of the model. *Model complexity*, for my purposes, can be defined as simply the number of free parameters (i.e., parameters that we need to estimate from data). Complexity of the model fitted to time-series data has an enormous effect on estimates of all dynamical quantities, such as signal/noise ratio, or trajectory stability. In our initial approach to nonlinear time-series analysis with RSM models (Turchin and Taylor 1992), we used a model of fixed complexity, model (7.8). We were well aware that by applying a model of standard complexity we risked misclassifying the qualitative dynamics of some case studies (Turchin and Taylor 1992:304). In fact, the contrast between our results and those of Hassell et al. (1976), who employed a much simpler first-order model (and, correspondingly, found much simpler dynamics), speaks eloquently for the need for model selection. A model with too small a number of lagged densities will not be able to approximate the dynamics of a high-order system. It will assign most of variation to "noise," and will more likely than not classify the model as stable. A graphic example of this bias in favor of stability is the misclassification by Hassell et al. (1976) of the larch budmoth data as the most stable data set in their collection, although this is perhaps the best example of a high-amplitude cyclic dynamics in ecology (see chapter 9). The converse problem is that a model with too high p will have many parameters and will "fit the noise" instead of fitting the endogenous feedbacks. An overfitted model will often produce a spurious chaotic result.

In the absence of a priori information that would help us to select an appropriate model structure, we have to use time-series data themselves. One approach that seems to work well is cross-validation. The basic idea is to fit a variety of models on a part of data (the fitting set), to predict the data points that were not used for fitting (the validation set), and to then select the model that predicts the validation data the best. Cross-validation guards against selecting an overly complex (overfitted) model, because such models will be fitting to the noise rather than the signal, and will thus produce poor predictions on the validation set. To make efficient use of limited data, cross-validation is done repeatedly on different divisions of the data into fitting and validation sets. For example, the data may be divided into k equally sized subsets, and cross-validation is done k times with one of the blocks as the validation set and the remainder as the fitting set; the sum of squared prediction errors from the entire process is used to measure the model's accuracy. The NLTSM approach uses an extreme version of such *sequential-blocks cross-validation*, in which $k = 1$ (see p. 193).

Cross-validation becomes impractical with even moderately large data sets. For splines one can use instead an analytic approximation to cross-validation that is valid for large sample sizes and can be computed from a single fit to the entire data set (Wahba 1990)—this is called "generalized cross-validation" (GCV). For neural nets, the Bayesian information criterion (BIC) seems to be reliable for avoiding spurious overfitting on short, noisy time-series data (Nychka et al. 1992).

Complexity of a phenomenological time-series model, such as (7.7), has two components. The first one is the question of what is the optimal process order, or embedding dimension (p). Process order has a very strong influence on the model complexity: typically, the number of fitted parameters grows exponentially with p. The second questions is how flexible should the surface be. In the RSM approach, functional flexibility is controlled by the parameter q, or the polynomial degree. Unlike p, the polynomial degree does not have a ready biological interpretation (the functional shape of the relationship between r_t and lagged densities is determined jointly by q, θ_i, and a_i). I now discuss approaches to selecting model complexity, starting with the process order.

Choice of the process order, p, is conceptually straightforward, but often difficult in practice with short data sets. Cheng and Tong (1992) showed that a consistent estimate of the correct order can be obtained by a cross-validation procedure, fitting successively models with $p = 1, 2, 3, \ldots$, and choosing the value of p at which the mean square prediction error is minimized. Their analysis assumes that the data are fitted with a fully nonparametric model (they consider specifically the kernel regression model), so their procedure inherits a hunger for data in higher dimensions. The analysis and simulation study in Cheng and Tong (1996) suggest that the data requirements are not catastrophically high, for example, that 1,000 data points are sufficient to yield reliable results even for $p = 10$. However, as many as 200–300 data points may still be necessary for $p = 2$ or 3. In ecological applications, these data requirements will rarely be met outside the laboratory.

If the data are too sparse for nonparametric estimation of p, one option is to impose some mechanistic constraints that reduce the number of parameters in the higher-dimensional models. For example, if a suitable parametric model (or set of candidate models) can be identified for r_t as a function of lagged population densities, we can similarly fit the model at a series of p values, estimate prediction accuracy by cross-validation, and select the p giving the most accurate predictions. While this is not guaranteed to find the correct p, it does give a value that can be regarded as optimal for the available data, according to a sensible criterion for model selection. A more conservative approach is to first fit a model with $p = 1$ (only direct density dependence), and test for significant effects of other lags, for example, by linear regression of residuals on lagged population densities.

NLTSM Implementation　　As indicated above, the NLTSM approach employs sequential-blocks cross-validation. To understand the logic of this approach, consider first cross-validation as it was originally proposed. The procedure is to split the data set into the part on which the model is fitted, and the part where the model is "validated," that is, where values predicted by the fitted model are compared with the data. This could be repeated for all possible choices of p, and the p that allows the best prediction is the final choice. This approach, however, is wasteful of data. For example, suppose we have only twenty observations. We can use ten of them to fit the model, but this is really not enough to explore different possibilities. Increasing the allocation to the fitting part, however, leaves us with too few data points for the validation part. An approach that is much more frugal with data points works like this. We form a fitting data set by omitting one data point, fit the model on this reduced data set, and predict the omitted point. The difference between the predicted and observed values is saved. Next, we omit another point and repeat the procedure. The process is repeated by omitting each data point in turn, and, at the end, we calculate a measure of prediction accuracy, the sum of squared differences between the predicted and actual data. Finally, we repeat the process for all different model specifications (e.g., different values of p), and choose the model form that produces the least MSS. This is a computer-intensive procedure, but it allows us both "to have the cake and eat it," since the fitting part at any given time is all but one of the data points, and the validation part is all data points. Cross-validation allows us to squeeze the data set dry of information, an important advantage when data are sparse.

Cross-validation for p proceeds largely as described above. For each choice of p, there are several models with different q and θ. We fit all combinations of p up to 3, q up to 2, and θ_i from the set $\{-1, 0.5, 0, 0.5, 1\}$. Ideally, θ should be allowed to be different for each lag, since the response surface can have very different nonlinear properties in different directions, or axes, corresponding to N_{t-1}, N_{t-2}, and so on. Allowing θs to vary among lags, however, would result in more parameter combinations at higher lags and a larger pool of prediction error estimates. For example, for five choices of θ and two choices of q, there are ten possible models for $p = 1$ (5 θ_1s times 2 qs), but fifty possibilities for $p = 2$ (5 θ_1s times 5 θ_2s times 2 qs). Thus, we would be comparing a R^2_{pred} minimized over ten choices to a R^2_{pred} minimized over fifty choices. A minimum over fifty choices is likely to be less than a minimum over ten choices, even if there is no real difference (i.e., the second lag does not add to prediction accuracy). In other words, such an approach may bias model selection in favor of higher p.

To avoid this bias, we are forced to use the same θ for all lags, which leads to the same number of possibilities in each p class. This penalizes higher lags, but it is better to be overly conservative than to find a spuriously high p, with all the consequent problems. After p has been chosen, we can use cross-validation to select q and θ_is.

At this point we can allow θ to vary from lag to lag, because we are comparing R^2_{pred} values minimized over the same number of possibilities. To summarize, cross-validation is done in two steps: first determining p, second determining q and θs. This two-step procedure avoids a bias in favor of selecting a higher-dimensional model.

Finally, there is the question of what quantity to predict. There are several possible choices: r_t, N_t, or $Y_t = \log N_t$. NLTSM uses Y_t. In applications, we usually need to predict the future population levels rather than the population change (r_t), and log-transformed population density often has a better statistical distribution, compared with N_t.

The accuracy of one-step-ahead prediction of Y_t is measured by R^2_{pred}:

$$R^2_{\text{pred}} = 1 - \frac{\sum_{t=1}^{n} (Y_t^* - Y_t)^2}{\sum_{t=1}^{n} (\bar{Y} - Y_t)^2} \tag{7.9}$$

where Y_t^* is the prediction, and Y_t is the "predictee" (the data point omitted). I call this quantity the **coefficient of prediction** (by analogy with the coefficient of determination used in regression analysis). R^2_{pred} tells us how much better a model does compared with using the mean of time series as a simple-minded forecaster. The closer R^2_{pred} to one, the better is the accuracy of one-step-ahead prediction. If R^2_{pred} is close to 0, however, then the model is not increasing our ability to forecast the population density, and $R^2_{\text{pred}} < 0$ indicates that the model is fitting the noise, and thus predicting more poorly than the mean. The value of the coefficient of prediction is a useful diagnostic in its own right. When it is around 0, the dynamics are either simple, with fluctuations driven by noise, or so complex that we cannot characterize them given the amount of data and present technology. If R^2_{pred} is substantially greater than 0, then dynamics are complex, and are governed by a sufficiently low-dimensional attractor so that we can capitalize on that for prediction purposes.

7.3 SYNTHESIS

In this chapter I reviewed general approaches to nonlinear time-series analysis of ecological data. I also described a particular, fully developed approach (NLTSM) based on the response surface method of fitting the relationship between the realized per capita rate of change and lagged densities (equation 7.7), and sequential-blocks cross-validation for selection of model complexity. The advantage of using NLTSM is that it can be completely automated, allowing rapid processing of both empirical data series and model output. In other words, the software accepts time-series data as input, and generates a set of numeric probes as the output. I will use these outputs in two ways: (1) in the initial phenomenological analysis of empirically observed population fluctuations, to generate and, perhaps, reject hypotheses about mechanisms underlying dynamics; and (2) in quantitative comparisons between data and trajectories generated by mechanistic models purporting to explain the observed dynamics. Incidentally, the software is available (free of charge and free of responsibility) from the author's website.

CHAPTER 8

Fitting Mechanistic Models

While the previous chapter focused on exploring the structure of density dependence, without worrying too much about the mechanistic content, in this chapter we shall consider more mechanistic approaches to analyzing time-series data. Recollect that phenomenological methods employ lagged population densities as **state variables**. The distinguishing feature of the approaches reviewed here is that we, at the very least, can postulate the ecological nature of state variables that drive the dynamics. The functional forms of dependencies between state variables may be completely known, in which case we need to estimate only the numeric values of parameters. Alternatively, we may have to decide among several candidate functions, and we have to use the data to help us with this choice. Finally, we may have to resort to **semimechanistic** methods that represent functional relationships with flexible nonparametric curves or surfaces subject, perhaps, to some qualitative constraints. For example, we might require that increased predator density must have a negative effect on prey, without specifying the functional form of the relationship.

This chapter is organized as follows. First, I revisit the issue of model selection that was already raised in section 7.2.4, but now discuss it in the context of mechanistic models (section 8.1). Next, I consider the various approaches to an exploratory analyses when we have access to **ancillary data** (section 8.2). Ancillary data document fluctuations of variables other than the population density of the focal species that, we suspect, may have an effect on focal species population. The topic of ancillary data leads us naturally to fitting models using the method of "one-step-ahead prediction," which can be applied when we have time-series data on all the state variables of the fitted model (section 8.3). Finally, I tackle the most difficult topic: fitting mechanistic models when we are missing information about some state variables. I review two general approaches: trajectory matching (section 8.4) and nonlinear forecasting (section 8.5).

Fitting mechanistic models to data is already a huge field in ecology, and one that is growing very rapidly. Given space limitations, I must focus only on issues that are of direct relevance to the main subject, complex population dynamics. One particular aspect of the problem that I cannot give justice here is the statistical issues. Fortunately, there exist an excellent book by Hilborn and Mangel (1997) that fills the gap. Another useful book is Quinn and Deriso (1999).

8.1   MODEL SELECTION

When fitting models to ecological data, we rarely find ourselves in the situation where the functional form of the model is completely known, so that the only task is the estimation of its parameters. More typically, the theory can suggest several alternative models, and we have to contrast each of them with the data in order to determine which one is "the best." When comparing alternative models that have the same number of free parameters (a *free parameter* is one whose value is not fixed a priori and must be estimated using the data), we can simply pick the one that explains the greatest proportion of variance in the data. This approach, however, does not work when we must choose among models of variable complexity, because it is clear that a model with many free parameters will generally capture a greater proportion of variance in the data than a simpler model, even if the actual process that generated the data is better described by the simple model. In other words, we need to guard against the problem of overfitting the data (see section 7.2.4).

One situation in which a standard statistical approach is available is when one of the alternative models is *nested* within the other. A simple model is nested within a more complex one if its functional form is a special case of the more complex model, obtained by setting one or more parameters to some constant (typically, 0). For example, the hyperbolic functional response is nested within the Beddington response (see table 4.1 in section 4.1.1), because we can obtain it by setting the parameter w to 0 in the Beddington form. When alternative models are nested, we can use a likelihood ratio test (almost any standard statistical text treats this topic; for an exposition in an ecological context, readers can consult Hilborn and Mangel 1997: chapter 7). However, in most situations of interest to an ecological analyst, alternative models are not nested within each other. This is usually the case when we wish to compare models based on different ecological mechanisms.

The most general approaches to model selection are cross-validation and the information criteria. I have already discussed cross-validation in the context of phenomenological time-series analysis (see section 7.2.4). I think that cross-validation is the way to go in a serious analysis of data, because it is based on the ability to predict out-of-sample data, the most stringent test a model can be subjected to (short of collecting more data). However, implementing cross-validation is laborious, because it requires programming, and there are also several conceptual issues that need to be resolved in order for cross-validation to work properly (this is discussed in section 7.2.4). Unfortunately, I know of no widely available statistical software package that implements the sequential-blocks cross-validation. By contrast, the information criteria, discussed below, are easily calculated using the standard output of statistical software (in fact, many packages routinely print them out). The information criteria, therefore, provide a "quick and dirty" approach to model selection.

The best-known tool for model selection, although perhaps a flawed one, is the Akaike information criterion (AIC). AIC is equal to $[-2 \times \ln(\text{maximized likelihood}) + 2 \times (\text{number of independent parameters estimated})]$ (Chatfield 1989:197). Another widely used index, the Bayesian information criterion (BIC) substitutes $p + p \ln n$ in place of $2p$ (p is the number of estimated parameters) in this formula (Chatfield 1989:197). A useful recent book on the use of information criteria is by Burnham and Anderson (1998). These authors advocate AIC as the main tool for model selection. However, AIC has a serious problem: it is overly liberal. For this reason, and because BIC enjoys the property of consistency, Nychka et al. (1992) suggest that we should use BIC.

8.2 ANALYSIS OF ANCILLARY DATA

Analysis of ancillary data is very straightforward when models are formulated in the discrete-time framework. For example, suppose that we have time-series data on the focal species, N_t, and its parasitoids, P_t, collected once a year. We can use the following general host-parasitoid model in the analysis:

$$N_{t+1} = N_t \exp[r(N_t) - a(N_t, P_t)P_t] \qquad (8.1)$$

where $r(N_t)$ is the density-dependent function, and $a(N_t, P_t)$ is the parasitoid attack rate. Writing this function in terms of realized per capita rate of change (and shifting subscripts, as usual), we have the following general model:

$$r_t = r(N_{t-1}) - a(N_{t-1}, P_{t-1})P_{t-1} + \epsilon_t \qquad (8.2)$$

Note that I explicitly added an exogenous noise term, ϵ_t, to the right-hand side. We are now in position to explore the effect of parasitism on the dynamics of the focal species using nonlinear regression (offered by any serious statistical software package). Usually we will not know which particular functional forms to use, so we can try a number of different ones, taking them "off the shelf" (for density-dependent functions, see table 3.1; for functional response functions, see table 4.1). For example, using the Ricker form and hyperbolic response, we have the following specific model:

$$r_t = r_0 \left(1 - \frac{N_{t-1}}{k}\right) - \frac{aP_{t-1}}{1 + ahN_{t-1}} + \epsilon_t \qquad (8.3)$$

Fitting a variety of combinations of $r(N)$ and $a(N, P)$ functions to data, we observe how much variance the inclusion of the parasitism term helps to explain. A more formal comparison of models can be performed using an information criterion such as BIC, which penalizes models with a large number of parameters.

Thus, the general approach to the analysis of ancillary data is to fit time-series data with models of the form

$$X_t = f(X_{t-1}, Y_{t-1}, Z_{t-1}, \ldots, \epsilon_t) \qquad (8.4)$$

where X_t is the *response* variable whose dynamics we are investigating. The potential *predictor* variables are, first, the variable we are investigating lagged by one year, X_{t-1}, and, next, other variables $(Y_{t-1}, Z_{t-1}, \ldots)$ that we have information about. The time subscripts are shifted (t and $t-1$ instead of $t+1$ and t) to emphasize that we are not *modeling* (predicting the future given the present state of the system) but *analyzing* data (understanding the present given the past). Finally, ϵ_t represents, as usual, the effect of stochastic factors. The form of the function f is partly, but not completely, data driven. The idea is to model mechanistic relationships between various variables (this is not phenomenological time-series analysis). Thus, we want to try a variety of functional forms suggested by ecological theory (logistic population growth, functional responses, etc.). We make a decision on which model is best supported by data based on formal criteria such as AIC or BIC, and on the basis of whatever diagnostic tools the software package supports.

The preceding discussion assumes that we are investigating a system that is well described by discrete-time models. If we are dealing with a system operating in continuous time (data, though, always comes in discrete bits), then we can follow one of two alternative approaches. The first one is to discretize the continuous models that we can write for the system (see section 3.1.2). After that, we follow the steps outlined above. The second, more general approach is to smooth the N_t data and calculate the derivative of N at each point t for which we have data. This approach gives us the response variable, which we can place on the left-hand side of the differential model that we have for the system. Thus, the logic is the same as with fitting discrete models, but we have to perform an extra step of estimating derivatives. For examples see Ellner et al. (1998) and Kendall et al. (1999).

An extended example of ancillary data analysis will be considered in chapter 9. There I use ancillary data analysis to test two rival explanations for the larch budmoth cycle, based on two different driving state variables: budmoth–plant quality and budmoth–parasitoid hypotheses.

8.3 ONE-STEP-AHEAD PREDICTION

Ancillary data analysis focuses on how the dynamics of the focal species are affected by various environmental factors for which we have data. The next logical step is to extend this approach by also fitting dynamical models to these other factors. Thus, suppose that fitting some form of equation (8.2) suggests that parasitism data help to explain a large portion of variation in r_t of the focal species. In that case, we should perform analysis of the second equation of this model, for example,

$$P_t = N_{t-1}\{1 - \exp[a(N_{t-1}, P_{t-1})P_{t-1}]\} \qquad (8.5)$$

An added complexity is that, in exploring the effect of different functional forms on the degree of fit, we must ensure that we use the same forms in both equations.

The general (deterministic) model underlying this approach to fitting data is

$$N_t = f(N_{t-1}, X_{t-1}, Y_{t-1}, \ldots)$$

$$X_t = g(N_{t-1}, X_{t-1}, Y_{t-1}, \ldots) \qquad (8.6)$$

$$\ldots$$

I call it the "one-step-ahead prediction" approach, because our fitting criterion is the accuracy with which this year's values of N_t, X_t, and so on are predicted on the basis of the values of the same variables last year. This approach is clearly a relative of the phenomenological time-series model such as equation (7.6) (see section 7.2.1). The difference is in the state variables: real ecological variables in equations (8.6) and "reconstructed" in equation (7.6). This relatedness suggests that we can use many methods developed for phenomenological time-series analysis in the analysis of equations (8.6). Thus, we can use cross-validation to determine which of the ancillary variables are relevant, just as we determine the order of the phenomenological model (section 7.2.4). We can also use flexible functional forms, such as response surfaces or neural nets, to fit the relationships between different variables, and then check whether the fitted curves or surfaces correspond to the theoretically suggested (quantitative or qualitative) shapes. This is the *semimechanistic* approach to which I referred above.

Models fitted with the one-step-ahead approach should always be iterated on the computer to check whether their long-term dynamics match the observed behavior of the studied system. This comparison constitutes an important diagnostic test of the fitted model, because a good ability to predict one step ahead is no guarantee of being able to generate correct long-term dynamics. For example, in the empirical example of larch budmoth dynamics, our analyses could not detect any effects of self-limitation (adding such terms did not noticeably increase the proportion of variance explained). Even though the fitted model explained near to 90% of variance in r_t, when iterated it generated diverging oscillations (section 9.3.2). To fix this problem, we had to add a self-limitation term, even though it was not supported by the regression analysis. A general message here is that fitting well on a short timescale is no guarantee of correct long-term behavior. Some of the considerations related to choosing the prediction interval, discussed in the context of phenomenological analysis (section 7.2.2), may be well worth checking on.

There are few examples of application of the one-step-ahead analysis, perhaps due to the practical difficulty of collecting information on all important state variables. One empirical example is considered in the context of the larch budmoth (chapter 9). Another excellent example deals with the flour beetle in the laboratory (Dennis et al. 1995, Costantino et al. 1995).

8.4 TRAJECTORY MATCHING

Model fitting based on one-step-ahead prediction requires time-series measurements for all state variables in the model. Such a happy situation is rare; usually we have to fit models when dynamics of some important variables are unknown. Two general methods have been developed to deal with this situation: trajectory matching and nonlinear forecasting, which is a generalization of the one-step-ahead prediction approach. The two approaches differ fundamentally in the assumptions they make about the source of stochasticity in the data. As we discussed on several occasions, there are two fundamental kinds of "noise": **process noise**, which reflects the action of exogenous variables affecting vital rates, and **measurement noise**, which reflects our inability to measure variables such as N_t very precisely. It is very difficult, perhaps impossible, to estimate the variances associated with both kinds of noise simultaneously (Hilborn and Mangel 1997). Put simply, trajectory matching and nonlinear forecasting differ in that the first method sets the variance of process noise to zero and estimates the measurement error, while the second does the reverse.

The logic underlying trajectory matching is simple, and can be illustrated by the example of fitting a parasitoid-host model to data. Let us suppose that we have only a set of measurements of host data, $\{N_t\}$, where $t = 1, 2, \ldots, n$. The model has a vector of parameters (e.g., r_0, k, and a if we are fitting the Beddington model; see equation 4.32) that we are interested in estimating. For any particular choice of parameter values and initial host and parasitoid densities (N_0 and P_0, respectively) we can solve the model forward for n steps, and obtain the model-predicted sequence $\{N_t^*\}$ (the star superscript denotes values generated by the model). We also obtain a sequence of parasitoid densities, which we have to discard since we have no data against which to compare it. To calculate the degree of fit between the model predictions and data, we employ some measure such as the **coefficient of prediction**, R^2_{pred} (section 7.2.4). We have now defined a mapping from a set of parameters and initial values (r_0, k, a, N_0, and P_0) to R^2_{pred}. The next step is to use some standard software for function minimization, and ask it to find the set of parameters and initial values that will minimize $1 - R^2_{\text{pred}}$ (i.e., maximize the coefficient of prediction).

Because of its conceptual simplicity, and because it can be programmed quite easily, trajectory matching is often used in ecological applications (e.g., Harrison 1995; Hunter and Dwyer 1998). A popular exposition of the statistical issues can be found in Hilborn and Mangel (1997). A particularly sophisticated application to an ecological problem is by Wood (2001). Nevertheless, I cannot recommend this approach wholeheartedly.

The main problem with the approach is in its assumption that system dynamics are completely deterministic, and that the difference between model predictions and data is entirely due to measurement error. As I stressed repeatedly in this book, this assumption is simply not tenable for ecological systems. Ecological systems are "phase-forgetting," using the terminology of Nisbet and Gurney (1982). Repeated influence by exogenous factors, coupled with trajectory divergence due to endogenous dynamics (if the system is chaotic), means that the system "forgets" its initial conditions after repeated iterations, certainly after several complete oscillations. Yet, the trajectory-matching method assumes that, knowing the initial state of the system, we can predict the last data point in the time series! Note that the longer the data series, the worse would be this problem. Paradoxically, therefore, longer series make our estimates worse rather than better.

Furthermore, trajectory matching also has a tendency to fit more chaotic models than necessary. Suppose we are given a set of highly variable randomly generated data, and we can use a nonlinear model with parameters that allow chaotic dynamics. Then what can happen is that the fitting routine might push the model into the chaotic regime, where each slight variation in the initial values and parameters results in wildly different trajectories. If the fitting routine is good enough, by sorting through many such trajectories, it will be able to find some that will fit the data quite well. Needless to say, the parameter estimates from such an exercise would be completely useless. This is not a thought experiment; I once got a manuscript for review where authors did just that, but with a real data set.

In summary, one must be extremely careful about using the trajectory-matching approach. With a dynamical system that retains memory of its initial conditions for a long time (at least, for a time comparable to the length of the data series), this approach can give meaningful results. And, as we shall see shortly, the alternative approaches have their problems. Hilborn and Mangel (1997) suggest that data should be analyzed both ways: in one approach setting process noise to zero, and in the other setting measurement noise to zero. If both approaches agree, then we can probably believe the results. In any case, the ability of the approach to yield correct parameter estimates should always be checked by subjecting simulated data to the same analysis that was used on real data.

8.5   FITTING BY NONLINEAR FORECASTING

The nonlinear forecasting method is a generalization of the one-step-ahead prediction approach for the case when we are missing information about some state variables. The basic approach was suggested by Tidd et al. (1993), and further developed by Turchin and Ellner

(2000a). Recollect that in the one-step-ahead method we assume that data are generated by a model such as

$$N_t = f(N_{t-1}, X_{t-1}, \epsilon_t)$$
$$X_t = g(N_{t-1}, X_{t-1}, \varepsilon_t)$$

(8.7)

where N_t and X_t are state variables (for simplicity, I am assuming a two-dimensional system), f and g are some functions, and ϵ_t and ε_t represent exogenous variables. Since we have data only on N_t, we cannot use model (8.7). The Takens theorem, however, tells us that there is another function, call it F, that gives us the value of N_t, given lagged densities:

$$N_t = F(N_{t-1}, N_{t-2}, \ldots N_{t-p}, \epsilon_t)$$

(8.8)

Equations 8.7 and 8.8 are equivalent in the sense that they are different representations of the same dynamical rule.

In some special cases we can go algebraically from model (8.7) to model (8.8) (e.g., the Nicholson-Bailey model; see equation 2.13 in section 2.5). In general, however, a closed-form solution is not available, and in order to predict $\{N_t\}$ from a mechanistic model, we must use nonparametric regression techniques to numerically construct the function F that holds for the model (Turchin and Ellner 2000a). Here is how this works in practice. We first iterate the mechanistic model to produce a long time series of simulated data $\{N_t^*\}$ (star superscripts again denote values generated by simulating the mechanistic model, and curly braces indicate *a set*, in this case the whole time series). This is the *atlas* generated by the model. We then treat the atlas as if it were a data set and apply nonparametric regression to produce an estimate of F. The F constructed from the model can then be applied to data values, forecasting N_t from $N_{t-1}, N_{t-2}, \ldots$, in the data. A variety of nonparametric regression methods are available for estimating F, but we used kernel regression, because properties of kernel regression make it possible to reestimate F relatively quickly when model parameters are adjusted and a new atlas is generated (Turchin and Ellner 2000a).

As usual, we quantify the accuracy of atlas-based forecasts by calculating the coefficient of prediction (equation 7.9). We are now again in the situation where we can predict future population densities using the mechanistic model, when given parameter values. We now invert this process and ask, what values of the parameters maximize the forecasting accuracy? In other words, we find the parameter values for which the model does the best job of predicting the observed data, and take those as estimates of the true parameter values. This method, which we call NLF (for nonlinear forecasting), is an example of simulated quasi-maximum likelihood and yields parameter estimates that are consistent and asymptotically normal as sample size increases (see Turchin and Ellner 2000a for technical details).

One of the outputs of the NLF method is the coefficient of prediction characterizing the model with best-fit parameters. This raises the following conceptual issue: how should we judge the success of an attempt at predicting future population densities? One approach that seems to make sense is, first, to determine how well the data predict themselves, and then compare this with the accuracy of atlas-based forecasts from the mechanistic model (Turchin and Ellner 2000a). To make the results comparable, we again used kernel regression, treating the real data exactly the same as the simulated data, with one exception: when predicting N_t from N_{t-1}, N_{t-2}, we omitted from the data set the "predictee," N_t, and a window of temporal neighbors before and after time t. This is, of course, the cross-validation approach already discussed in section 7.2.4. As a quantitative measure of forecasting accuracy, we employed the coefficient of prediction, using the notation R^2_{data} to denote the prediction R^2 from the data predicting themselves via kernel regression, and R^2_{atlas} for the data being predicted using an atlas derived from a mechanistic model.

To recapitulate, R^2_{data} provides us with a measure of how predictable the data are in the absence of any knowledge about the mechanisms driving population dynamics. However, we must keep in mind that R^2_{data} is biased toward underestimating the true predictability, because the forecasts are based on a limited and imperfectly measured set of data. A more accurately measured or, most important, a longer empirical time series would increase our ability to make accurate predictions. Despite this limitation, R^2_{data} provides a useful benchmark for evaluating the accuracy of predictions made by the mechanistic model.

Three possible outcomes can occur when we compare R^2_{data} and R^2_{atlas}:

1. If $R^2_{\text{data}} \approx R^2_{\text{atlas}}$, then the model has successfully passed the test. This outcome does not necessarily mean that the

model is correct, because another model, perhaps based on different mechanisms, may do as well. Additionally, it does not mean that the model fully captures the predictability in the data, because we expect that R^2_{data} underestimates the true predictability. Nevertheless, this outcome raises the stakes by requiring any alternative models to perform at least as well.

2. If on the other hand $R^2_{\text{data}} > R^2_{\text{atlas}}$, then it is likely that the model failed to capture all the predictability in the data, suggesting that either the model is based on incorrect mechanisms, the model equations contain some incorrect functional forms, or model parameters have been estimated poorly. In general, this outcome means that we should go back to the drawing board.

3. Finally, an outcome of $R^2_{\text{data}} < R^2_{\text{atlas}}$ is a strong endorsement of the model (with the caveats listed above still applicable, however). It suggests that independent data, on which the model and parameter estimates are based, provide an even better prediction accuracy than the data themselves. If we have a reasonably long time series, then this outcome provides strong support for the mechanistic model.

In summary, the nonlinear forecasting method offers a general solution of fitting mechanistic models that include dynamical noise in situations where data do not provide a complete specification of the state variable vector. The specific application of the method to vole data will be discussed in chapter 12. Technical issues, as well as results of testing the approach with artifical data, are discussed in Turchin and Ellner (2000a). Looking forward, we believe that nonlinear forecasting is generalizable to almost any kind of the model: "if you can simulate it, then you can fit it." In particular, nothing prevents us from adding measurement noise to the simulated data, and then treating it as the atlas for forecasts. We should not expect to be able to *estimate* the parameters of both dynamical and measurement noise from data, but we can test the effect of measurement noise if its variance is known from some external information (this is in fact the case for the vole data). This is an important advantage of nonlinear forecasting over trajectory matching.

What are the drawbacks associated with nonlinear forecasting? There is a practical disadvantage: nonlinear forecasting requires a significant amount of coding, and also it takes a long time to run. On a more fundamental level, nonlinear forecasting relies on the presence of strong signal in the data. If such a signal is lacking, then the method will fail. Essentially, this means that one cannot gainfully employ the method when dynamics of the studied system are simple. In the worst case of perfect regulation, there are only two numbers that completely specify the dynamics of the system: the mean and variance of fluctuations. Clearly, one cannot fit even a simple model with just three parameters (one of which must be noise variance) to such data series. Crudely speaking, one parameter will set the mean, another the variance, and ... we have run out of degrees of freedom to estimate the third. Of course, this is a generic problem of any methodology for fitting models to time-series data, including trajectory matching. Finally, a potential problem affecting any one-step-ahead approach, including nonlinear forecasting, is that care must be exercised in choosing the appropriate prediction interval τ_{pred} (see the discussion in the context of phenomenological model fitting, section 7.2.1).

PART III
CASE STUDIES

CHAPTER 9

Larch Budmoth

9.1 INTRODUCTION

If there were a beauty contest for complex population dynamics, then population oscillations of the larch budmoth (LBM), *Zeiraphera diniana*, in the Swiss Alps would be a credible contender for first place (figure 9.1). Not only are these oscillations remarkably regular, but the moth population swings through a stunning range of densities during a typical cycle, covering five orders of magnitude! It is, therefore, not surprising that the larch budmoth was featured as one of the best examples of complex population systems in a recent news article in *Science* (Zimmer 1999). However, the ecological mechanisms that drive this remarkable oscillation have not been unambiguously identified, although a number of hypotheses have been advanced.

Food Quality   Larch trees suffering greater than 50% defoliation lack nutrient resources to grow high-quality needles during the following spring. Needles grown after the LBM peak are short (< 20 mm, compared with normal length of > 30 mm) and have a high raw fiber content of about 18% (compared with the normal 12%), while the protein content falls from 6% to 4%. Low quality of food (as measured by high raw fiber, and indexed by low needle length) strongly depresses larval survival and female fecundity in bioassays (Benz 1974; Omlin 1977). Furthermore, poor needle quality persists for several years after an outbreak. This "quality transmission" effect imposes delayed density dependence on LBM population growth rates, and can theoretically lead to cycles (see section 4.4.4).

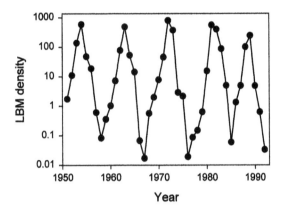

FIGURE 9.1. Population oscillations in the larch budmoth at Sils (Upper Engadine Valley, Switzerland). Moth density is the number of larvae per kg of larch branches.

Pathogens   Theory suggests that the interaction between pathogens and their hosts can exhibit oscillatory dynamics (section 4.5.2). In 1957, after the first cycle that was studied intensively, it seemed obvious to everybody that a granulosis virus disease played a critical role in suppressing the LBM outbreak (Baltensweiler and Fischlin 1988). Unfortunately, the virus incidence decreased during the next outbreak, and then the virus disappeared completely. As a result, the pathogen hypothesis fell out of vogue, at least among the field-workers. Nevertheless, Anderson and May (1980) used LBM as their prime example of how an epidemiological model may explain population cycles in a forest insect.

Parasitoids   General theory suggests that parasitoids may play an important role in population dynamics of forest insects, and therefore LBM parasitoids were intensively studied from the beginning of the systematic research program on LBM oscillations (e.g., Baltensweiler 1958). Once the data on parasitism rates became available, however, the initial enthusiasm for the parasitoid hypothesis waned. Parasitism rates at population peak are typically low, around 10–20% (Baltensweiler and Fischlin 1988), suggesting that parasitoids play a minor role in *limiting* LBM densities, that is, in preventing further LBM increase. The parasitism rate reaches a high of around 90%

during the collapse stage, but this high is reached only during the second (or even third) year after the peak. Accordingly, Delucchi (1982) concluded that control of LBM by parasitoids alone is not possible.

Polymorphic Fitness Hypothesis   There are two races of larch budmoth with distinct differences in color and ecological traits: a dark morph that feeds primarily on deciduous larch, and a light morph that feeds primarily on evergreens (*Pinus cembra* and *Picea abies*). The frequency of the dark morph tends to increase together with density increase, and decrease after the outbreak collapses (Baltensweiler 1993a: figure 1). Baltensweiler (1977, 1993a) proposed the following explanation of this pattern. During population increase, dark morphs increase because they are characterized by faster development time and higher survival than light morphs. During population collapse, the dark morphs decrease faster than light morphs, because they rely primarily on larches for food, and the quality of larch foliage declines after defoliation. Once the effects of the previous defoliation on host quality dissipate, dark morphs begin increasing faster than light ones, and the cycle repeats itself. Baltensweiler (1993a) argued that this polymorphism plays a key role in explain the LBM cycle. In particular, he suggested that it helps explain why the LBM population at low densities switches immediately from the decline to the increase phase. However, as we shall see, the abrupt switch from decline to increase is not a pattern that needs a special explanation: it arises naturally in several trophic models considered later in this chapter. Furthermore, the polymorphic fitness hypothesis is not an elemental mechanism, because it invokes plant quality as the primary factor causing population collapse (without prolonged decrease in plant quality, the population density of dark morphs would not decrease, and no cycle would ensue). In fact, it is not clear how the polymorphic fitness hypothesis adds to the explanation of the primary question (why LBM populations oscillate). It is rather an explanation of why morph frequencies change regularly during the LBM cycle.

Other Hypotheses   Several other theoretical possibilities need to be briefly discussed. First, the natural history of the LBM-larch system is such that food *quantity* is an unlikely factor to explain LBM oscillations. Mortality of the host trees due to defoliation is less than 1%

(Baltensweiler and Fischlin 1988). Although the length of needles is reduced after a severe defoliation, the total amount of needle biomass is decreased by only about twofold. Such a small variation in food availability cannot drive a second-order population cycle in which the ratio of peak/trough densities is around 100,000. Food, however, should have strong first-order effects on LBM dynamics, since most LBM outbreaks are accompanied by widespread defoliation of host trees (Baltensweiler and Fischlin 1988), leading to high starvation mortality of larvae during peak years. Second, the role of specialist predators of LBM, other than parasitoid wasps, is poorly studied, but it is believed that they play a minor role. However, some parasitoids act as functional predators because they host-feed on LBM larvae. Third, maternal effects constitute a theoretically well-established intrinsic mechanism of second-order oscillations (section 3.4.1). However, there are no data that would suggest that this mechanism operates in the LBM, and, as far as I know, there are no advocates of the maternal effect hypothesis as the explanation of LBM cycles.

Hypotheses: Summary   The influential review of Baltensweiler and Fischlin (1988) concluded that food quality change induced by previous budmoth feeding was the most plausible explanation for this insect's cycles. During the 1990s various reviews of insect population dynamics (e.g., Bowers et al. 1993; Ginzburg and Taneyhill 1994; Den Boer and Reddingius 1996; Hunter and Dwyer 1998; Berryman 1999) generally concurred with the "received view" that budmoth cycles are driven by the interaction with food quality. Baltensweiler and Fischlin also stated that empirical studies of LBM parasitoids "generally indicate that parasitism merely tracks the larch budmoth population; that is, budmoth fluctuations regulate the numbers of parasitoids and not vice versa" (Baltensweiler and Fischlin 1988:344). Indeed, data do not support an important role of parasitoids in stopping LBM increases. However, this observation does not mean that parasitoids do not play an important role in driving LBM cycles: stopping population increase is a first-order effect, while population oscillations result from the action of second-order factors. This consideration suggests that the parasitism hypothesis should remain a viable contender for the explanation of LBM cycles, in addition to the plant quality hypothesis.

The rest of this chapter is organized as follows. I begin with phenomenological analyses of time-series data to establish the basic pattern of density dependence exhibited by larch budmoth populations. Next, I focus on the two main hypotheses for LBM cycles: plant quality and parasitism. For each hypothesis, I start with analyses of appropriate time-series data, and then discuss dynamical models. Finally, I consider a model of joint effects of plant quality and parasitism.

9.2 ANALYSIS OF TIME-SERIES DATA

The best data on LBM dynamics come from the Upper Engadine Valley (Switzerland). Population census started in 1949 and with minor modifications continued until 1977 (Baltensweiler and Fischlin 1988). The valley was divided into twenty sites, and data were collected at each site separately (Auer 1977; data tabulated in Fischlin 1982: table 10). The twenty sites oscillated in close synchrony, and thus we can average them into a single series, hereafter called "Engadine." After 1977, sampling the Upper Engadine Valley continued on a reduced scale. At one site, Sils, data were collected in an uninterrupted sequence from 1951 to 1992 (Baltensweiler 1993b). We call this data series "Sils" (depicted in figure 9.1). Furthermore, there are several shorter data sets on larch budmoth dynamics from other valleys in the Alps: Lungau, Goms, Val Aurina, and Briançonnais (Baltensweiler and Fischlin 1988: Figure 4). These data can serve as replicates of the main series (table 9.1).

The first striking feature of LBM time-series data is the sheer amplitude of oscillations (figures 9.1–9.2). Values of the amplitude index, S, range between 1.1 and 1.5 (table 9.1). Of the case studies that I discuss in this book, only lemming systems rival this amplitude. The second feature of LBM data is the very regular periodicity (see the ACF plot in figure 9.2). The dominant period is practically the same in all series.

What about the process order? PRCF estimated for the Engadine series is characterized by a strong spike at lag 2 (figure 9.2). In fact, all series are characterized by highly significant PRCF[2]. Additionally, the PRCF for Briançonnais and Sils exhibit significant spikes at lag 3. Cross-validation results are evenly split between p estimates

TABLE 9.1. LBM primary data: summary of results of nonlinear time-series analysis. Quantities: number of data points, n; measure of amplitude (SD of log-transformed densities), S; dominant period, T; the autocorrelation at the dominant period, ACF[T]; estimated process order, p; polynomial degree, q; the coefficient of prediction of the best model, R^2_{pred}; and the estimated dominant Lyapunov exponent, Λ_∞

Location	n	S	T	ACF[T]	p	q	R^2_{pred}	Λ_∞
Sils	42	1.35	9	0.67**	3	2	0.79	0.05
Engadine	31	1.53	9	0.79**	2	2	0.91	−0.01
Briançonnais	20	1.25	8	0.68**	3	2	0.94	0.45
Goms	21	1.28	9	0.90**	2	2	0.94	0.23
Val Aurina	20	1.08	10	0.63**	2	1	0.83	−0.06
Lungau	19	1.15	9	0.76**	3	1	0.85	−0.01

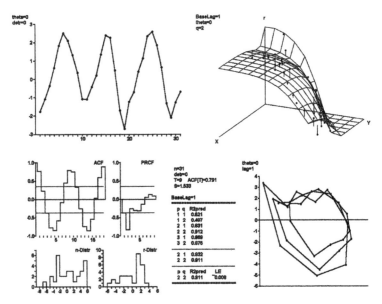

FIGURE 9.2. Graphical output of the NLTSM analysis of Engadine data. *Upper left:* time plot; *upper right:* response surface; *lower right:* phase plot.

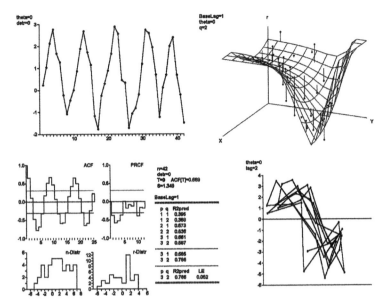

FIGURE 9.3. Graphical output of the NLTSM analysis of Sils data. *Upper left:* time plot; *upper right:* response surface; *lower right:* phase plot.

of 2 and 3 (table 9.1). Thus, we can conclude that the evidence for second-order dynamics is overwhelming. There are indications of even higher-order dynamics, but the case for that is not as compelling.

Regulatory process in the LBM is characterized by a high degree of nonlinearity. The estimates of the polynomial degree, q, are mostly 2 (except for the two shortest series), suggesting highly nonlinear relationship between r_t and lagged densities. Examination of the response surface fitted to the Engadine data (figure 9.2) suggests that the non-linearity mainly affects the N_{t-2} axis. This particular pattern of non-linearity is easy to understand. For low Y values, lagged population density is very low, and r_t is essentially at r_0. However, when Y approaches the peak levels, r_t finds itself on an increasingly precip-itous slope. In other words, this type of nonlinearity arises because of a very large amplitude of LBM oscillations. Visual inspection of response surfaces fitted to other data sets indicates that this is a generic feature of the structure of LBM regulation. In two or three series, there are even more extreme nonlinearities, for example, the Sils data (figure 9.3).

Finally, additional indications of complex dynamics are the high predictability, as indicated by R^2 values usually in excess of 0.8, and estimates of the dominant Lyapunov exponent clustering around zero. In fact, the mean ($\pm$ SE) $\Lambda_\infty = +0.11$ (± 0.08) is positive, although not statistically different from zero.

Taken together, the results of nonlinear time-series analysis suggest the following probe values for LBM dynamics: S values in excess of 1.1; dominant period of 9 with ACF at this lag around 0.5; second- or even higher-order dynamics; R^2_{pred} around 0.9; and Λ_∞ between -0.05 and $+0.25$. These are the quantitative patterns of dynamics that a good mechanistic model for LBM cycles needs to match.

9.3   HYPOTHESES AND MODELS

9.3.1   Plant Quality

The plant quality hypothesis is currently the "reigning" explanation for larch budmoth oscillations. In this section I evaluate the previous analytical and modeling efforts, as well as attempt some improvements of previous analyses. The ultimate goal is to contrast predictions of models based on the plant quality hypothesis with predictions of models based on the alternative, the parasitism hypothesis.

Needle Length as an Index of Plant Quality   As an indirect measure of plant quality dynamics, we have a long time series of average needle lengths at Sils during 1961–1992 (Baltensweiler 1993b). Needle length provides us with a good index of plant quality because it is related to both raw fiber and protein content of larch needles (Omlin 1977), which directly affect LBM survival and reproduction. Previous analyses of the interaction between plant quality and LBM dynamics emphasized the raw fiber content of larch needles as the main food quality index (raw fiber content measures the physical toughness of needles). However, there are no time-series data available for this index, unlike the average needle length. How good an index of food quality is the needle length? We can answer this question with the bioassay data of Benz (1974: table 8). Benz fed LBM larvae foliage

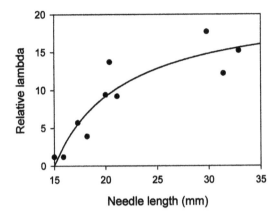

FIGURE 9.4. Effect of average needle length on the relative multiplicative rate of change, λ' (calculations based on data from Benz 1974: table 8).

from larch trees with known defoliation history, and measured larval survival and pupal weight. Because female pupal weight is linearly related to fecundity, we can translate the measured pupal weights into expected fecundity using the relationship estimated by Benz (1974: figure 2). Multiplying larval survival by the calculated fecundity, we then obtain a measure related to the multiplicative rate of population increase, λ' (the prime is to remind us that this measure is not the true λ, because it does not include the effects of egg and adult mortalities). Plotting λ' against the needle length index, we observe that there is a well-defined relationship between these two quantities (figure 9.4), with a high coefficient of determination ($R^2 = 0.86$). Interestingly, the alternative index, raw fiber quality, explains a somewhat lower percent of variance in λ' ($R^2 = 0.66$). Thus, the somewhat surprising conclusion is that needle length appears to be a better index of effects of food quality on LBM rate of population change than the index based on raw fiber content. Clearly, food quality is a complex variable, whose effect on LBM survival and fecundity is mediated by physical (e.g., toughness as measured by raw fiber content) and nutritional (e.g., protein content) properties of needles, as well as, perhaps, tree chemical defenses, such as resin content (Benz 1974). However, the result that the average length of needles is an accurate *explanans* of LBM rates of population change is encouraging, because we can

now proceed, using this index in subsequent analyses with greater
confidence.

Analysis of the Ancillary Data   The general approach to the analysis
of ancillary data is discussed in section 8.2. The first step is to investigate the interrelations between LBM density and plant quality (as
indexed by the needle length index). The basic model used is equation (4.44) from section 4.4.4. Rewriting the first equation in terms
of r_t, we have

$$r_t = r_0 \frac{L_{t-1} - a}{b} - \frac{r_0}{k} N_{t-1}$$

where N_t and L_t are LBM density and average needle length in year t.
Parameters a and b are needed to rescale L_t, measured in mm, into
the unitless Q_t. The equation is overparameterized, so in practice we
fit the model

$$r_t = \beta(L_{t-1} - a) - \gamma N_{t-1} \qquad (9.1)$$

to estimate parameters a, β, and γ. The dynamical equation for needle
length is

$$L_t = \delta + \alpha L_{t-1} - \frac{c N_{t-1}}{d + N_{t-1}} \qquad (9.2)$$

where the extra parameter δ again reflects the need to rescale the
needle length into unitless plant quality.

   Fitting equation (9.1) to the Sils data on LBM density and needle lengths, we find that the effect of L_{t-1} is statistically significant
(although the density-dependent effect—the logistic term—is not).
However, the model explains a disappointing 31% of variance in the
data. Trying different functional forms, or adding the current year's
plant quality (L_t), does not increase R^2.

   The dynamics of L_t, by contrast, are well explained by LBM density and the previous year's plant quality ($R^2 = 0.76$). The effect of
"memory," represented by the autoregressive parameter α, is highly
significant, and by itself explains about 47% of variance. The effect
of LBM by itself explains about 37% of variance. It appears that both
factors are needed in the model.

   To summarize, the surprising result from the analyses of ancillary data is that an index of plant quality explains, at best, a measly

31% of variance in the LBM rate of change. The low level of predictability yielded by plant quality contrasts unfavorably with R^2 of around 90% characterizing phenomenological response-surface models (section 9.2) or ancillary analyses utilizing parasitism data (see section 9.3.2 below). While the regression analyses described above do not constitute a "proof" that plant quality is an unimportant variable in the LBM dynamical system, they considerably weaken the case for plant quality as the primary factor responsible for LBM oscillations. However, in order to pass the final verdict, we need more information. Specifically, we need to know whether a model based on the plant quality hypothesis with empirically supported parameters is capable of mimicking the observed LBM dynamics (since that is what we shall be doing with the parasitism hypothesis below). I first review the previous most credible effort of building such a model (Fischlin 1982), and then offer a somewhat different, and hopefully improved, approach.

The Fischlin Model   The model of Fischlin (1982; see also Fischlin and Baltensweiler 1979) has two state variables, LBM density and food quality, indexed by the raw fiber content of needles. Fischlin used the data of Benz (1974) and Omlin (1977) to fit three linear regressions to small larva survival, large larva survival, and adult fecundity as functions of raw fiber content. The overall relationship between the net replacement rate and food quality is a curve similar to the one depicted in figure 9.4.

Dynamics of food quality were modeled as follows. A severe defoliation in year t drastically lowers food quality in year $t + 1$ (raises food's raw fiber content). Further defoliation would keep food quality at low levels. However, after LBM density collapses, plant quality does not jump immediately to its maximum value. Return of quality after defoliation to its maximum value is regulated by a key parameter, the recovery rate. Lacking direct estimates, Fischlin simulated his model for a large set of postulated values for this parameter, and chose the one that mimicked the observed LBM dynamics best.

Using the description in Fischlin (1982), I simulated the dynamics of LBM density and the food quality index implied by the Fischlin model. In order to be able to directly compare the predictions of the Fischlin's model to the one I develop below, I translated raw

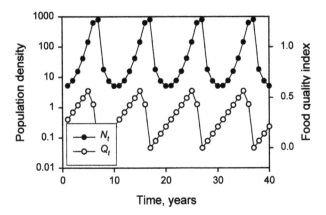

FIGURE 9.5. Dynamics of LBM density and food quality index predicted
by the Fischlin model.

fiber content into unitless food quality, Q_t. (The maximum raw fiber
content, corresponding to the lowest food quality, maps into $Q_t = 0$,
while the minimum raw fiber content maps into $Q_t = 1$).

It is clear that the LBM dynamics generated by the Fischlin model
match *qualitatively* the observed LBM dynamics; that is, both predicted and observed dynamics are second-order oscillations of about
the same period (figure 9.5). Quantitatively, however, there is a certain
degree of mismatch between model predictions and data. First, modeled LBM density peaks at somewhat higher values than observed.
Second, and more important, the Fischlin model does not capture the
behavior of the LBM system during the troughs. Note that the Fischlin
model predicts a "soft landing" characterized by a gradual turnaround
of the LBM trend (figure 9.5), while in the data the declining trend
changes abruptly to the increasing trend (figure 9.1). Additionally,
predicted trough density is one or two orders of magnitude too high.
The third, and perhaps most important, difference between predictions and data is the very gradual increase in food quality generated
by the Fischlin model. This pattern is a result of assuming a low
value of the recovery rate parameter, discussed above. Thus, Q_t inches
up with small incremental steps, and then collapses before reaching
the maximum value during the next LBM outbreak. This is clearly a
problematic prediction for the model, because empirical observations
suggest that plant quality recovers fully 3–4 years after the outbreak. I

conclude that, although the Fischlin model was an important development showing that a mechanism based on plant quality can generate
the correct qualitative dynamics in the LBM system, the model fails
to accurately match the quantitative aspects of the data.

A Model Based on the Needle Length Index   An alternative approach
to modeling the LBM–plant quality hypothesis is to switch to the
average needle length as the food quality index. There are two important reasons for preferring this approach: (1) we have a lengthy time
series for this index (figure 9.6), and (2) the reanalysis of Benz (1974)
data suggests that needle length is a better index of the effects of plant
quality on LBM rate of increase (figure 9.4). The needle length index,
L_t, varies between a minimum of 15 mm and a maximum of around
30 mm. I reparameterize the needle length data into a Q_t index as
follows: $Q_t = (L_t - 15\text{mm})/15\text{mm}$.

In developing the model for LBM–food quality interactions, I will
follow the discrete-time plant quality framework introduced in section 4.4.4 (see equation 4.44). That model, however, assumed a linear
relationship between the quality index and the realized per capita rate
of change, while the data from the LBM system suggest a curvilinear
relationship (figure 9.4). Fortunately, we can use the data directly to
model the effect of Q_t on the budmoth per capita rate of population
change. After trying several two-parameter relationships, I found that
the one fitting the data best appears to be a negative exponential function (this is a purely phenomenological approach, as we do not have
any mechanistic basis for postulating a functional form). The fitted
curve was

$$r_t' = a\left(1 - \exp\left[-\frac{Q_{t-1}}{\delta}\right]\right) \qquad (9.3)$$

where $a = 3.8$ is the saturation level, or maximum r_t', occurring at the
best food quality ($Q_t = 1$), and $\delta = 0.22$ is the parameter determining
how fast r_t' approaches the saturation plateau.

There are two things still missing in this model. First, it assumes
that there is no mortality in the adult and small larva stages. We
can remedy this omission by replacing the saturation level, a, with
the average per capita rate of population change observed when
plant quality is at its maximum. A good choice for this parameter

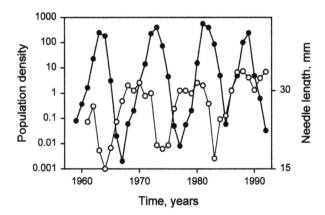

FIGURE 9.6. Observed dynamics of LBM density and the average needle length (the food quality index) in Sils.

is $r_0 = 2.5$, corresponding to about tenfold increase in N_t per year (because this is the average rate at which the LBM density climbs out of the trough). Second, the model lacks a self-limitation term that is due to larvae overeating their food supply and starving. As in section 4.4.4 I will assume the Ricker-type self-limitation term. This assumption appears to be consistent with the results of ancillary data analyses in the previous section. Adding these two ingredients, we have the N_t equation of the LBM–plant quality model:

$$N_{t+1} = N_t \exp\left\{ r_0\left(1 - \exp\left[-\frac{Q_t}{\delta}\right]\right) - \frac{r_0}{k}N_t \right\} \qquad (9.4)$$

where k is the carrying capacity, as usual.

For the Q_t equation, we simply use the second equation in model (4.44). Nonlinear regressions of Q_t on Q_{t-1} and N_{t-1}, using a variety of functional forms, suggest that this three-parameter equation provides a good description of the data, explaining about 75% of variation in Q_t (see figure 9.7).

In summary, the empirically based model for LBM–food quality interaction is

$$N_{t+1} = N_t \exp\left\{ r_0\left(1 - \exp\left[-\frac{Q_t}{\delta}\right]\right) - \frac{r_0}{k}N_t \right\}$$

$$Q_{t+1} = (1 - \alpha) + \alpha Q_t - \frac{cN_t}{d + N_t} \qquad (9.5)$$

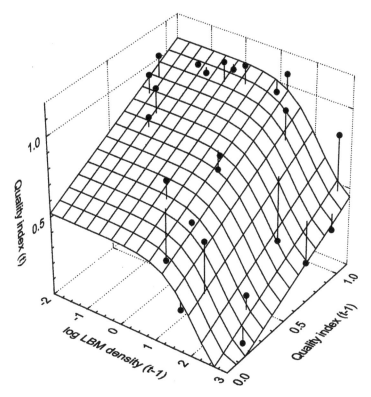

FIGURE 9.7. The relationship between plant quality (Q_t) and LBM density (N_{t-1}) and plant quality (Q_{t-1}) last year (Sils data).

The regression-based parameter estimates (mean $\pm$ SE) are $\delta = 0.22 \pm 0.05$, $\alpha = 0.5 \pm 0.1$, $c = 0.7 \pm 0.2$, and $d = 150 \pm 150$. Additionally, we have guesstimates (means $\pm$ some reasonable range) $r_0 = 2.5 \pm 0.2$ and $k = 500 \pm 200$.

Numerical exploration of model dynamics for the parameters within the ranges defined by mean $\pm$ SE indicates that model (9.5) is readily capable of generating population trajectories resembling data (figure 9.8). Comparison of figures 9.8 and 9.6 suggests that the output of model (9.5) matches tolerably well both the period and the amplitude of the observed LBM oscillations. Additionally, the model mimics the quantitative pattern of the quality index dynamics reasonably faithfully, including the amplitude of variation and the timing of declines and increases (compare figures 9.6 and 9.8).

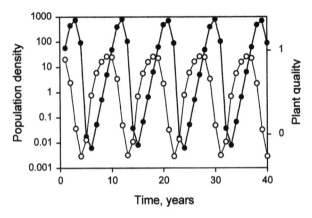

FIGURE 9.8. Predicted dynamics of LBM density and the food quality index. Equations: model (9.5). Parameters: $r_0 = 2.3$, $k = 600$, $\delta = 0.17$, $\alpha = 0.5$, $c = 0.9$, and $d = 100$. No process noise ($\sigma = 0$).

In summary, the model of LBM–plant quality interactions with biologically reasonable parameters is capable of matching the empirically observed quantitative patterns in the time-series data. Does it mean that we have found the explanation for the LBM oscillations? Unfortunately, there remains one serious problem for this hypothesis: lack of empirical evidence for a strong feedback from plant quality to LBM rate of population change. Perhaps it is in general difficult to detect this feedback in a model such as (9.5). To investigate this possibility, I added process noise to the model (choosing the maximum value of σ for which dynamics are not too "wild"), and analyzed the model output using an identical approach to the one applied to the real data. I found that even a simple linear regression, $r_t = a + bQ_{t-1} + \epsilon_t$, captured approximately three-quarters of the variance in the simulated data, despite the strong nonlinearity in the actual relationship between r_t and plant quality. Fitting the data with a power model for Q_{t-1}, which allows for nonlinearities in the relationship between the two variables, yielded an R^2 of more than 0.9. Note that, in order to make this exercise more realistic, I did not use the true model in these regressions (using the correct functional form results in $R^2 = 0.99$).

There is, therefore, a key mismatch between the predictions of the plant quality hypothesis and data: lack of detectable feedback from

Q_t to LBM rate of change. It gets even worse. If we examine the last documented LBM outbreak, which peaked in 1989, we notice that the plant quality index hardly declined, with needle lengths remaining above the 30 mm threshold throughout the whole period. As discussed by Baltensweiler (1993b), a sequence of unusual weather in 1989–1991 was conducive to high egg mortality. As a result, budmoth population never reached the level at which widespread defoliation occurs (the 1989 peak density was only 240 larvae per kg of larch branches, while previous peak densities observed at Sils were 490, 590, 800, and 560 larvae/kg). Correspondingly, no defoliation resulted in no decline in plant quality. Yet, the LBM population collapsed during 1990–1992! In other words, we have here a natural experiment suggesting that a large decrease in plant quality is not necessary for driving LBM cycles (since a population decline after peak sustained for 3–4 years is a necessary condition for LBM oscillation).

9.3.2 Parasitism

Our investigation of the parasitism hypothesis employs a structure paralleling the one used in assessing the plant quality hypothesis. Therefore, first I discuss some characteristics of the Engadine data on LBM fluctuations and parasitism rates. Next, I subject the data to the ancillary analysis. Finally, I develop an empirically based LBM-parasitoids model that attempts to mimic the observed LBM dynamics.

Ancillary Data: Parasitism Rate Although more than hundred species of parasitoids are associated with the larch budmoth, there are two groups that are particularly important in affecting LBM dynamics (Delucchi 1982). The first group consists of a complex of three eulophid species (*Sympiesis punctifrons*, *Dicladocerus westwoodii*, and *Elachertus argissa*). The second group are several ichneumonids, of which the most important is *Phytodietus griseanae*. Eulophids attack primarily the third (and to a lesser extent, the fourth) LBM instar, while the ichneumonid attacks primarily the fifth instar. LBM parasitoids and their effect on the moth dynamics have been extensively studied (Baltensweiler 1958; Aeschlimann 1969;

Renfer 1974, 1975; Herren 1976, 1977; Delucchi and Renfer 1977; Delucchi 1982). Combining the results obtained by these authors, we have an almost continuous sequence of parasitism rates from 1952 to 1976 (with only 1968 missing). Our approach in constructing the total parasitism data set essentially followed previous compilations by Baltensweiler, Auer, and Delucchi, with the exception that we traced the origin of all the cited data, and excluded those that turned out not to be based on actual observations.

It is worth discussing the problems encountered in compiling time-series data on LBM parasitism rates, because they are symptomatic of forest insect systems in general. It is often difficult to obtain an unbiased estimate of parasitism rates, because most simple approaches tend to miss some instances of parasitism. This is a general problem in studies of insect population dynamics. It is ironic that one of the initial reasons that insect ecologists have focused on parasitoids so intensively is the ease of quantifying parasitism rates (an additional consideration, of course, is that parasitoids are a ubiquitous and highly important component of insect communities; see Godfray 1994). In fact, obtaining unbiased estimates of parasitism in most systems is as laborious as measuring predation rates. This general problem is illustrated with the larch budmoth.

Because routine parasitism data are obtained by collecting LBM larvae during regular census, that is, during their third instar, such data will not only miss parasitism by the ichneumonids (who attack fourth and fifth instars) completely but also underestimate parasitism due to eulophids (who also attack fourth instar larvae). Additional collections of fifth-instar caterpillars are, therefore, necessary in order to correct for this bias. However, a single collection of fifth instars will still miss the parasitism that could occur during the period between collection and pupation. In other words, any method of measuring parasitism rate that is based on collecting larvae and rearing them in the lab underreports the true rate to an unknown degree. This problem is compounded by another feature of ichneumonid biology: *P. griseanae* tends to impose substantial direct mortality on LBM larvae via hostfeeding (Renfer 1974). In cage experiments, wasps killed and fed upon 30% of the larvae (Delucchi and Renfer 1977). In short, the available parasitism data underestimate the true contribution of parasitoids to LBM mortality to an unknown degree, and this fact has to be taken into account in the analysis.

That is the bad news. The good news is that the data provide reasonable estimates of the parasitoid numbers in the next generation. There is a curious asymmetry here: slight errors in parasitism rate have little effect on the estimate of parasitoid density, while having a disproportionately large effect on estimates of parasitoid impact. This idea is best illustrated with a simple numerical example. Suppose that the true parasitism rate is 95%, and therefore host survival is 5%. Our estimate of parasitoid density next year, P_{t+1}, is the product of the parasitism rate and host density. If host density is $N_t = 100$ larvae per larch tree, then $P_{t+1} = 95$ wasps per tree (assuming no overwinter mortality; including mortality does not affect the argument). An error of 1% in either direction, for example, a parasitism rate of 96%, would affect this estimate by about 1% (96 wasps instead of 95). The estimate of host density next year, on the other hand, is the product of larvae surviving parasitism and average fecundity, λ. Suppose $\lambda = 10$. The host density next year, therefore, $N_{t+1} = 10 \times (1 - 0.95) \times 100 = 50$. The same error in parasitism rate (96% instead of 95%) affects this estimate as follows: $N_{t+1} = 10 \times (1 - 0.96) \times 100 = 40$. That is, a 1% error in parasitism rate imposes a 20% error in host density estimate!

The take-home message here is that the parasitism rate may not be a very good predictor of next year's host density, due to measurement errors. On the other hand, it gives a good estimate of next year's parasitism density. If there is a tight coupling between parasitoid numbers and host mortality, then, paradoxically, we may be able to make better forecasts of host density two years ahead, using a host-parasitoid model, than one year ahead, using the simple demographic model.

Analysis of Ancillary Data   Our general model is the Beddington[2] (see equation 4.33 in section 4.3.2), but we rewrite it in terms of the proportion parasitized in year t, Π_t:

$$\Pi_t = 1 - \exp\left[-\frac{aP_{t-1}}{1 + ahN_{t-1} + asP_{t-1}}\right] \qquad (9.6)$$

The parasitoid density last year, P_{t-1}, is not directly observed, and therefore we need to estimate it by multiplying the host density during the previous year, $t - 2$, by that year's parasitism rate: $P_{t-1} = \Pi_{t-2}N_{t-2}$. In other words, we are indirectly sneaking in the delayed density dependence! Note that our estimate of P_{t-1} does not incorporate the (unknown) overwintering mortality. Thus, P_t is actually a relative index that is linearly related to the true parasitoid density, but with an unknown proportionality constant. This has no effect on the estimate of the proportion of variance resolved by parasitism; however, the estimate of parameter a is affected by the same proportionality constant.

Results of nonlinear regression suggest that the parasitism rate is very well resolved. Almost 90% of variance in the ichneumonid parasitism rate is explained by a simple two-parameter model (see table 9.1). A similar R^2 is obtained for the eulophid parasitism rate, but at the expense of an extra parameter. The total parasitism rate is less well modeled (see table 9.1). This probably reflects the fact that "total" parasitism is a compilation of several heterogeneous sets of observations.

Similarly high coefficients of determination are obtained when modeling LBM dynamics as a function of parasitoid density. Essentially all measures of parasitism (eulophids, the ichneumonid, and total parasitism) resolve a high proportion of variance (table 9.2). What is particularly impressive is that a very simple model, with only one predictor variable, P_{t-1}, manages to capture such a high proportion of variance in r_t (see figure 9.9). Interestingly, adding eulophids as an extra predictor variable after the ichneumonid does not increase R^2 (and vice versa). The reason, most likely, is that the two parasitoids are highly cross-correlated, acting essentially as a **dynamical complex**. Thus, both eulophid and the ichneumonid parasitism rates appear to provide basically the same information for predicting r_t.

The best model suggested by this analysis (high R^2 with fewest parameters) is the one based on the parasitoid interference functional response,

$$r_t = \frac{aP_{t-1}}{1 + awP_{t-1}} + \epsilon_t \tag{9.7}$$

The degree of predictability achieved by this parasitoid model is quite impressive. However, as I remarked above, the parasitoid density used in regressions, P_{t-1}, is calculated by multiplying N_{t-2} with Π_{t-2}.

TABLE 9.2. Results of ancillary analyses: parasitism rate. Υ_t, Σ_t, and Π_t: proportion parasitized by inchneumonids, eulophids, and total. r_t: LBM realized per capita rate of change. N_t, I_t, E_t, and P_t: population densities of LBM, ichneumonids, eulophids, and total parasitoids. P_q: polynomial of degree q

Variable	Model	R^2
Ichneumonid	$\Upsilon_t = 1 - \exp[-aI_{t-1}/(1 + awI_{t-1})]$	0.871
Eulophid	$\Sigma_t = 1 - \exp[-aE_{t-1}/(1 + ahN_{t-1} + awE_{t-1})]$	0.870
All parasitoids	$\Pi_t = 1 - \exp[-aP_{t-1}/(1 + ahN_{t-1} + awP_{t-1})]$	0.712
LBM	$r_t = r_0 - aI_{t-1}/(1 + awI_{t-1})$	0.861
LBM	$r_t = r_0 - aE_{t-1}/(1 + awE_{t-1})$	0.810
LBM	$r_t = r_0 - aP_{t-1}/(1 + awP_{t-1})$	0.865
LBM	$r_t = r_0 - aP_{t-1}/(1 + ahN_{t-1} + awP_{t-1})$	0.880
LBM	$r_t = a + b\log N_{t-2} + c(\log N_{t-2})^2$	0.646
LBM	$r_t = P_5[\log N_{t-2}]$	0.688
LBM	$r_t = P_2[\log N_{t-1}, \log N_{t-2}]$	0.877

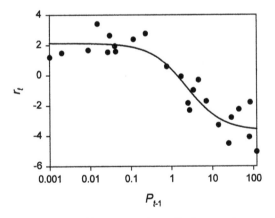

FIGURE 9.9. The relationship between the realized per capita rate of LBM change and parasitoid density (Engadine data). Note that parasitoid density is plotted on a logarithmic scale in order to resolve the relationship for small P_{t-1}; this changes the J shape of the relationship into an S shape.

Thus, one might ask, how much of the effect, which we ascribe to the parasitoid impact, is due simply to delayed density dependence? To answer this question, I fitted r_t with a variety of three-parameter functions of N_{t-2} (i.e., the same number of parameters as in the best parasitism model). The best R^2 (obtained with a quadratic polynomial in log N_{t-2} (see table 9.2) was far below R^2 yielded by equation (9.7). In fact, increasing functional flexibility by fitting a fifth-degree polynomial still results in R^2 of only 0.687, suggesting that this predictor variable by itself cannot approach the level of resolution achieved by P_{t-1}. The reason why is apparent when we examine the phase plot of Sils data in figure 9.3, where r_t is plotted against N_{t-2} (N_{t-2} is log-transformed, as indicated by the legend "theta = 0" in the upper left corner of the graph). It can be seen in this phase plot that the observed LBM trajectory exhibits noticeable cycling—the ascending and descending phases do not go through the same region of the phase space. Although by plotting the trajectory in the $r_t - N_{t-2}$ instead of $r_t - N_{t-1}$ phase space we have considerably "flattened" the reconstructed attractor, we did not manage to collapse it to a one-dimensional curve. In fact, as we know from the phenomenological analysis in section 9.2, we need a three-dimensional phase space to resolve all the variation in the Sils data.

To summarize, a simple but theoretically sound model based on the parasitism hypothesis resolves close to 90% of variation in the LBM r_t. This parasitoid effect is no simple artifact of "sneaking" the delayed density dependence "through the back door." Knowing parasitism rates allows us to predict the rate of LBM population change with much greater precision than is possible on the basis of only N_{t-2}. Furthermore, the performance of the parasitism hypothesis should be contrasted with the disappointing results yielded by analyses of the index of plant quality, which explains less than a third of variance in LBM rate of change. Incidentally, this result is not due to some subtle difference between the Sils and the Engadine data sets: for example, the Engadine parasitism rate predicts Sils data much better than the Sils plant quality index does. Taken together, the regression analyses of ancillary data are consistent with the hypothesis that the primary factor responsible for LBM oscillations is the larch budmoth's interaction with parasitoids.

Modeling the Parasitism Hypothesis   The analysis of ancillary data suggests a simple model of LBM-parasitoid interactions. It is based on the Nicholson-Bailey model, to which we add host limitation (in the Ricker form) and parasitoid interference functional response (since that is the functional form suggested by the regression analysis):

$$N_{t+1} = N_t \exp\left[r_0 \left(1 - \frac{N_t}{k} \right) - \frac{aP_t}{1 + awP_t} \right]$$

$$P_{t+1} = N_t \left\{ 1 - \exp\left[\frac{aP_t}{1 + awP_t} \right] \right\}$$

The regression analysis also suggests parameter values $a = 2.5 \pm 1$ and $w = 0.17 \pm 0.02$ (means $\pm$ SE estimated when fitting the LBM r_t as a function of P_{t-1}). We have already estimated r_0 and k above (thus, $r_0 = 2.5 \pm 0.2$ and $k = 250 \pm 50$). Simulating the model within these parameter ranges, we find that it produces high-amplitude oscillations for all reasonable values of parameters. For the median parameter values, however, the period is a bit short, $T = 7$ yr. It is necessary to reduce w to 0.15 in order to lengthen the period to 8 yr, and to 0.13 (2 SE from the point estimate—still within the realm of the possible) in order to lengthen the period further to 9 yr. The model output matches other probes, such as the amplitude and the cross-correlation function between N_t and Π_t, quite well. In particular, proportion parasitized peaks on average 2 yr after the LBM peak, similarly to the pattern observed in the data.

9.3.3   Putting It All Together: A Parasitism–Plant Quality Model

The preceding analyses of data and models suggest an interesting conclusion. On one hand, the model with plant quality as the only mechanism driving second-order oscillation fails to match data patterns as well as the LBM-parasitoid model. On the other hand, short-term experiments suggest that there is a strong effect of changes in plant quality on LBM survival and reproduction. This raises an important question: should we be satisfied with the parasitism-only explanation of the LBM dynamics, or do we really need a multifactorial model, combining plant quality and parasitism effects? One way to

address this issue is to investigate the dynamics predicted by the multifactorial model, and contrast its ability to match empirical patterns
with the two simpler alternatives.

Combining the effects of plant quality and parasitism is quite
straightforward, now that we have invested so much effort in building models for each component separately. The equations of this
parasitism-quality model are

$$N_{t+1} = N_t \exp\left\{ r_0\left(1 - \exp\left[-\frac{Q_t}{\delta} \right]\right) - \frac{r_0}{k}N_t - \frac{aP_t}{1 + awP_t} \right\}$$

$$P_{t+1} = N_t\left\{ 1 - \exp\left[\frac{aP_t}{1 + awP_t} \right]\right\} \tag{9.8}$$

$$Q_{t+1} = (1 - \alpha) + \alpha Q_t - \frac{cN_t}{d + N_t}$$

Parameter estimates are the same as before: $r_0 = 2.5 \pm 0.2$, $k =
250 \pm 50$, $a = 2.5 \pm 1.0$, $w = 0.17 \pm 0.2$, $\delta = 0.22 \pm 0.05$, $\alpha =
0.5 \pm 0.1$, $c = 0.7 \pm 0.2$, $d = 150 \pm 150$. Simulating the model
within these parameter ranges, we find that the model does very well
for parameters essentially at, or very near, their median values. In
particular, with slight modifications (specifically, $r_0 = 2.3$, $c = 0.9$,
and $d = 100$; note that we are staying within 1 SE of the median
estimate), the model output matches the periodicity, amplitude, and
cross-correlations between LBM and parasitism or plant quality index.
Note that this is an improvement on the parasitism-only model, which
required a rather low value of $w = 0.13$.

In order to compare the quantitative patterns predicted by the
parasitism–plant quality model with data, we need to add to the model
some terms representing the action of exogenous variables. I added
three such terms: (1) process noise with parameter $\sigma = 0.2$; (2) observation noise with $\sigma_{obs} = 0.2$; and (3) a small amount of immigration,
$i = 0.01$ moths per kg of branches per year (modeled as a random
variable uniformly distributed between 0 and 0.02). The values for
these parameters were chosen by the method of trial and error (the
model output does not appear to be very sensitive to the specific
values).

A typical series predicted by the parasitism–plant quality model is
shown in figure 9.10. The graphical output in this figure is arranged
in exactly the same way as in figure 9.3 to aid visual comparison.

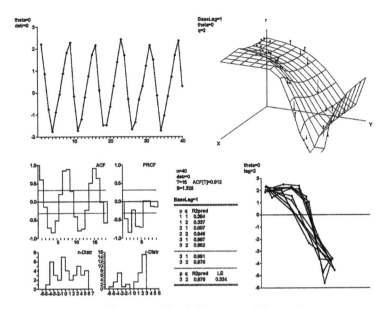

FIGURE 9.10. Nonlinear analysis of the output of the parasitism-quality model. *Upper left:* time plot; *upper right:* response surface; *lower right:* phase plot.

The model output appears to be somewhat less noisy than data (this is particularly clear when comparing the phase plots), suggesting that higher noise parameters are called for. The period and amplitude are matched by the model very well. But what is particularly interesting is that the model also manages to match the shape of the response surface, in particular the slight upturn observed at high values of Y (the delayed density-dependence axis).

9.4 SYNTHESIS

Theoretical and empirical analyses in this chapter suggest the following conclusions. First, a simple model of LBM–plant quality interaction with biologically plausible parameters predicts population dynamics that are quite similar to the observed pattern (second order, correct periodicity and amplitude). However, the plant quality hypothesis has weaknesses: although it predicts that there should be a strong feedback effect from plant quality to the LBM rate of change, analysis of real data does not reveal it. Additionally, lack of plant quality decline during the last recorded cyclic collapse is hard to understand, if plant quality is the main factor driving LBM oscillations. Second, an empirically based model of LBM-parasitoid interaction is also capable of mimicking the observed LBM dynamics. Unlike the rival explanation, however, the parasitism hypothesis is supported by regression analyses of the feedback relationship from parasitism rates to the LBM rate of change. To get the appropriate period, however, we have to "stretch" some parameters values. Finally, a tritrophic model combining both hypotheses does the best job in matching the observed dynamics for biologically reasonable parameter values. Whether this improvement is compelling enough to accept the more complex hypothesis over the two simpler alternatives cannot, probably, be resolved on the basis of existing data. What seems clear, however, is that previous authors have incorrectly rejected the parasitism hypothesis. Both empirical and modeling analyses in this chapter suggest that parasitoids play a key role in driving LBM cycles.

CHAPTER 10

Southern Pine Beetle

10.1 INTRODUCTION

The southern pine beetle, *Dendroctonus frontalis*, belongs to the family of scolytid bark beetles. Its generic name, *Dendroctonus*, can be loosely translated as "tree death." This is an apt name for this beetle, because it is the most important agent of mortality for several pine species, most notably the loblolly (*Pinus taeda*), in the southern United States, Mexico, and parts of Central America (Flamm et al. 1988). The estimated damage due to the southern pine beetle (SPB) over the last three decades is well over $1 billion (see Price et al. 1992).

Pine trees protect themselves from insects and fungi by exuding resin. As long as a pine continues to produce resin, SPBs are unable to utilize its tissues for feeding and reproduction. This beetle, however, has evolved a remarkable strategy to overcome the tree's defenses. Pioneering beetles (individuals initiating attack) emit a congregation pheromone that attracts other conspecific beetles. As more beetles bore into the tree, they release more pheromone to attract additional beetles, resulting in a positive-feedback process known as *mass attack*. As beetles congregate on the tree, they literally drain it of its resin resources, nullifying the tree's ability to defend itself (Hodges et al. 1979). Around two thousand beetles are needed to overcome the defenses of a healthy pine tree (Goyer and Hayes 1991). As the mass attack progresses, and the larval resource—inner bark of the tree (phloem)—starts to fill up with beetles, they begin to release a repelling pheromone that eventually inhibits congregation at the tree

(Payne 1980) and shifts the attack focus to adjacent hosts. Successful attacks almost invariably result in the death of attacked trees because the destruction of phloem by beetles prevents the tree from moving photosynthates down to its roots (this is known as "girdling"). One result of mass attack spilling over onto adjacent trees is that the spatial pattern of SPB attack is very patchily distributed. Typically, there is an area containing tens or hundreds (and on rare occasions much greater numbers) of dead trees, surrounded by normal forest. Such compact areas of trees killed by the SPB are known as *spot infestations*, or simply "spots."

Southern pine beetles do not go into a diapause, and there is some SPB activity (new trees coming under attack) occurring year-round. However, while it takes only about a month for one complete SPB generation in summer, during winter months it may take up to four months, depending on weather (mainly temperature; see Thatcher and Pickard 1964). There are about six SPB generations per year (Reeve et al. 1995). Because the SPB has fast generation times, growth of spot infestations begins to be fueled primarily by internally generated beetles within a few months after initiation. Other aspects of SPB biology are covered in a very useful compendium by Thatcher et al. (1980). Reviews emphasizing population dynamics are Flamm et al. (1988) and Reeve et al. (1995, 2002).

10.2   ANALYSIS OF TIME-SERIES DATA

Although direct measurements of SPB density exist (and will be discussed in a later section), they are too short for meaningful analysis with phenomenological models. The longest series come from indirect measurements of SPB activity, such as the Texas Forest Service records of SPB activity in East Texas since 1958. The Texas Forest Service conducts aerial surveys at 3–6 week intervals from May to October (Billings 1979). Spot infestations of more than ten trees are located from the air, and later visited on the ground to ensure that tree mortality is due to SPB activity. Currently, the surveys cover some 44,000 square km in thirty eight East Texas counties (before 1973, however, spots were recorded only for Southeast Texas). Adding together all spots detected by aerial surveys in one year gives us an index of SPB activity. Clearly, the relationship between this index and SPB density must be nonlinear. The nonlinearity may arise as a result of most beetles being found in small infestations below the detection threshold during periods of low SPB activity. Additionally, during outbreaks, the average spot size is larger than during troughs. As a result, the spot-based index probably exaggerates the degree of SPB fluctuations (this supposition is confirmed by the comparison between SPB densities and SPB spots during the 1990–1994 outbreak in Louisiana; see figure 10.7). Nevertheless, the relationship between spot index and SPB density, while nonlinear, is probably monotonic. Thus, these data provide useful material for phenomenological analyses (it is clearly not a good idea to fit mechanistic models to these data). One other problem with the data is that during the period of 1979–1981, the Texas Forest Service temporarily discontinued aerial surveys, because SPB activity was extremely low. Following the usual procedure (section 7.1.3), I substituted these zeros with the smallest nonzero number of spots observed.

Results of the nonlinear analysis of these data are very instructive (figure 10.1). Examining the time plot, we are struck by how irregular—even noisy—the population trajectory appears to be. ACF suggests oscillatory dynamics, but none of the autocorrelations are significantly different from 0. The PRCF, however, indicates a second-order process, thus providing the first indication that fluctuations in SPB activity may not be simply a result of some exogenous noisy driver. The real surprise comes from the response surface fitted to these data. First, the cross-validation suggests $p = 2$ (thus supporting the PRCF result). Second, the coefficient of prediction is very respectable 0.71. Finally, the estimated Lyapunov exponent is 0.16, suggesting that SPB dynamics are quasi-chaotic or weakly chaotic. It is interesting to note that when I first analyzed these data in late 1980s, using the records up to 1987 (Turchin et al. 1991), I obtained similar results to the ones described above.

Simulations of the fitted RSM model without noise indicate that the purely deterministic component of SPB dynamics is characterized by a stable point equilibrium. Adding noise, however, has a greater effect than simply creating random fluctuations around this point equilibrium, as would be expected in a linear model. Note that the response surface fitted for the SPB data (figure 10.1) is highly nonlinear in the

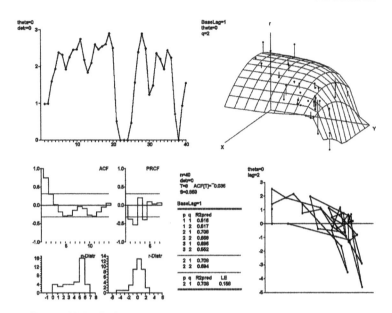

FIGURE 10.1. NLTSM analysis of SPB data. *Upper left:* time plot; *upper right:* response surface; *lower right:* phase plot.

Y direction ($Y = \log N_{t-2}$ quantifies delayed density dependence): it is very flat in the lower range of Y and then steeply declines for high Y. Without stochasticity, or with low amounts of noise, the trajectory never enters the region where the relationship between r_t and Y is characterized by a steep slope. Adding more substantial amounts of noise ensures that the population density recurrently enters this region. Such events are then followed by a sustained collapse of density (because Y is the delayed density dependence, density will continue collapsing for at least two years after a peak). Thus, adding noise to this nonlinear dynamical system changes its dynamics from stable (with Λ strongly negative) to quasi-chaotic (with $\Lambda \approx 0$). The response surface model, estimated for the SPB data, therefore provides another example of the general notion of **noise-induced chaos** discussed in section 5.3.3.

The most robust conclusion from the time-series analysis is the one pointing to the importance of delayed density dependence in the SPB dynamics. Note that each of the major peaks was followed by two years of decline. Taken together with the PRCF, cross-validation, and

RSM results, this observation suggests that SPB dynamics have a very strong second-order component, despite weak periodicity. Thus, what we apparently have here is an aperiodic (or very weakly periodic) second-order dynamical system with high signal/noise ratio. Note, further, that in the analysis of spot numbers I used the base time delay of 1 year, while SPB have about six generations per year. This result—that the dynamical process driving SPB oscillations is second-order on the temporal scale of years, while SPB generation time is on the order of a few months—provides an important clue for the inquiry into causes of SPB outbreaks.

10.3 HYPOTHESES AND MODELS

10.3.1 General Review of Hypotheses

Exogenous Factors   When I started working on SPB dynamics in 1987, the prevailing idea in the field was that its oscillations were driven by climatic variables (Wyman 1924; Craighead 1925; Beal 1927; 1933; St. George 1930; King 1972; Kroll and Reeves 1978; Kalkstein 1981; Michaels 1984; Michaels et al. 1986). Most of these papers linked SPB outbreaks to fluctuations in rainfall, but some also argued for the importance of cold temperatures in winter (e.g., McClelland and Hain 1979). The impact of rainfall on SPB population was postulated to be mediated by the physiological condition of host trees. In particular, it was thought that drought conditions would weaken pine trees, and make them more susceptible to SPB attack. Subsequent research showed that the actual relationship between drought and susceptibility to SPB attack is more complex than was previously envisioned (Lorio 1986; Lorio et al. 1990). Very severe drought, indeed, brings trees to the brink of death, at which point they are easily attacked by bark beetles, such as the SPB and *Ips* spp. However, moderate water stress actually increases the ability of pines to defend themselves against the SPB, because it limits growth more than photosynthesis, resulting in an increase in energy allocated to secondary metabolism (Reeve et al. 1995). In sum, the relationship between moisture stress and resistance to the SPB is curvilinear: resistance increases at moderate levels and declines at extreme levels of stress. Thus, new ideas and experimental results (Dunn and Lorio 1993; Wilkens et al. 1998) undermine the mechanistic basis of the moisture stress hypothesis.

Even more important, fluctuations in such weather factors as the amount of rainfall are by definition an exogenous factor in population dynamics that cannot drive second-order oscillations. One remaining possibility that would save the climate hypothesis is that SPB are driven by a *second-order exogenous process*. For example, annual rainfall amounts could oscillate in a second-order manner, and the SPB populations could simply follow along. We tested this possibility by analyzing the relationship between several climatic variables (temperature, moisture, etc.) and the SPB per capita rate of population increase (Turchin et al. 1991). We found no effect of weather on SPB population change. Furthermore, there were no indications that variation in climatic variables follows a second-order dynamical process. Our results, therefore, contradicted the voluminous previous literature that had no trouble finding statistical connections between weather and SPB outbreaks. Examination of the statistical methods used by previous authors suggested that their results were most likely spurious (Turchin et al. 1991). One common problem was a tendency to run hundreds of analyses and then select only those that yielded "significant" results. For example, one paper performed more than 500 regressions, and found that 42 of them were significant at the 0.05 level. Another study tried eleven independent variables in fitting a data set consisting of eleven observations. It was hardly surprising that the "best" model, employing four predictor variables, managed to explain >90% of variation.... Finally, almost all studies ignored the first law of population dynamics by using a measure of current population numbers as the response variable in the analyses (i.e., N_t, rather than r_t as I advocate). Ignoring this basic fact of population dynamics can lead to embarrassing predictions. Thus, one article made the following prediction of the course of the SPB epidemic in Hardin County, Texas, in 1979: 0 spots in June, 1,254 spots in July, and 0 spots again in August. The highest number of spots ever observed in that county was 836 in 1985, and it took three years to build up to that level from 3 spots observed in 1982. Given the inertial nature of SPB dynamics, it is biologically impossible for this population to increase from 0 to 1,254 spots in one month.

The only study that avoided these pitfalls (Michaels 1984) found that weather exerts a very weak influence on SPB rate of change. Michaels's final regression included seven predictor variables, but explained only 25% of variance in r_t. This is quite typical: climate usually accounts for less than 30% of variance in insect population dynamics (Martinat 1987).

In summary, while the exogenous factors hypothesis is initially appealing, given the pattern of apparently irregular SPB fluctuations, it fails to account for the high signal/noise ratio and the strong evidence for second-order dynamics revealed by the phenomenological time-series analysis.

First-Order Factors   There are several ecological factors potentially affecting SPB rate of change that we expect to act in a first-order manner. These factors are intraspecific competition, generalist predation (particularly from such avian predators as woodpeckers), and perhaps interspecific competition (e.g., from other bark beetles). If SPB dynamics are indeed dominated by some second-order mechanism or mechanisms (and the evidence for this seems quite strong), then first-order mechanisms cannot be "primary movers" of SPB oscillations. However, this does not mean that these factors are irrelevant to our understanding of SPB dynamics. First-order factors may play a key role in imposing an upper limit on population density, and perhaps they can be responsible for some short-term fluctuations around some equilibrium or mean density.

There are, thus, two key things we need to know about these mechanisms. First, we need to test empirically whether a mechanism such as intraspecific competition indeed acts in a first-order manner. Second, we need to empirically measure its strength and other attributes in order to be able to model it. These empirical issues will be addressed in section 10.4.

Second-Order Factors: Overview   Having discussed exogenous and first-order endogenous factors, our next step is to consider ecological mechanisms that are in principle capable of acting in a second-order manner. The theory (part I) supplies many candidate processes: maternal effects, interaction with host (quantity or quality), and specialist natural enemies (pathogens, parasitoids, and predators). One of these hypotheses, maternal effects, we shall not be able to evaluate empirically. Although we know that there is a substantial amount of variation among individual beetles (e.g., in their energy reserves upon emergence; Kinn et al. 1994), we lack systematic data to evaluate the potential of this mechanism to explain SPB cycles. Furthermore, theoretical considerations (discussed below in section 10.5) suggest that the maternal effect hypothesis is not a particularly likely explanation of SPB oscillations.

Another ecological mechanism, pathogens, also does not appear to hold much promise for explaining SPB dynamics. Pathogens (particularly, viral diseases) commonly attack another large class of forest insect herbivores, lepidopteran defoliators (Dwyer et al. 2000). Because caterpillars become infected by ingesting viral particles as they feed on foliage, viral diseases can easily spread through a defoliator population. In bark beetles, by contrast, each larva feeds in its own mine, without coming in direct contact with other larvae. This feature of bark beetle biology is probably responsible for lack of known viral pathogens in the southern pine beetle (Berisford 1980). A pathogen, in order to be able to transmit itself between bark beetle larvae, would need an ability to cross from one larval gallery to another. Some organisms associated with the SPB are, indeed, capable of growing through the wood on their own. For example, the bluestain fungus *Ophiostoma minus* negatively affects the survival of SPB larvae (Barras 1970; Reeve et al. 1998). However, this fungus affects SPB dynamics not by infecting beetles but by decreasing the amount and quality of resources available to feeding larvae. In other words, it is a competitor, not a natural enemy. Accordingly, we expect that competition from the bluestain fungus may act as a first-order regulatory factor, rather than a second-order factor that can promote oscillations (this assumption will be tested with empirical data in section 10.4).

I argue, therefore, that considerations of the SPB natural history allow us to rate certain hypotheses, for example, microparasites, as inherently unlikely to explain second-order oscillations in this organism. Another hypothesis that we can remove from the candidate list, using similar qualitative reasoning, is the interaction between the SPB and induced plant defenses. As discussed in section 4.4.4, previous feeding by herbivores may induce the plant to increase the amount of physical and/or chemical defenses, which in turn will negatively affect the herbivore's rate of population change. This mechanism cannot operate in the SPB system, because these beetles are "parasitoids" of pines rather than their parasites, since successful reproduction by the SPB almost invariably implies that their host was killed. Unless dying pines can communicate their distress to trees that will be attacked next (which seems quite far-fetched), therefore, we lack a mechanistic basis for inducible defenses.

In summary, we are left with three mechanisms whose potential contribution to SPB oscillations warrant a closer theoretical scrutiny: interactions with hosts (food quantity), parasitoids, and predators. In the following paragraphs I review what we know about each of these mechanisms in the context of the SPB, and advance simple models whose purpose is to investigate certain quantitative features of SPB oscillations implied by each mechanism, with a particular focus on the oscillation period.

10.3.2 Interaction with Hosts

There are two empirical observations that create serious difficulties for any explanation of SPB oscillations based on interactions between this beetle and the availability of food (host trees). First, even during the most severe outbreaks, the SPB kills only 1–2% of available hosts (Price et al. 1992). Second, major SPB outbreaks occur at intervals of 8–10 years. It is difficult to imagine how such relatively short-period oscillations could occur when one of the interacting species is a slowly growing tree. Even in the southern United States, where stand rotations are quite short, it still takes at least 15–20 years for pines to mature to the point where they become a good habitat for the SPB.

To test this intuition, I developed the following simple model of SPB-host interaction. The starting point of the model is

$$H_{t+1} = \mu_t H_t + \rho_t \tag{10.1}$$

Here H_t is the density of mature host trees in year t. *Mature* host trees are defined as those trees that are at least τ years old. These trees are old enough to be susceptible to SPB attack. Additionally, in order not to introduce an extra parameter, I assume that τ is also the minimum age of reproduction. The function μ_t is the survival rate from year t to year $t+1$, and ρ_t is the recruitment of new trees to the population of mature trees. The subscript t reminds us that these quantities are not fixed parameters but dynamic functions, with arguments to be specified shortly.

To write down the recruitment function, I assume that each mature tree every year produces β seedlings. Seedling establishment rate is density dependent, and the proportion of seedlings successfully establishing equals the proportion of space unoccupied by mature trees: $1 - H_t/k$, where k is the tree density at which all space is occupied by trees (and seedling establishment rate is 0). Finally, γ is the proportion of seedlings surviving from establishment to mature trees (assumed to be density independent). Thus, the recruitment rate is a product of the following quantities: mature tree density, number of seedlings produced per mature tree, proportion successfully establishing, and proportion surviving to maturity:

$$\rho_t = \beta H_{t-\tau}\left(1 - \frac{H_{t-\tau+1}}{k}\right)\gamma \tag{10.2}$$

Note the time subscripts: because seedlings are produced τ time units before they mature into trees, seedling production rate is proportional to $H_{t-\tau}$. On the other hand, the establishment rate depends on the tree density *next* year; thus, the $t-\tau+1$ subscript. Substituting this formula for ρ_t in equation (10.1), we have

$$H_{t+1} = \mu_t H_t + \lambda H_{t-\tau}\left(1 - \frac{H_{t-\tau+1}}{k}\right) \tag{10.3}$$

where I replaced the product $\beta\gamma$ with λ, since we do not need two separate parameters for this combination. We are not yet done: equation (10.3) allows the density of mature trees to exceed the maximum density k. We fix this problem by a simple expedient of killing all trees in excess of k:

$$H_{t+1} = \min\left\{k, \mu_t H_t + \lambda H_{t-\tau}\left(1 - \frac{H_{t-\tau+1}}{k}\right)\right\} \tag{10.4}$$

Now we are ready to specify the form of the survivorship function, μ_t. Because the SPB is functionally a parasitoid of pines (section 10.1), the simplest possible approach is to use the Nicholson-Bailey form:

$$\mu_t = \exp[-aN_t]$$

Although I could use a more complex and realistic functional response, I am content with this simple term because I am primarily interested in the dynamical consequences of disparate temporal scales on which the two interacting species operate. The equation for the SPB follows directly from this assumption:

$$N_{t+1} = sH_t\left(1 - \exp[-aN_t]\right)$$

Next, I scale the SPB and tree densities to reduce the number of model parameters: $H_t' = H_t/k$ and $N_t' = N_t/(sk)$. Parameter a is also rescaled: $a' = ask$. Substituting these relations and dropping primes, I have the following model of SPB-pine interaction:

$$H_{t+1} = \min\left\{1, H_t\exp[-aN_t] + \lambda H_{t-\tau}(1 - H_{t-\tau+1})\right\}$$
$$N_{t+1} = H_t\left(1 - \exp[-aN_t]\right)$$
$$(10.5)$$

The model has three parameters: τ is the pine generation time, and λ is the maximum per capita reproductive rate of pines. The interpretation of a can be seen if we rewrite the second equation in (10.5) as follows:

$$\frac{N_{t+1}}{N_t} = \frac{H_t(1 - \exp[-aN_t])}{N_t}$$

For periods when host density is at its maximum ($H_t = 1$) and SPB density is very low ($N_t \ll 1$), we can further simplify this equation, employing the approximate relationship $\exp[x] \approx 1 - x$ for x near zero:

$$\frac{N_{t+1}}{N_t} \approx \frac{1 - (1 - aN_t)}{N_t} = a$$

Thus, for small N_t and large H_t, the SPB replacement rate is a. In other words, parameter a can be interpreted as the multiplicative rate of increase of beetles at low densities.

Numerical investigation of this model showed that it is very prone to unstable dynamical behavior. Setting $\tau = 20$ yr, which seems to fit the biology of loblolly pines, I simulated model dynamics for all combinations of parameter values of $\lambda = 2, 3, \ldots, 10$ and

$a = 1.1, 1.2 \ldots 10$. (The lower bound on both λ and a is 1, since otherwise populations would not be able to replace themselves.) I sampled values of a more thoroughly because this parameter had a stronger effect on dynamics than λ. The dynamical behavior of model (10.5) is very complex: for different values of parameters it can generate cyclic, quasiperiodic, and chaotic attractors. Additionally, multiple coexisting attractors are possible. None of the parameter combinations yielded stability, while many led to SPB extinction (typically, after some very violent fluctuations). As a increases, it becomes increasingly more difficult to obtain bounded dynamics. A thorough characterization of the dynamics of this model, however, is not what we are after. We are interested in determining whether it is capable of generating oscillations characterized by short periods of around 8–10 years, as observed in the SPB. Plotting the dominant period against the parameter a, and frequency distribution of periods characterizing those series where the SPB does not go extinct, we see that there appear to be three or four typical periodicities in this model (for $\tau = 20$) (figure 10.2). The first peak in the frequency distribution is periodicities between 26 and 32 years, which I interpret as **generation cycles**, because this period is near the fundamental lag of $\tau = 20$. These generation cycles are somewhat longer than 20 years, for the following reasons. After a destructive SPB outbreak it takes 20 years for pines to grow back to densities where the SPB replacement rate can go above 1. An additional 6–12 years are, then, required for the SPB population to "climb out of the trough," at which time another outbreak occurs, and the whole "cycle" repeats (actually, all generation "cycles" were chaotic, as far as I could determine, so we are talking about statistical rather than mathematical periodicity here).

Within the range 40–80 years ($2 - 4\tau$) there is one major group of periods in the range 42–48 years, and another subsidiary peak around 60 years (figure 10.2b). These dynamics fit my definition of **first-order oscillations** (see section 2.5). Finally, there is a large group of periods located between 80 and 140 years ($4 - 7\tau$) which can be interpreted as **second-order oscillations**.

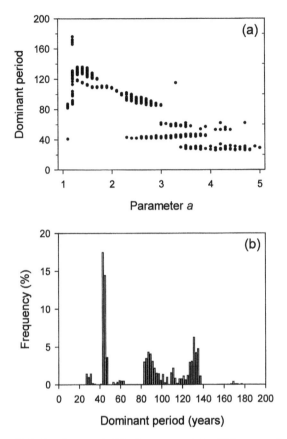

FIGURE 10.2. Dominant period in the output of the SPB-tree model: (a) periods as a function of parameter a, (b) frequency distribution of dominant periods. Results shown are for $1 < a \leq 5$ only, because for $a > 5$ few parameter combinations result in SPB persistence.

Similar results were obtained with $\tau = 30$ yr, although dynamics, in general, were more violent (a smaller proportion of parameter combinations yielded bounded oscillations), and all dominant periods for bounded cases were longer than for $\tau = 20$.

Returning to the main question that motivated this theoretical exercise—can the SPB-tree interaction drive the observed SPB oscillations?—we now see that the answer is no, it cannot. Even generation cycles, which are characterized by the shortest periods, are still too long: around 30 years versus the observed 7–9 years.

And generation cycles occurred for $<5\%$ of parameter values; first-and second-order oscillations were much more common.

To summarize, the explanation based on the SPB–food quantity interaction encounters very grave theoretical difficulties. Granted, the SPB-tree model advanced above is overly simplistic, but it is difficult to see how one could in principle devise a model that would get around the main theoretical difficulty: it takes a long time for tree stands to regenerate after a destructive outbreak. The only possibility that I see is to bring food quality into the model. As I discussed above, the possibility of induced defenses is contradicted by the biology of the system. However, it is conceivable that hosts vary in their quality, and that SPB outbreaks collapse after depleting *high quality*, rather than total resources. This scenario is, in fact, a plausible explanation for dynamics of some other bark beetles in the *Dendroctonus* genus, such as the mountain pine beetle (*D. ponderosae*) (Berryman 1976). In the mountain pine beetle (MPB) system, bark beetles primarily attack large pines with thick phloem (population rate of change in small trees is, in fact, negative). When the supply of large trees is exhausted, MPB population declines, even though many (small) trees still remain. This hypothesis, however, does not fit the biology of the SPB. The SPB is much less discriminating than the MPB. Within spot infestations, all trees (practically speaking) are destroyed, while few trees outside the infestations are attacked (primarily, trees struck by lightning). Thus, the major difficulty of the SPB-host hypothesis is to explain why a huge proportion of apparently suitable trees remain unattacked by the end of the SPB outbreak. Furthermore, note that MPB population cycles are characterized by long intervals (30–60 years) between outbreaks. This observation is consistent with the hypothesis that MPB cycles are driven by their interaction with host trees (and thus consistent with the model advanced above). After a destructive outbreak, practically all old trees are killed. It then takes 10–20 years for smaller trees to grow to the point where they become a suitable resource for the MPB. Then, some time has to pass before the MPB populations can build up to the point where they start inflicting serious mortalities on pines. This scenario can probably be modeled by a slight modification of model (10.5), and provides a plausible explanation of MPB dynamics.

10.3.3   Interaction with Parasitoids

Arthropods are among the principal natural enemies of the SPB (Berisford 1980). These enemies fall into two general classes: parasitoids (including several wasp species in the braconid and pteromalid families) and predators (primarily coleopteran, but also including a dipteran and some predacious mites) (see Berisford 1980 for a general overview).

We can assess the potential of parasitoids to drive SPB oscillations using the same logic that underlined the SPB-tree model. The SPB parasitoids tend to have rather fast development times, so that a generation is completed in around 2–3 weeks in summer (Berisford 1980). Since the dominant periodicity of SPB oscillations is around 8 years, each oscillation requires about 50 SPB generations and >50 parasitoid generations. Can a host-parasitoid model exhibit such long-period oscillations? As a first step to addressing this question, let us assume that the Nicholson-Bailey framework offers a reasonable approximation to the SPB-parasitoid dynamics. Since, in actuality, parasitoids have faster generation times than those assumed in the model, we would expect that any answer we obtain would be conservative with respect to finding long cycles. The simplest model for host-parasitoid interactions (that is not limited to diverging oscillations of the Nicholson-Bailey model) is the model of Beddington et al. (1976a):

$$N_{t+1} = N_t \exp\left[r_0(1 - N_t) - aP_t\right]$$

$$P_{t+1} = N_t(1 - \exp[-aP_t])$$

where N_t is host density (scaled so that the carrying capacity $k = 1$), and P_t is the parasitoid density. Parameters r_0 and a have their usual interpretations: the intrinsic rate of host population growth and the parasitoid attack rate, respectively. Historical data on SPB density suggests that the intrinsic rate of population increase is $r_0 \approx 1.8$ yr^{-1} or ≈ 0.3 gen^{-1} (Reeve et al. 2002). We have no data on a, but we know that $a > 1$, in order for parasitoids to be able to increase, and a cannot be too high, because then violent oscillations in parasitoid density cause it to go extinct. In order to determine the

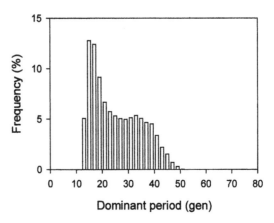

FIGURE 10.3. Frequency distribution of dominant periods found in the SPB-parasitoid model.

spectrum of dominant periods characterizing the Beddington model, therefore, I investigated the range of $r_0 = [0.20, 0.21, \ldots, 0.40]$ and $a = [1.1, 1.2, \ldots, 10.0]$. Excluding parameter combinations for which the parasitoid goes extinct, and those for which the dynamics are stable, I plotted the frequency distribution of dominant periods (figure 10.3). We see that the bulk of dominant periods lies substantially below 50 gen (the 90% range, excluding the shortest 5% and the longest 5%, is 13–41 gen). In fact, it is necessary to balance the parameter values just right, pushing r_0 to the lower bound of 0.2 and $a = 5.7 \pm 0.1$, in order to obtain $T = 50$ gen.

To summarize, it is possible, but difficult, to get oscillations of the correct period (≈ 50 SPB generations) in the Beddington model. In order to obtain the right period, one needs to push the value of r_0 quite low (perhaps unrealistically low) and select just the right value of the parameter a. This value of a also happens to lie on the boundary in the parameter space where the parasitoid persists (increasing a slightly causes the parasitoid to fluctuate violently and eventually go extinct). Although this result is not as strong as our rejection of the SPB-tree hypothesis, it still throws doubt on the viability of the SPB-parasitoid model.

10.3.4   The Predation Hypothesis

A number of studies (reviewed in Reeve et al. 1995) suggest that the clerid beetle *Thanasimus dubius* may be a particularly important natural enemy of the SPB. It is one of the first species to appear on trees mass-attacked by the SPB, because it is attracted by SPB pheromones and volatiles emitted by the damaged host tree. Adult clerids capture adult SPBs participating in mass attack, while larval clerids attack SPB larvae under the tree bark. Population oscillations of clerid beetles are shifted in phase with respect to SPB oscillations (Reeve et al. 2002). As a result, clerids achieve high densities during the years of SPB population decline. For example, in 1993–1994 clerid density on mass-attacked trees averaged 1.25 beetle per dm^2 of bark (Reeve 1997). Variation in clerid density has a strong explanatory effect on SPB population change (Billings 1990). For example a simple regression model, $r_t = a + bN_{t-1} + cP_{t-1}/N_{t-1}$, where N_t and P_t are SPB and clerid densities, explains 70% of variance in SPB realized rate of change, r_t (Reeve et al. 2002).

Adult clerids can inflict a substantial amount of mortality on attacking SPBs. Reeve (1997) showed that up to 60% of adult SPB can be killed by clerids before the bark beetles enter the galleries within the tree bark. We lack comparable experiments to quantify the impact of clerid larvae on SPB larvae within trees. However, exclusion experiments have shown that the whole complex of SPB natural enemies can reduce SPB survival severalfold: from 32% in treatments where enemies were excluded to 5.6% where they were not (Linit and Stephen 1983). Linit and Stephen (1983) attributed a substantial proportion of this mortality to clerids.

A particularly interesting feature of *T. dubius* biology is its tendency to undergo an extended period of development inside the host tree (Reeve et al. 1996): while some clerids emerge from mass-attacked trees about half a year after the trees were attacked, the majority of clerids emerge 1, 1.5, or even 2 years later. This puzzling but well-documented feature of clerid biology (see Reeve et al. 1996) has profound implications for hypotheses attempting to explain SPB oscillations.

Reeve et al. (2002) developed the following model to investigate the dynamical consequences of extended development in the clerid predators. The model is based on the Nicholson-Bailey framework (as are the SPB-tree and SPB-parasitoid models, allowing for direct cross-comparisons between the three hypotheses). For the SPB equation, we simply use the Beddington model:

$$N_{t+1} = N_t \exp\left[r_0\left(1 - \frac{N_t}{k}\right) - aP_t \right]$$

Here the units of t are SPB generations (of which there are six per year).

Upon completing development, new clerids do not immediately emerge to attack SPB, but instead join the population of "diapausers," whose density is Q_t. The equation for Q_t is

$$Q_{t+1} = Q_t + \chi N_t(1 - \exp[-aP_t])$$

In other words, every generation Q_t is incremented by the number of newly developed clerids. Parameter χ specifies how many new clerids are produced per each killed SPB (since each clerid larva needs to consume several SPB larvae to complete development, $\chi < 1$). Meanwhile, the population of adult predators is governed by the following equation, assuming density-independent dynamics:

$$P_{t+1} = P_t - \delta P_t$$

where δ is the proportion of adult predators dying each SPB generation. The last ingredient in the model is the connection between the diapausing and free-flying clerids. We assume that twice per year (or every three SPB generations) a certain proportion of diapausers, v_0, leaves the tree and becomes free-flying adults.

The model has six parameters, but two of them, k and χ, can be scaled out. The equations of the scaled model are

$$N_{t+1} = N_t \exp[r_0(1 - N_t) - aP_t]$$

$$Q_{t+1} = Q_t - v_t Q_t + N_t(1 - \exp[-aP_t])$$

$$P_{t+1} = P_t - \delta P_t + v_t Q_t \tag{10.6}$$

$$v_t = \begin{cases} 0 & \text{for } t = 1, 2, 4, 5, 7, \ldots \\ v_0 & \text{for } t = 3, 6, 9, \ldots \end{cases}$$

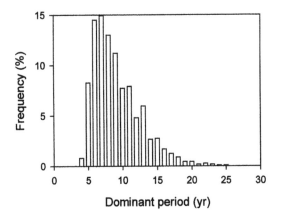

FIGURE 10.4. Frequency distribution of dominant periods found in the SPB-clerid model (for the parameter ranges described in the text).

For the SPB intrinsic rate of increase, we use the same range of values as was justified above: $r_0 = [0.2, 0.4]$. We lack data to directly estimate the clerid attack rate, so we will use a very broad range for this parameter: $a = [1, 10]$. The proportion of clerids breaking diapause is around 0.5, with the range $v_0 = [0.3, 0.7]$ (Reeve et al. 2002). Finally, the death rate of free-flying clerids can be crudely estimated by observing how fast their numbers decline after episodes of spring and fall emergence. Our estimate lies in the range $\delta = [0.5, 0.9]$.

Numerical solution of the SPB-clerid model for all combinations of the four parameters showed that this model behaves as a typical second-order difference model. The mode of the distribution is at 7 years, and the bulk of periods (around 90%) falls within the range of 6–13 years (figure 10.4).

This is a striking result, especially when contrasted with the predictions of models based on the two rival hypotheses. While the SPB-tree model could not generate the correct period for any biologically reasonable values of its parameters, and the SPB-parasitoid model could do it only for the most extreme values, the SPB-clerid model generates approximately correct periods for almost any biologically reasonable combination of its parameter values. Our conclusion from this theoretical exercise is that the likelihood of the clerid model is much higher than the likelihood of the parasitoid or (especially) the tree models.

Strictly speaking, our argument in favor of the clerid *model* cannot be extended more broadly to argue in favor of the clerid *hypothesis*. Without checking many different models for rival hypotheses, we cannot know whether under some of them the observed period may not, indeed, be quite probable. However, general theory of population dynamics suggests that it would be difficult to construct a model to produce either an unusually long period (which we would need for the parasitism hypothesis) or an unusually short period (needed for the tree hypothesis). Furthermore, Reeve et al. (2002) explored the effect of using a more realistic functional response than the linear one used in model (10.6), and found qualitatively the same result. Thus, given what we know about population dynamics, it seems likely that if we were to perform a survey of many different kinds of models for the three hypotheses, we would still come to essentially the same conclusion.

Two other caveats need to be kept in mind. First, there may be other hypotheses, in addition to the ones we have explicitly considered here, and there is no guarantee that one of them would not outperform the clerid hypothesis. Nevertheless, finding that a model based on the clerid hypothesis explains the observations better than the rivals is clearly a step forward. Second, all the hypotheses that we have considered in this section are monocausal, yet the world is likely to be complex. Thus, the results of the theoretical exercise we have worked through should not be interpreted as denying any dynamical role for parasitoids or hosts. The observation that the monocausal clerid hypothesis explains the observed period well enhances our belief in the proposition that the *primary* factor driving SPB oscillations is SPB interaction with the clerid predators. However, it does not preclude the possibility that parasitoids may play an important secondary role in, for example, providing some first-order feedback instrumental in stabilizing SPB-clerid cycles. In fact, our experimental results appear to support such a role for some natural enemies of the SPB, as will be discussed in the next section.

10.4   AN EXPERIMENTAL TEST
OF THE PREDATION HYPOTHESIS

10.4.1   Rationale

Theoretical arguments and comparisons to dynamical patterns observed in time-series data, discussed in the previous section, strongly suggest (to me) that the predation hypothesis is a better explanation of the SPB oscillations than the rival hypotheses that I considered. I suspect, however, that most ecologists would not be particularly swayed by "just" logic, and would require an experimental test. How can we test the predation hypothesis? The most drastic proof would be to exclude predators and to see whether SPB cycles would stop. This approach, of course, is completely unrealistic, simply because of the large spatial scale on which such an experiment would have to be conducted. The median dispersal distance for the SPB was measured to be around 0.5–1 km (Turchin and Thoeny 1993). Even a 1 km^2 chunk of pine forest is too small a scale on which to try to "stop the cycle," because of the highly patchy distribution of SPB attack, which is organized in spot infestations. During a major outbreak, there are only about 0.3 large spots per 1 km^2 area (data from historical records for East Texas), so we would need at least 10×10 km experimental units to capture areawide SPB dynamics (rather than local spot growth and collapse).

Although it is physically impossible to experimentally stop the SPB cycle, given present knowledge and realistic resources, there are other empirical ways to test the predation hypothesis. First, it is possible to exclude the SPB natural enemies on a local scale (one tree, or a portion of a tree). By comparing SPB survival and reproduction in predator exclosures versus controls, this experimental manipulation allows us to measure the predation impact on SPB population change. Second, both the general theory and the specific model based on the predation hypothesis (equation 10.6) make quantitative predictions about the pattern and the magnitude of predation impact, and these predictions can be empirically tested.

Before we consider these predictions, it is worth reiterating the following point. Demonstrating that predators impose a substantial (or even overwhelming) mortality at any particular point in the population cycle does not tell us whether predators are in fact the mechanism driving the oscillations. If predators kill *the same* proportion of SPB every year, then some other factor must be driving the cycle. What we need to determine is how the predator impact changes with time, or more precisely with cycle phase. In other words, we need to measure predation impact throughout the complete increase-peak-decline cycle.

We can distinguish three broad outcomes that such a long-term predator exclusion experiment can have. These outcomes correspond to the hypotheses that predators are

- an exogenous (or zero-order) factor
- a first-order endogenous factor
- a second-order endogenous factor

(figure 10.5). The graphs in figure 10.5 were constructed by, first, postulating a particular time course of a single SPB oscillation (the dotted lines). Second, assuming for simplicity that fecundity does not change systematically with the cycle phase, the numerical course of the outbreak is completely determined by the dynamics of survival rate. The survival rate that produces the oscillation is plotted in figure 10.5 as the solid curves. Note that the dotted and solid lines do not vary between cases. The only thing that varies is the survival rate of beetles protected from predators (the dashed lines), corresponding to three general scenarios of how predators may affect SPB dynamics.

In the first case (figure 10.5a), there is no dynamical feedback between prey density and predation impact. As I pointed out above, the average predator-induced mortality may be very high and still predators might have no dynamical impact, simply reducing the intrinsic rate of population increase to a lower value. Random (phase-independent) fluctuations in predator-imposed mortality will affect prey density in a stochastic manner, but cannot drive a regular oscillation. In the second case (figure 10.5b), predators respond to changes in prey population without a significant lag time. The dynamical role of predators, therefore, is stabilizing rather than causing oscillations. Generalist predators may act in this manner, reducing the amplitude of oscillations or preventing diverging oscillations. Only in the third

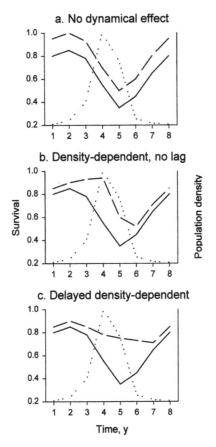

FIGURE 10.5. Dynamical effects of predation. Dotted line: SPB density. Solid line: SPB survival in the presence of predators. Dashed line: SPB survival when predators are excluded. (After Turchin et al. 1999.)

case (figure 10.5c), when acting in a delayed density-dependent manner, are predators actually causing the oscillation. Note that the three scenarios represent extremes of a continuum, since it is possible for the predator community to act in a mixed manner. For example, a mixture of generalist and specialist predators would act in a manner intermediate between cases (b) and (c).

The predation hypothesis for SPB cycles predicts that the empirical outcome of the predator exclusion experiment should resemble

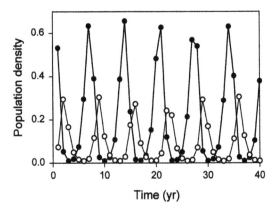

FIGURE 10.6. Dynamics predicted by the SPB-clerid model (10.6) for the reference set of parameter values. Solid circles: SPB density. Hollow circles: clerid density. Note that although the model has six time steps per year, SPB and clerid densities were averaged within each year, to provide a direct comparison with data such as shown in figure 10.7.

case (c). In fact, we can obtain an even more specific and quantitative prediction, by solving the SPB-clerid model developed in the previous section (equation 10.6) for best parameter values, and calculating the expected predation impact with respect to the phase of the cycle. To arrive at a set of such "best" values, I used the data-suggested estimates of parameters $r_0 = 0.3$, $v = 0.5$, and $\delta = 0.7$. Parameter a must be estimated indirectly. I selected $a = 5$, which gives the closest match to the observed period and amplitude of SPB oscillations. Dynamics predicted by the SPB-clerid model with these parameters are shown in figure 10.6. The comparison between the predation impact values predicted by this model and the data in the predator exclusion experiment will be considered below (see figure 10.11). Note that this is a very strong test of the hypothesis, because there is no circularity in obtaining predictions: the data from the predator exclusion experiment were not used in estimating model parameters.

Experimental Method   Our empirical approach for testing the predictions of the predation hypothesis relied on using cylindrical cages to exclude SPB natural enemies from a portion of a tree trunk (Turchin et al. 1999). The cage design is explained in Reeve et al. (1998). In each iteration of the experiment, we selected a stand of predominantly loblolly pine within Kisatchie National Forest, and located experimental trees within the stand. The trees were of approximately the same diameter, and were separated by about 100 m. Some trees were designated as "cage" trees, and others as "control" trees. On cage trees we installed 2 m–long cylindrical exclosures made from polyethylene screening. The cage contained a central 1 m–long experimental area, and two 0.5 m buffer zones, which acted as barriers to the movement of insects (both SPB and its natural enemies) into the experimental area. These buffer zones were not completely successful in excluding predators. However, we suspect that predators who entered the central zone did so toward the end of the SPB development period (when the beetles construct pupation chambers in the outer bark), and thus did not affect SPB survival very much. In any case, the possible semipermeability of our predator exclosures does not affect our main conclusions, although it raises the possibility that our measures may underestimate the actual predator impact.

The control trees did not have cages installed on them, but in all other respects were treated in the same way as the cage trees. After exclosures were installed on cage trees, all experimental trees were baited with SPB congregation pheromone to induce attack. A total of two thousand adult SPB were added to exclosures, generating attack density that was similar to that outside cages. We monitored the temporal course of the mass attack on the cage tree (outside the exclosure) and matched by varying the frequency and the size of SPB additions within the cage.

Several measurements were made to estimate SPB reproduction and survival on cage trees within exclosures and in areas above and below exclosures, and on control trees. First, we estimated the density of successful attacks and eggs by taking bark samples (see Reeve et al. 1998 for details). Second, when brood development was complete, we cut sections of trunks, placed them in individual rearing cans, and recorded the numbers of emerging SPB. Our two primary measures of SPB performance were (1) survival from egg to emerging adult, estimated by dividing the numbers of emerging SPB by the numbers of eggs laid; and (2) the ratio of increase λ', estimated by dividing the numbers of emerging SPB (or the offspring generation) by the numbers of successful attacks (or the parent generation). The prime

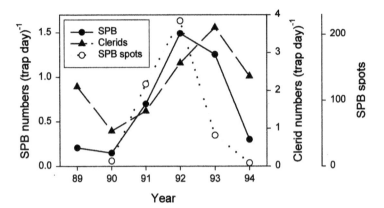

FIGURE 10.7. Population densities of SPB (solid circles) and clerids (solid triangles) during 1989–1994, measured by a network of pheromone-baited traps within Kisatchie National Forest (KNF), Louisiana. Also shown are the number of spots detected in KNF (hollow circles).

reminds us that the ratio of increase is not the true multiplicative rate of population change, because it does not take into account mortality of adult beetles between emergence and successful attack.

Each experimental iteration described above was replicated (1) spatially, by using two stands separated by at least 3 km; and (2) seasonally, by conducting one study in late spring–early summer and another in late summer–fall. Finally, the whole study was repeated for five years. The 5 yr period covered a complete cycle, including increase (1990–1991), peak (1992), and decrease (1993–1994; see figure 10.7).

10.4.2   Results

The first thing we need to check is how well our experimental manipulation succeeded in mimicking patterns of SPB attack and reproduction inside cages. Figure 10.8 compares the densities of successful attacks and egg densities inside cages (protected from predation) versus outside cages (exposed to predation; this category combines measurements taken on control trees with those taken above and below exclosures on cage trees). Fluctuations in attack density inside cages paralleled those occurring naturally, and there

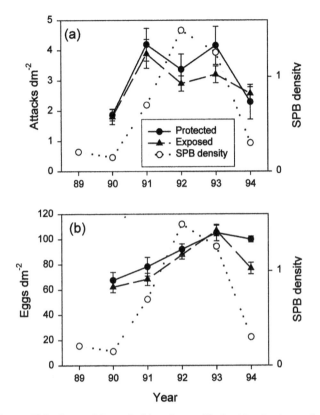

FIGURE 10.8. Successful attack (a) and egg (b) densities in the predation exclusion experiment. Solid circles: measurements inside cages ("protected"). Solid triangles: measurements outside cages ("exposed"). Hollow circles: the temporal course of the SPB cycle, as measured by a network of pheromone traps (from figure 10.7).

were no statistically significant differences between the two variables in any year (figure 10.8a). The same pattern held for egg density trajectories, apart from the last year (1994), when egg density inside cages was significantly higher than that outside cages. Because SPB survival from egg to adult is strongly density dependent (this will be discussed below), we expect that our estimate of predation impact in 1994 may be underestimated (lower survival inside cages decreases the difference between it and survival outside cages). The general message from the data in figure 10.8 is that our experimental

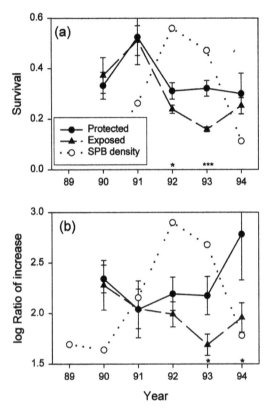

FIGURE 10.9. SPB survival (a) and log-transformed ratio of increase, $r' = \log \lambda'$ (b) in the predation exclusion experiment. Solid circles: measurements inside cages ("protected"). Solid triangles: measurements outside cages ("exposed"). Hollow circles: the temporal course of the SPB cycle, as measured by a network of pheromone traps. $* = P < 0.05$. $*** = P < 0.001$.

technique was successful in creating similar starting conditions for SPB populations inside versus outside cages. In one instance where initial conditions diverged (egg densities in 1994), our measure of predation impact should be conservative; that is, we expect that it will underestimate the true impact.

Turning to the main results of the experiment, I begin by considering the predation impact on SPB survival (figure 10.9). The survival on control trees versus cage trees outside exclosures (note that both are *exposed* treatments) was not significantly different in all but one iteration of the experiment (fall 1992). For this reason, we treated these replicates as a single category, "exposed." The *P*-values reported below are, thus, based on within-year comparisons between exposed and protected replicates (each tree serving as an experimental unit) employing t-tests.

In 1990 and 1991, during the increase phase of the cycle, the survival of protected (inside cages) SPB broods did not differ from that of broods exposed to predation, indicating negligible predation impact during the increase phase. Predators imposed detectable mortality during the peak year (1992; $P < 0.05$), but numerically the strongest (and statistically most significant, $P < 0.001$) effect of predation was observed during the first year of decline (1993), when SPB survival rate was halved (from 0.32 to 0.16) by predators. The survival difference in the second year of decline (1994) was not significantly different from zero.

We observed a qualitatively similar pattern in the effect of predators on the SPB ratio of increase (figure 10.9b). However, this measure of predation was statistically significant during both decline years (1993–1994), and not during the peak year. As in survival rate, the ratio of increase during the two increase years (1990–1991) was indistinguishable between the exposed and protected replicates.

As I noted above, egg densities were significantly different between the exposed and protected treatments in 1994, raising the possibility that the predation impact on brood survival in that year was masked by this difference. Furthermore, experiments that manipulated SPB density within trees documented a strong effect of it on both survival and the ratio of increase (Reeve et al. 1998). Analysis of density dependence in the vital rates of SPBs inside predation exclosures yielded similar results (e.g., figure 10.10 shows density dependence in the ratio of increase). Because different replicates varied in initial SPB densities, it is possible that this variation obscures some of the patterns in the data. To check on this possibility, we reanalyzed the data in which effects of density were removed. For example, we defined a new variable, the adjusted survival rate:

$$S_{i,j}^* = S_{i,j} - \beta E_{i,j}$$

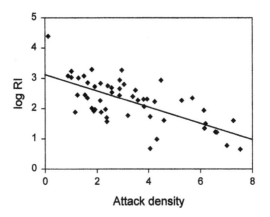

FIGURE 10.10. The relationship between the ratio of increase (emerging beetles/attacking beetles) and the density of attacking beetles inside cages. Linear regression results: $F_{1,51} = 42.8$, $P < 0.00001$, $R^2 = 0.456$.

where $S_{i,j}$ is the observed survival in replicate i of treatment j ($j = 1$ is protected, $j = 2$ is exposed), $E_{i,j}$ is the egg density in this replicate, and β is the slope of linear regression of $S_{i,1}$ on $E_{i,1}$. Analyzing the adjusted survival rates $S^*_{i,j}$ using the same approach as before (t-tests within each experimental year), we found that for years 1990–1993 the qualitative results were unchanged, but for 1994 the difference between protected and exposed was borderline significant ($P < 0.07$). This result suggests that the difference between survival inside versus outside cages is detectable even two years after the peak.

Performing similar reanalysis of the ratio of increase, we found that, as before, during the increase years there were no differences between the exposed and protected treatments. However, the difference in the peak (1992) year became significant ($P > 0.05$), and the results during the years of decline were also strengthened (P-values of 0.001 and 0.03, respectively). In short, we get much crisper results after tuning out variance due to different initial densities in replicates.

Taken together, these results suggest that the predator complex acts primarily as a second-order process, with an admixture of a weaker first-order impact. This qualitative result is consistent with the predictions of the predation hypothesis. What about a quantitative comparison? Recollect our earlier discussion of the fully specified SPB-clerid model. The expected impact of predation as a function of

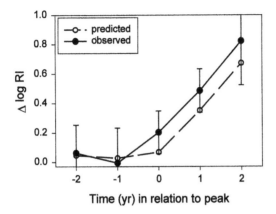

FIGURE 10.11. Comparison of predicted and observed predation impact in relation to cycle phase. Hollow circles: predicted by the SPB-clerid model. Solid circles: observed in the predator-exclusion experiment. Predation impact, $\Delta r'$, is the difference between log-transformed ratios of increase inside versus outside cages.

the phase of the cycle, predicted by the SPB-clerid model, is plotted in figure 10.11. The SPB-clerid model predicts negligible predation impact during the increase and peak years, and strong impact during the two years of decrease, especially in the second year, when $\Delta r' = 0.7$. Note that, since $\exp[0.7] = 2$, predators impose a 2-fold decrease on SPB survival. This may not appear to be a strong impact, but remember that there are six SPB generations per year, so a 2-fold survival differential per generation translates into $2^6 = 64$-fold differential per year. This survival differential is strong enough to drive substantial cycles in SPB density (as seen in figure 10.6).

The observed pattern in $\Delta r'$ is very similar to the SPB-clerid model predictions (figure 10.11). As predicted, the greatest impact of predators occurs during the two years of population decline. Most important, the observed impact matches numerically the predicted magnitude of predation impact (in fact, it is even greater than predicted, but not significantly so, because the predictions lie within 1 SE of observed means). The importance of this match is that it suggests that the predation impact measured in the experiment is numerically strong enough to drive the oscillation *by itself*.

The observed impact deviates from the predicted in one way: it begins to increase already during the peak year. While evidence for increased predation impact during 1992 is not as strong as that for the decline years (1993–1994), it is substantial enough to suggest that something more is going on than is predicted by the SPB-clerid model. This is perhaps not surprising, given the extreme simplicity of the model. In particular, an argument was made earlier that another important group of natural enemies of the SPB, parasitoid wasps, should act largely as a first-order factor. Additionally, some arthropod predators that, unlike the clerid beetle, do not have an extended diapause could contribute to the first-order effect seemingly present in the data. Finally, vertebrate predators such as woodpeckers should also act in a nondelayed density-dependent manner. Thus, it is likely that this diverse group of SPB natural enemies contributes to first-order regulation of SPB density.

I should also note that our predator exclusion method excluded not only natural enemies but also some potential competitors, such as sawyer beetles or the bluestain fungus. However, neither sawyer beetles nor bluestain fungi exhibit any signs of second-order control (Reeve et al. 2002). Thus, these associates most likely add to the first-order regulation in the SPB (although numerically their effect may be swamped by the much stronger direct density dependence in the SPB).

Another caveat is that we should not be too hasty in attributing the second-order component of predation impact, demonstrated by the predator exclusion experiment, entirely to the effect of clerids. After all, as discussed in the previous two paragraphs, our experimental cages excluded not only clerids but also many other organisms potentially detrimental to SPB reproduction and survival. Our evidence implicating clerids as the main factor driving SPB oscillations is indirect: because of their extended development, clerid populations behave in the dynamically appropriate way to drive SPB oscillations, as demonstrated by the SPB-clerid model. We do not know enough about other enemies of the SPB, particularly other insect predators, to either support or refute their potential role in helping to drive SPB cycles. Our direct experimental evidence implicates the whole complex of SPB natural enemies, rather than any particular species.

The final caveat is that our experiment examined only one oscillation in only one (broad) area. Ideally, it would be very useful to

repeat this experiment in another state during the next oscillation. Fortunately, we have some additional observations that suggest that the
1990–1994 cycle in Louisiana was not unusual. Stephen et al. (1989)
conducted a longitudinal study of the 1975–1978 SPB outbreak in
Arkansas, in which they measured SPB egg and late-stage immature
densities, as well as densities of its natural enemies. As SPB density
increased from 1975 to 1976, the survival rate from eggs to late-stage
immatures decreased. However, the most substantial drop in survival
occurred after the peak, during the decline year of 1978. This pattern
parallels the one we observed in our long-term predator exclusion
study.

10.5 SYNTHESIS

SPB populations in the southern United States exhibit a rather
irregular-looking pattern of outbreaks, occurring at intervals of 6–9
years. Time-series analysis suggests that despite this apparent irregularity, SPB outbreaks are driven by a strong second-order dynamical
process. Theoretical investigation of three likely hypotheses (interaction with hosts, parasitoids, and specialist predators) suggest that only
models based on the last hypothesis (predation) can generate oscillations of appropriate periodicity for biologically plausible parameter
values. Most important, the predation hypothesis successfully passes
the severe test of a long-term predator exclusion experiment. It
correctly predicts not only the observed second-order pattern of
predator-imposed mortality but also the magnitude of predation
impact. The experimental results do not preclude the possibility that
other second-order factors than specialist predators may contribute
to oscillatory dynamics in the SPB. However, the observation that
the experimentally observed predation impact is numerically strong
enough to drive SPB oscillations of correct period and amplitude
suggests that if such other mechanisms are present, then their effect
is weak.

CHAPTER 11

Red Grouse

Periodic dynamics are not common in bird populations (Kendall et al. 1998: table 1). A major exception to this general pattern is birds of the grouse family (Tetraonidae, order Galliformes) (Middleton 1934; Williams 1954). Population cycles have been reported in Scottish rock ptarmigan (Watson et al. 1998); black grouse, capercaillie, and hazel grouse (Lindén 1989); ruffed grouse and prairie grouse (Keith 1963); and red grouse (Potts et al. 1984; Williams 1985). Most nontetraonid examples of oscillations in bird populations appear to be exogenously driven (e.g., owls feeding on cyclic voles or arctic geese periodically suffering from lemming predators; see chapter 12).

Red grouse (*Lagopus lagopus scoticus*, a subspecies of the willow grouse) is the best-studied tetranoid bird. Because red grouse is the favorite game bird of the British sportsman (Hudson 1992), we have a multitude of long-term data indexing grouse population dynamics. Equally important is the willingness of British foundations to support empirical research on grouse ecology. As a result, we may be in good position to resolve the question of what ecological mechanisms are responsible for grouse oscillations, having time-series data, short-term data on grouse ecology, parameterized mathematical models, and field experiments testing the models.

In my review of red grouse cycles, I follow the standard sequence: I start with phenomenological modeling of time-series data, then review the hypotheses and mechanistic models based on them, and finally discuss the experiments. The literature on red grouse cycles is voluminous; fortunately, there are two recent reviews that provide very useful guides to this complex topic (Moss and Watson 2000; Hudson et al. 2002).

11.1 NUMERICAL PATTERNS

The main fodder for phenomenological analysis of red grouse fluctuations comes from bag data—records of the number of grouse shot per year on a hunting estate. There is an approximately linear, although rather noisy, relationship between the log-transformed grouse density and the log of number shot plus 1 (Hudson 1992: figure 4.2). The bag data, however, exaggerate the degree of fluctuation observed in grouse populations. Thus, judging by figure 4.2 in Hudson (1992), as bird density increases by one order of magnitude (between approximately 30 and 300 birds km^{-2}), the log of number shot plus 1 increases by two orders of magnitude (100-fold). These data, therefore, suggest that the estimate of S (standard deviation of log-transformed numbers) indicated by the analysis of bag data inflates the true amplitude by a factor of about two.

Williams (1985) Analysis First analyses of grouse bag data were done by Middleton (1934) and MacKenzie (1952), although these investigators did not employ very sophisticated statistical approaches. Another study of tetraonid dynamics using qualitative data is by G. R. Williams (1954). Using modern methods of time-series analysis, Jennifer Williams (1985) reanalyzed the data tabulated by Middleton and MacKenzie. Because data show temporal trends (see, e.g., figure 11.1), Williams detrended them using polynomials of up to sixth order. Generally, I advocate against routine detrending. However, bag data are particularly prone to trends (Hudson 1992). British society underwent abrupt structural shifts, particularly during the twentieth century, on top of which one should add the impact of two world wars. As a result, the management of game estates evolved, perhaps changing the nature of red grouse dynamics (this will be discussed below), and the "measuring apparatus" did not always faithfully reflect fluctuations in grouse numbers. Thus, some approach to dealing with this nonstationarity is necessary, although fitting polynomials of sixth order seems excessive.

ACFs estimated by Williams (1985) for the fourteen time series indicate that there is strong evidence for periodicity in four data

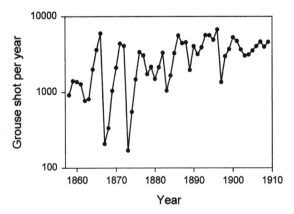

FIGURE 11.1. Annual numbers of red grouse shot on a Lanarkshire estate. (Data from Middleton 1934.)

sets (ACF statistically significant at the dominant period; see section 7.1.3). In five other series, there is weaker evidence of periodicity (ACF significantly less than zero at half the dominant period), and in the last five series ACF is not statistically significant at any lags. The most common dominant period is 6 years (six cases), but there are also single instances of 4-, 5-, and 10-year periodicities. In summary, Williams's analysis of the Middleton and MacKenzie data suggests that the oscillatory tendency in the grouse bag data is real, but while some populations undoubtedly exhibit cycles, others fluctuate in a much less regular manner.

NLTSM Analysis of the Middleton Data   To place the red grouse cycles in the general context of oscillatory dynamics, I submitted the Middleton data to the NLTSM analysis (I focused on the Middleton data because MacKenzie presented data only as averages over a whole region, which may hide important differences in fluctuations between individual estates). Unlike Williams (1985), however, I did not detrend the data. Visual examination of the series suggested that not only the mean of fluctuations drifts with time, but the character of dynamics themselves may also change from the first to second half (figure 11.1). Thus, I used the alternative approach of splitting the series (see section 7.1.2). Splitting each series in half yields segments of 26–30 years long, which should be rather optimal for NLTSM

TABLE 11.1. Red grouse bag data from Middleton (1934) and Thirgood et al. (2000a): summary of results of nonlinear time-series analysis. Number of data points, n; measure of amplitude (SD of log-transformed densities), S; dominant period, T; the autocorrelation at the dominant period, ACF$[T]$ ($**$ = significantly > 0 at T; $*$ = significantly < 0 at $T/2$); estimated process order, p; polynomial degree, q; the coefficient of prediction of the best model, R^2_{pred}; and the estimated dominant Lyapunov exponent, Λ_∞

Location/segment	n	S	T	ACF$[T]$		p	q	R^2_{pred}	Λ_∞
Middleton data									
Cumberland 1	29	0.45	6	0.67	**	2	2	0.64	0.12
Cumberland 2	28	0.37	—	—		2	1	0.06	−0.91
Aberdeenshire 1	28	0.38	6	0.83	**	2	2	0.37	0.53
Aberdeenshire 2	28	0.20	7	0.35	*	3	1	0.16	−0.30
Lanarkshire 1	26	0.38	6	0.65	**	3	2	0.51	0.45
Lanarkshire 2	26	0.16	8	0.37	*	1	0	0	−∞
Thirgood et al. data									
Moor F	24	0.45	7	0.48	**	2	2	0.18	0.22
Moor G	24	0.47	7	0.48	**	3	1	0.41	−0.12
Hudson et al. data									
Gunnerside	15	0.26	—	—		2	1	0.09	−0.16
Watson et al. data									
Kerloch	18	0.19	8	0.09	*	3	2	0.87	0.57
Rickarton 1	20	0.41	—	—		1	1	0.38	−0.58
Rickarton 2	21	0.37	10	0.20	*	2	1	0.65	−0.41

analysis, especially since, at 6 years per cycle, we end up with 4–5 oscillations per piece (section 7.1.2).

The summary of NLTSM results is presented in table 11.1. We see that there is very strong evidence for dynamic nonstationarity. Almost all probes change in a systematic fashion. First pieces are classi-fied as characterized by strong periodicity (with the dominant period

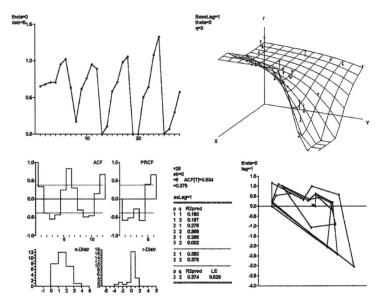

FIGURE 11.2. Graphical output of the NLTSM analysis of grouse bag data: first 30 years of Aberdeenshire (Middleton 1934). *Upper left:* time plot; *upper right:* response surface; *lower right:* phase plot.

of 6 years), strong nonlinearities ($q = 2$), high signal-to-noise ratio (R^2 around 0.8), and, rather surprisingly, positive Lyapunov exponents. An example of analysis is shown in figure 11.2.

Second segments have much weaker (and in two cases statistically not significant) periodicity, more linear and less predictable dynamics, and are characterized by trajectory stability (negative Lyapunov exponents). In fact, cross-validation suggested that none of the models predict Lanarkshire 2 better than the mean (thus, the best model for that series is simply its mean with variance around it). However, in the other two cases cross-validation selected second- or higher-order models, hinting that some oscillatory process, although weaker than during the first period, may still underlie grouse population fluctuations.

In line with other probes, the amplitude declines from first to second halves (table 11.1). Note, however, that the amplitude of oscillations is generally rather low. Because S estimated for bag data is probably twice the S of densities, the real S, therefore, is probably around 0.2 during the first period, and 0.1 during the second one. Such S values roughly correspond to fourfold and twofold peak/trough ratios (see section 7.1.3), respectively. In other words, even during the more oscillatory periods, grouse fluctuations are rather mild.

In summary, the NLTSM analysis of the Middleton data yielded quite interesting results. The general finding of second-order periodic dynamics supports the previous conclusions of Williams (1985). Furthermore, the analysis strongly suggests that grouse dynamics changed qualitatively around 1880, from more complex to simpler dynamical patterns. The insistence of NLTSM on estimating chaotic models ($\Lambda > 0$) for data observed during 1850–1880 is somewhat puzzling, given the generally low amplitude of grouse oscillations. This finding, however, is strengthened by several observations: generally high R^2 characterizing first halves (around 0.8) and the consistency between time-series patterns observed at such widely separated locations as Lanarkshire (southern Scotland) and Aberdeenshire (northeastern Scotland). On the other hand, the three time series cannot be taken as completely independent replicates. For example, the two most pronounced collapses (in 1866 and 1872) were synchronized across all three sites.

Analysis by Potts et al. (1984) and Hudson (1992)   Potts et al. (1984) and Hudson (1992:138–144) obtained long-term bag records from 63 estates in England and 110 estates in Scotland. The longer time series extended back into the 1870s. Many series showed a declining trend, and other signs of non-stationarity, such as low bag numbers during World War I and II (Hudson 1992: figure 24.1). These trends were removed by smoothing (see Potts et al. 1984:22 for details). Smoothed time series were subjected to the autocorrelation analysis. Apparently, autocorrelations were calculated on raw (not log-transformed) data, which is generally not a good idea in the analysis of population dynamics (see section 7.1.3).

Very few of the ACFs yielded significant positive autocorrelations at dominant periods, but a substantial proportion were characterized by significantly negative ACFs at the estimated half-period. Thus, the strength of periodicity in red grouse dynamics is not very strong.

This result is in agreement with my reanalysis of Middleton data, which showed that periodicity substantially declined after 1880 in the three Scottish data sets. Around 60% of English and 70% of Scottish data were deemed periodic, on the basis of significantly negative ACF at half-period.

Autocorrelation analysis suggested that English and Scottish populations were characterized by slightly different periodicities. The modal periodicity for English data was 4 yr (range: 4–6 yr), while for Scotland it was 5 yr (most series within the range of 3–8 yr, but a few exhibiting 10-yr or longer cycles). The cycle period tends to increase toward northwest within the British Isles, and wetter and colder sites tend to be more likely to have periodic oscillations than drier, warmer sites.

NLTSM Analysis of Thirgood et al. (2000a) Data   Thirgood et al. (2000a: figure 5) reported grouse bag numbers at three moors in South Scotland. On one of the moors, Langholm, control of predatory raptors was relaxed in 1990, while on the other two, moors F and G, management of predators continued throughout the complete period covered by data. I will focus on the two moors (F and G) where conditions were stationary.

NLTSM analysis of the two moors reveals a pattern that is very similar to that exhibited by Middleton's series before 1880 (figure 11.3; table 11.1). The amplitude and periodicity of fluctuations are almost identical. Second-order (or higher) response surface models are selected for both data series. Most interestingly, note that both data sets are characterized by similar shapes of fitted response surfaces: the surface tends to be flat both along the X and Y axes, but then declines nonlinearly and precipitously for jointly high values of N_{t-1} and N_{t-2}. This shape is characteristic of all five series, and is probably a result of the tendency of bag numbers to collapse precipitously, and often in one time step, after spending several years at high levels. This is a feature that is probably also responsible for the chaotic or quasi-chaotic dynamics detected in the data sets (Λ positive or near 0). However, we should remember that this particular feature may have nothing to do with the biology of red grouse cycles, being, rather, a result of the measuring apparatus. For

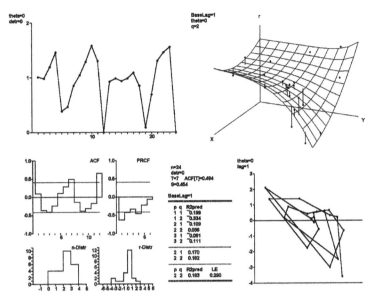

FIGURE 11.3. Graphical output of the NLTSM analysis of grouse bag data: Thirgood et al. (2000a), moor F. *Upper left:* time plot; *upper right:* response surface; *lower right:* phase plot.

example, when grouse numbers on an estate decline, the owners may cancel the shooting in an attempt to preserve the breeding stock.

NLTSM Analysis of Watson et al. (1984) Data The final data that I analyze here—Kerloch and Rickarton—come from two areas in northeastern Scotland intensively studied by Robert Moss, Adam Watson, and coworkers. The longer Rickarton series is clearly nonstationary, so I split it into equal-size segments. Like data analyzed above, Kerloch and Rickarton have a mild amplitude of oscillations ($S = 0.2$–0.4). The first Rickarton segment is characterized by very noisy dynamics (at least, NLTSM was unable to find any signal in these data). Kerloch and Rickarton 2, however, show very similar patterns (see Kerloch data in figure 11.4). The evidence for second-order dynamics is quite strong (note the strongly negative PRCF[2]). The ACF exhibits weak evidence of periodicity (it is significant at half the period). The response surface model selected for these data captures a high proportion of variance in r_t, and the associated Lyapunov

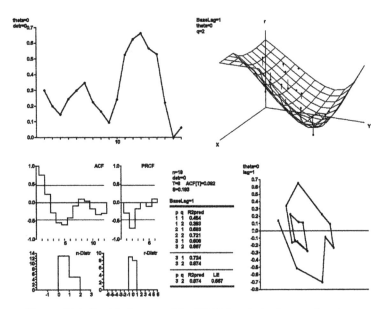

FIGURE 11.4. Graphical output of the NLTSM analysis of grouse bag data: Kerloch (Watson et al. 1984). *Upper left:* time plot; *upper right:* response surface; *lower right:* phase plot.

exponents suggest chaotic or quasi-chaotic dynamics (again, this is an interesting result, given the mild amplitude of oscillations).

Despite these similarities, however, Kerloch and Rickarton 2 exhibit a clearly different structure of density dependence compared with other data sets, which are characterized by a slow buildup to the peak, a tendency to linger there, followed by a precipitous collapse. Thus, the oscillations are markedly asymmetric (figures 11.1–11.3). By contrast, Kerloch and Rickarton 2 are characterized by very symmetric cycles (figure 11.4). The difference between the asymmetric and symmetric oscillations is also apparent in the shape of the response surface (compare the upper right panels of figures 11.1–11.3 and figure 11.4). This interesting difference has been previously commented on by Moss et al. (1993). Systematic differences in the shape of oscillations may provide important clues about the ecological mechanisms responsible for population cycles (section 4.2.3). I shall return to this point below.

11.2 HYPOTHESES AND MODELS

11.2.1 Overview

Current research has focused on two rival hypotheses for red grouse cycles: intrinsic factors and interaction with macroparasites. Before dealing with these hypotheses in depth, I need to say a few words about other oscillatory mechanisms suggested by population theory and why they do not appear as viable hypotheses for red grouse cycles. The first obvious factor to consider is food. Red grouse is a specialist herbivore on heather (Hudson 1992). However, birds are selective feeders and consume only a small proportion of total food biomass even during periodic population peaks (Moss and Watson 2000). Thus, any dynamical effects of food must be due not to total plant biomass but to availability of *high-quality* food, which may include food accessibility, physical form, proportion of digestible nutrients, and concentration of secondary compounds (Moss and Watson 2000). Indeed, food quality, in particular the content of digestible protein, declined at high grouse density in one intensively studied population (Moss et al. 1993). However, there is no evidence that heavier browsing of food plants in peak years results in lower-quality diet in subsequent years (Moss and Watson 2000). Since the critical delayed-density effect on food quality appears to be lacking, it is unlikely that food quality is the mechanism driving the red grouse cycles.

Turning to natural enemies, we note that the main predators of red grouse in unmanaged moorlands are red foxes and crows (Hudson 1992). In managed populations, grouse can be subject to predation by hen harriers and peregrine falcons (Thirgood et al. 2000b), unless they are (illegally) controlled by managers. None of these species is a specialist on red grouse (Moss and Watson 2000). Accordingly, predators also do not appear as likely agents to drive population cycles. However, predator density varies between sites (partly as a result of management practices), and generalist predators may be an important factor in explaining why some grouse population cycle, and others do not.

Another important source of grouse mortality is game shooting. The impact of hunting is somewhat mitigated by its taking place in the fall, before the winter population bottleneck, when most of the density-dependent mortality occurs (Hudson et al. 1992). Thus, especially during high-density years, hunters may be killing many birds that would die anyway during the following winter. The dynamical pattern of shooting mortality conforms well to the generalist predation paradigm. First, human numbers do not fluctuate as a result of changes in grouse densities (and, in any case, human numbers change very slowly on the temporal scale relevant to grouse cycles). As a result, hunting mortality can respond to elevated grouse numbers without a time lag. Second, the functional response of human hunters to grouse density is sigmoid: when grouse density declines below a certain threshold, hunting usually completely ceases. Taken together, these observations imply that generalist predation by humans will at most have a stabilizing first-order effect on grouse dynamics. If hunting impact is largely compensated by winter mortality, then shooting mortality will have only a weak effect on regulating grouse numbers.

Red grouse do have specialist enemies, but they are parasites, not predators. A particularly important macroparasite is the nematode *Trichostrongylus tenuis*. The population interaction between red grouse and this parasitic worm has been the subject of an intensive study by Hudson, Dobson, and coworkers. I review this research below. Next, I consider the rival hypotheses, advanced by Moss, Watson, and coworkers, focusing on factors intrinsic to grouse populations.

11.2.2   Parasite-Grouse Hypothesis

Adult nematodes inhabit the large cecum of red grouse (Hudson et al. 1992). Nematode eggs pass from the host in the bird's feces, and embryos begin developing when the temperature exceeds 5°C. Development from the egg to the infective larval stage (under optimal conditions) takes seven days. Infective third-stage larvae migrate to the growing tips of heather, where they are ingested by feeding grouse (Hudson et al. 1992). Most grouse older than two months are infested with worms, but parasite burdens (average number of worms per bird) vary substantially across individual birds. As we know from section 4.5.3, the frequency distribution of the number of parasites per host has an important effect on dynamics, because a greater heterogeneity of risk leads to a greater degree of dynamic stability. However, the degree of heterogeneity in red grouse is relatively low (the parameter k of the negative binomial distribution ranged between 1.2 and 5.8, with a mean of 2.85; Hudson et al. 1992). Such a low degree of contagion is probably due to the lack of acquired immunity (Hudson and Dobson 1996) and the virtually universal prevalence of infection in older birds.

To model the dynamics of the red grouse–nematode parasite system, Dobson and Hudson (1992) employed the theoretical framework developed by Anderson and May (section 4.5.3). The model proposed by Dobson and Hudson was essentially equations (4.49–4.51), but with a slightly different parameterization and adding direct density dependence in the grouse equation:

$$\frac{dH}{dt} = (a - b)H - gH^2 - (\alpha + \delta)P$$

$$\frac{dW}{dt} = \lambda P - \gamma W - \beta WH \qquad (11.1)$$

$$\frac{dP}{dt} = \beta WH - (\mu + b + \alpha)P - \alpha \frac{P^2}{H} \frac{k+1}{k}$$

Here a and b are per capita rates of birth and death, respectively (in the Anderson-May model, they enter as a single parameter, $r_0 = a - b$). Parameters α and δ reflect how per capita death and birth rate are affected by increased parasite load (see section 4.5.3 for the explanation of the term associated with α; the δ term is analogous). Other parameters are the same as in the Anderson-May model. However, the second term on the right-hand-side of the parasite equation is $(\mu + b + \alpha)P$, rather than $(\mu + \alpha)P$, because in my formulation of the Anderson-May model, I included the effect of b into μ. Finally, the term gH^2 in the first equation reflects self-limitation in the grouse population. The assumed mechanism of density dependence is grouse territorial behavior.

Dobson and Hudson (1992; Hudson et al. 1992) used a variety of data sources to estimate the parameters of model (11.1). Using the median values of parameters in their table 2 ($a = 1.8$, $b = 1.05$,

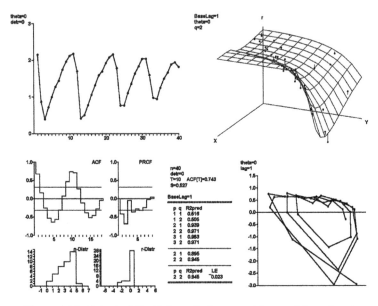

FIGURE 11.5. Output of the Dobson-Hudson parasitism model; analyzed with NLTSM. *Upper left:* time plot; *upper right:* response surface; *lower right:* phase plot.

$\lambda = 11$, $\gamma = 13$, $\alpha = 0.0005$, $\delta = 0.0003$, $k = 1$, and $\beta = 0.1$) and the value of density-dependence parameter from Hudson et al. (1998) ($g = 0.004$), I solved model (11.1) numerically. A small amount of process noise has been added to the simulation by periodically (once per year) perturbing grouse density. For these values of parameters, the parasitism model generates cycles with 8–9 year periodicity and amplitude $S \approx 0.5$ (figure 11.5). Although the period and amplitude of model output are greater than those in the majority of data sets analyzed in section 11.1, informal experimentation with parameter values suggests that it is easy to get shorter, less violent cycles in model (11.1). For example, increasing the strength of density dependence decreases the amplitude and shortens the period.

In the analysis of time-series data on grouse bags (section 11.1), we noted that the structure of density dependence in many populations has one stereotypical feature: grouse density approaches peaks slowly but collapses rapidly. The response surface fitted to this trajectory tends to be flat along both X and Y axes, but have a nonlinear "fold-down" in the part of the phase space where both X and Y

are high (see, e.g., Aberdeenshire data, plotted in figure 11.2). It is interesting that the output of the parasite-grouse model matches this feature of the data. This is a striking result. My informal numerical explorations of model (11.1) suggest that the "fold-down" is a robust feature of its output, obtaining as long as the parameters are such that the model is in the limit cycles regime. (When the model is characterized by oscillatory stability, adding dynamical noise, as usual, generates persistent pseudoperiodic oscillations. However, the response surface for such dynamics does not have a nonlinear fold-down, being instead quite flat.) Because the Anderson-May model was taken "off the shelf," we can be sure that this match between theory and data was not contrived by fiddling with model structure. In other words, the fact that the parasite-grouse model captures a recurring structural feature of grouse density dependence should greatly strengthen our degree of belief in this hypothesis.

11.2.3 Kin Favoritism Hypothesis

Development of theory attempting to explain red grouse cycles using intrinsic mechanisms has proceeded along two tracks: simple analytical models (Matthiopoulos et al. 1998, 2000b, 2000a) and complex individually based simulations (Mountford et al. 1990; Hendry et al. 1997). I reviewed the analytical models in section 3.4.2, and came to the conclusion that the theory has not yet developed to the point where we could see how assumptions about individual behavior and demography would cause population cycles (although work in progress by J. Matthiopoulos and coworkers may soon be able to fill this gap). For this reason I do not review those models further in this section. Instead I focus on the individual-based simulation developed by Hendry et al. (1997).

The basic idea underlying the kin favoritism hypothesis was already discussed in section 3.4.2. Strong philopatric behavior by new recruits causes them to settle near their fathers. The result of this behavior is formation of spatial clusters of related territory owners. Males behave aggressively toward nonkin, but tolerate closely related individuals (fathers, sons, brothers). One result of this behavior is that individuals finding themselves surrounded by kin are prepared to accept smaller territories. Furthermore, presumably because males do not fight relatives, larger kin clusters produce relatively more offspring per unit of area. Recruitment of young males into kin clusters is initially facilitated by reduced aggression. However, as kin clusters grow, eventually territory sizes approach an irreducible minimum. As a result, recruitment falls, and kin clusters can no longer be maintained. Because most males begin interacting with nonkin, aggression levels increase, minimum territories grow, recruitment continues to fall, and population density declines. Eventually, density becomes low enough so that recruitment can increase and new kin clusters can begin forming, and the cycle repeats itself.

Hendry et al. (1997) modeled this mechanism by simulating territorial behavior and recruitment rates of red grouse males inhabiting two-dimensional spatial "arenas." The heart of the simulation is the algorithm for acquisition of territories in the fall. A family group consisting of a father with any surviving sons (or a group of brothers if the father did not survive to the fall) is initially placed within the father's spring territory. The simulation then allocates territory space to each bird according to the concept of "pressure." The pressure that each cock exerts on its neighbors' territories depends on its fighting ability (chosen from a random distribution), its territory size (cocks with greater territories exert less pressure on their neighbors), and whether the two birds are kin or not. The simulation iteratively adjusts territories until the pressure from each side is equalized. If a male is left with a territory smaller than the minimum size (which is a function of how many relatives are in the kin cluster), then he is removed, and his territory is apportioned to neighbors by running the pressure algorithm again. The procedure is repeated until all males have territories exceeding the minimum size. The territory-acquisition algorithm is coupled to a module for winter survival and summer reproduction, to produce a complete model of grouse dynamics.

To determine the effect of kin favoritism, Hendry et al. (1997) first ran the "null" simulation in which the effect of kin tolerance was turned off (i.e., birds exerted the same territorial pressure on both kin and nonkin, and the minimum territory size was a constant independent of how many relatives were in the same kin cluster). In the null simulation the number of territories quickly approached an equilibrium of around 150 males per modeled area, and showed very little fluctuation around this value.

Adding kin favoritism to the null simulation also generated a rapid approach to an equilibrium. However, the equilibrium number of territories was around 300, or twice that in the null simulation. Additionally, there appeared to be a greater degree of fluctuation around the equilibrium. Hendry et al. (1997:31) concluded that kin-tolerant behavior destabilized the dynamics of the model grouse populations, because "kin-tolerant populations had much noisier dynamics than intolerant populations." This conclusion is not warranted by their modeling results, however. First, and most important, the question that we are trying to resolve is *what mechanisms explain second-order oscillations in red grouse populations.* The observation that kin-tolerant populations fluctuated with somewhat greater amplitude than kin-intolerant ones does not address this question in any meaningful manner, especially since both populations did not exhibit second-order dynamics. Second, "much noisier" is an exaggeration. Judging by eye, the numbers in kin-tolerant populations probably fluctuated within 10% of the mean (see figure 3a in Hendry et al. 1997; the authors unfortunately did not quantify the amplitude of fluctuations in their model output).

When Hendry et al. (1997) added dispersal to their simulation, they discovered that the model output exhibited oscillations with 2-year period and amplitude of about twofold. The mechanism leading to this oscillation is somewhat obscure, but my guess is that a high proportion of males emigrate during the years of high density, leading to an undershoot of the equilibrium next year. Whatever the mechanism, the oscillations exhibited by the model are clearly first-order in nature, and thus again fail to address the question posed in the previous paragraph. In fact, the results of the grouse simulation reviewed so far agree completely with the general lesson in chapter 3 that intrinsic-factors models typically generate first-order dynamics (either stability or first-order oscillations).

The models considered up to now have been firmly based on mechanisms of individual behavior and demography. The next two modifications considered by Hendry et al. (1997) departed from this approach by assuming that the phase of population cycle has a direct effect on individual behavior. The first modification assumed that individual males keep track of whether population declined or increased from last year. If population declined, males then increase their aggressiveness toward kin (a certain proportion start treating kin as nonkin). Conversely, a year of increasing density is accompanied by decreased aggressiveness toward kin. The second modification assumed that dispersal rate is a function of the cycle phase. Not surprisingly, both models incorporating these modifications were found to generate second-order oscillations for certain values of parameters.

I argue that the two last models, in which behavior is cycle phase-dependent, do not contribute to our theoretical understanding of grouse cycles. I am not convinced by the argument advanced by the authors that grouse would first perceive changes in population density between years, and second choose to modify their behavior toward kin on the basis of such observations. Even more tellingly, the logic of this modeling is circular in the following sense. Second-order oscillations occur as a result of some equivalent of inertia in population dynamics. For example, the inertia in predator-prey cycles during the collapse stage occurs because even though prey density is already low, there are still lots of hungry predators running around trying to kill the last few prey individuals. Some time has to pass before enough predators die off, releasing mortality pressure on prey population. This population inertia is not built in by fiat; instead, it is an epiphenomenon arising from behavioral and demographic characteristics of prey and predator individuals. By contrast, assuming that some individual characteristic is directly affected by whether population declines or increases is putting inertia in by fiat. There is no individual-based mechanism here. As a result, this procedure does not help us to understand how second-order cycles may arise. If we build second order into the model by assumption, then we should not be surprised that the model generates second-order oscillations.

The paper of Hendry et al. (1997) is often cited in the red grouse literature as a theoretical basis of the kin facilitation hypothesis. My reading of this paper is very different. First, a model including only kin favoritism exhibits very stable dynamics with minor oscillations around the equilibrium. Second, adding density-dependent dispersal causes the model to undergo first-order oscillations for certain values of parameters. Note that certain parameter values have to be rather

large in order for these oscillations to occur. For example, the parameter ΔTol, the proportion by which the minimum required territory
is decremented by each additional kin individual in the group, must
be 0.2 in order for any oscillatory behaviors to occur in the model
(Hendry et al. 1997: figures 3b and 5). Because the average brood
size assumed by the model is 3.6, many individuals will have their
territories reduced by more than 70%, and some by 100% (the latter
could happen if brood size is 5 and all individuals survive, a situation
that is possible under the assumptions of the simulation). It would be
interesting to know whether this is a realistic description of red grouse
natural history. However, the most important point is that models with
defendable structure among those advanced by Hendry et al. (1997)
cannot generate second-order cycles. This observation, taken with the
review of the analytic model of Matthiopoulos et al. (1998) in section 3.4.2 suggests to me that we still do not have a firm theoretic
basis for the kin favoritism hypothesis.

11.3 EXPERIMENTS

11.3.1 Density Manipulation

The red grouse cycle is an unusual system, because it was subjected
to two manipulative experiments (whereas most ecological systems
are lucky to get even one). The first experiment attempted to "stop
the cycle" by manipulating grouse density. This experiment was conducted on Rickarton moor in northeastern Scotland by Robert Moss,
Adam Watson, and coworkers (Moss et al. 1993; Moss et al. 1996).
The expected outcomes of the experiment were published prior to
conducting the experiment (Moss and Watson 1985).

Moss et al. divided the moor into two sections: the experimental,
from which they removed territorial males, and the control, which
was left unmanipulated. The experimental and control areas were 203
and 318 ha, respectively. There was a buffer zone about 500 m wide
between the two areas. Using a time-series model fitted to historical
data, the investigators predicted that grouse numbers would peak in
1982–1983. The main goal of the experimental manipulation was to
prevent the cyclic decline by not allowing the population to reach

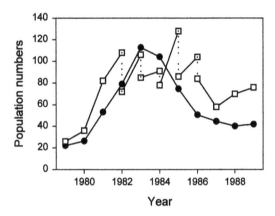

FIGURE 11.6. Population numbers of territorial grouse (cocks and hens) on the experimental (hollow squares) and control (filled circles) areas of the Rickarton moor. Grouse numbers in the experimental area are shown twice each spring, before and after experimental removals. Data from Moss et al. (1993: figure 4).

high (peak) density during the period of 1982–1986. To accomplish this, each spring the investigators reduced the numbers of territorial cocks to the 1981 (prepeak) level. The removal of cocks resulted in a similar number of hens lost from the population, so the experimental manipulation had an effect of keeping total density of territorial birds (both males and females) at the 1981 level.

Grouse density on the control area peaked in 1983 as expected, and exhibited a sustained decline during the next five years, until 1989 (figure 11.6). On the experimental area, by contrast, grouse density had a tendency to increase (and, therefore, had to be repeatedly thinned) until 1986. After that year, however, the experimental area also entered the decline phase. Thus, the experimental manipulation did not entirely "stop the cycle," instead delaying the decline phase by four years. Moss et al. (1996) speculated that the transient 1-year decline was caused by density-independent factors. Another possibility is the emigration of grouse to nearby areas (such as the control), where grouse densities were low and therefore territories could easily be established.

In conclusion, the hypothesis of Moss et al. that grouse cycles are driven by density-dependent factors is supported by their data. If the cycle were driven by exogenous, density-independent drivers, then the population in the experimental area should have declined in 1983 despite the density manipulation, four years before it actually did so. We must, of course, keep in mind that the study was not replicated. However, doing large-scale experiments is inherently laborious and expensive, so I would argue that even a single "replicate" is better than no data at all.

Moss et al. measured numerous population parameters throughout the complete cycle, which allows us to clarify the demographic machinery producing cyclic increases and declines. Of particular interest are the dynamics of the cecal parasite *Trichostrongylus tenuis*. To measure parasite prevalence, the investigators counted worm eggs in grouse feces for eight years (1982–1989) corresponding to the peak and decline of the cycle. They made two interesting observations. First, the worm loads at Rickarton were much lower than those observed at the northern England sites, where it is believed that grouse cycles are driven by the interaction with their parasites (Hudson et al. 2002). At Rickarton, average worm burdens varied between 100 and 2,600, depending on the year. By contrast, worm burdens in northern England sites typically vary between 1,000 and 10,000. Second, there was no apparent relationship between the average worm burden and rate of population increase. These observations suggest that the interaction between grouse and their worm parasites was not supported as the mechanism driving cycles at Rickarton.

11.3.2 Parasite Manipulation

The objective of the experiment by Hudson et al. (1998) was to test the parasitism hypothesis. Earlier, Hudson (1986) developed a procedure for reducing parasite prevalence in a grouse population by catching and treating grouse in spring with an anthelmintic drug. It is not practical to treat all birds in a population, so we need some quantitative predictions about what would be the effect on grouse dynamics of treating $x\%$ of a population. Hudson et al. (1998) addressed this question with a modified model (11.1), in which they added a second set of equations for treated grouse density and adult worm population within the treated hosts. They found that even if relatively small

proportions of grouse are treated, the amplitude of oscillations is substantially reduced. For example, if 20% of grouse are treated, then the
model predicts that population trajectory approaches the stable point
after 1–2 oscillations (Hudson et al. 1998: figure 3). This is a deterministic prediction. In the presence of exogenous noise, this model
would probably continue to oscillate, but at a reduced amplitude and
with weaker periodicity than the model with all birds untreated (control population).

Hudson et al. selected six independently managed moors for use
as replicates. Long-term data from these moors were used to predict the years of crushes in these populations (1989 and 1993). On
two moors, as many birds as possible were treated in both 1989 and
1993; on two other moors, birds were treated only in 1989; and the
last two moors were unmanipulated controls. Hudson et al. estimated
that they treated from 15 to 50% of grouse in manipulated populations. Thus, according to the model of grouse-parasite interaction,
the treated populations should exhibit a much reduced amplitude of
oscillations. This indeed appeared to be the case (figure 11.7). Note
that not only the magnitude of decline was reduced by anthelmintic
treatment, but also the height of the subsequent peak was reduced,
compared with controls.

There is, however, one potential problem with this experiment
(Lambin et al. 1999). Rather than estimate grouse density directly,
Hudson et al. relied on recording the number shot. As we discussed
above, the relationship between grouse density and bag statistics is a
nonlinear one; in particular, bag data amplify the variance of oscillations. An additional problem was that grouse were not shot in the
untreated areas during low years, unlike in treated areas. Assigning
zero values to replicates where grouse were not shot during certain
years tends to exaggerate the difference between untreated and treated
areas. In my opinion, this is a minor flaw that does not invalidate the
basic result of the experiment. Although it would certainly be better
to have direct estimates of density, if this was not practical (Hudson
et al. 1999), it is much better to have indirect statistics than nothing at all. However, I do want to criticize Hudson and colleagues
for overstating their results. Although their 1998 paper is titled "Prevention of Population Cycles by Parasite Removal," in actuality they

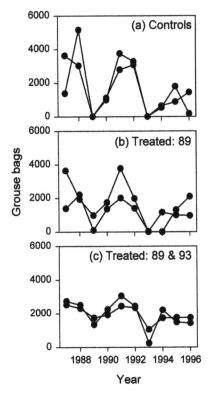

FIGURE 11.7. Population trajectories in the Hudson et al. parasite-reduction experiment.

failed to "stop the cycle." What they actually did was show that treating a proportion of grouse with anthelmintic substantially reduces the amplitude. This observation is in line with the quantitative prediction of the empirically based model for grouse cycles, and therefore these experimental results provide strong evidence in favor of the parasitism hypothesis.

In summary, the study of Hudson et al. (1998) is one of the best currently available experimental tests of a trophic mechanism for population cycles (May 1999). Among its remarkable features is its spatial scale (17–20 km^2), the degree of replication (six replicates over two separate cycles), and quantitative predictions obtained from an empirically based model. Clearly, the huge scale demanded trade-offs, such as reliance on indirect indices of population dynamics, and ability to treat with anthelmintic only a portion of the experimental population. Thus, although Hudson and colleagues failed to "stop the cycle," the match between their predictions and data, in my opinion, constitutes a successful confirmation of the parasitism hypothesis for red grouse cycles.

11.4   SYNTHESIS

Time-series analysis of grouse bag data indicates that grouse population dynamics are quite variable in time and space. As the contrast between the first and second pieces of Middleton data shows, at times grouse oscillate in a regular (quasi-chaotic) manner, while at other times fluctuations have a much weaker endogenous component, and can be characterized as stability with noise. Furthermore, the structure of density dependence documented at northeastern Scotland sites, studied by Moss, Watson, and colleagues, is quite different from that observed at northern England sites, studied by Hudson, Dobson, et al. (table 11.1).

The best hypothesis explaining population oscillations at the northern England sites currently appears to be the interaction of grouse with their worm parasites. The support for this hypothesis is as follows. First, a parameterized empirically based model of grouse-worm interaction predicts dynamics that, generally, match the observed ones. Although the model with independently estimated parameters predicted slightly longer periods than observed, the ability to predict correct period lies well within the biologically reasonable parameter ranges. What is particularly striking is the ability of the model to predict the correct structure of density dependence (slow buildup to the peak followed by a rapid collapse; the shape of the response surface). Second, the parasitism hypothesis has strong experimental support: grouse populations treated with an anthelmintic drug exhibit reduced amplitude of oscillations. Moreover, the experiment supports the quantitative predictions of the model that connect the proportion of birds treated with the degree of reduction in oscillatory tendency. Third, the free-living stages of the worm parasite require moisture for development to the infective stage. Thus, parasite transmission, and therefore propensity to cycle, should be reduced in drier red grouse habitats. Analyses of time-series data are in agreement with this prediction (Hudson et al. 2002). Finally, the kin selection hypothesis, currently the main alternative to parasitism, suffers from several problems. The first problem is its current lack of theoretical support. Second, unlike the parasitism hypothesis, the kin selection explanation has not yet been challenged experimentally (and survived the challenge).

The situation at the Scottish sites (Rickarton and Kerloch) studied by Watson, Moss, and coworkers is much less clear. The Rickarton experiment provides evidence in favor of the endogenous source of red grouse cycles, since the density manipulation succeeded in preventing a cyclic decline for at least four years. Furthermore, the data of Moss et al. on worm burden dynamics apparently contradicts the predictions of the parasitism hypothesis. This observation, coupled with different dynamical patterns of oscillations, suggests that mechanisms driving oscillations may differ between the two areas. What mechanism may drive oscillations at the Scottish sites, however, remains obscure. An explanation based on the kin selection hypothesis or, in fact, any intrinsic-factor hypotheses is currently unconvincing for reasons stated in the previous paragraph.

CHAPTER 12

Voles and Other Rodents

12.1 INTRODUCTION

Ecologists who are not working on small rodents may consider that the subject of population cycles in voles and lemmings remains as muddled as ever, if not increasingly more muddled. Small rodent ecologists appear to be in the business of proposing new hypotheses rather than rejecting old ones: the number of hypotheses has increased with time instead of being reduced to (ultimately) one plausible explanation. For example Batzli (1992) lists more than twenty distinct hypotheses, and more were added since the publication of his article (e.g., Boonstra 1994; Jedrzejewski and Jedrzejewska 1996; Inchausti and Ginzburg 1998).

In a recent article (Turchin and Hasnki 2001) we argued that this impression is incorrect: a certain degree of consensus has actually been emerging among small rodent ecologists during the 1990s, as manifested by discussions at two workshops held in Grand Forks and Oslo in 1996. Writing about the Grand Forks, North Dakota, workshop, George Batzli (1996) observed that "perhaps the most surprising outcome of the meeting was the degree to which consensus emerged regarding the most important issues." One area of agreement was that there are three hypotheses that are more likely than others to explain small rodent population cycles: predation, food, and maternal effects. All three hypotheses have now been translated into mathematical models, enabling us to directly and quantitatively assess their relative merits by contrasting their predictions with data.

In this chapter I review the progress toward elucidating the ecological mechanisms driving rodent population cycles. My main focus here is on *Microtus* and *Clethrionomys* voles, but I also devote a section to lemming population dynamics.

12.2 ANALYSIS OF TIME-SERIES DATA

Beginning with the formal discovery of rodent population oscillations by Elton (1924) and until the 1970s, population ecologists believed that all vole populations cycled (e.g., Krebs and Myers 1974). This mistaken impression was a result of two factors: lack of formal methods for distinguishing cycling from noncycling populations, and lack of long time-series data to apply these methods to. In fact, the lack of rigor in definitions led some ecologists to deny that any rodent populations exhibited such periodic dynamics (Cole 1951; Garsd and Howard 1981). The truth lies between the two extremes. While not all vole populations oscillate (Hansson and Henttonen 1988), some most certainly do. Motivated by Elton's paper, population ecologists, especially those based in northern Europe (Fennoscandia, Russia, and Poland), started collecting long-term data on rodent population fluctuations. As a result, it is now possible to attempt a synthesis of microtine rodent fluctuation patterns, especially in such well-studied areas as northern Europe.

12.2.1 Methodological Issues

Few of the data series I shall be surveying are based on absolute estimates of rodent population density. Such estimates are usually obtained by intensive mark-recapture programs that are very difficult to sustain beyond the length of research time for a typical Ph.D. thesis. Clearly, such data are too short even to attempt to quantify long-term rodent dynamics, yielding at most a single data point—an oscillation 3–5 years long. Furthermore, an observation of a single event of population increase followed by a decrease tempts investigators to claim that they have observed a "cycle," without having any knowledge whether it is really periodic. In order to properly analyze a time series, we need at least three oscillations (section 7.1.2). Given the average period of 3–5 years, this means that the shortest usable series would have a length of 9–15 years.

Because mark-recapture estimates are labor-intensive, the bulk of the data long enough for time-series analysis have been obtained using indirect methods. The most common and very useful relative index of population fluctuations is obtained by snap-trapping, for example, the small quadrat method widely practiced in Fennoscandia (Myllimäki et al. 1971). The index is usually expressed as the number of rodents trapped per 100 trap nights. This index is linearly related to the absolute population density (Hanski et al. 1994). Another indirect method that has been applied to the study of *Microtus arvalis* dynamics is counting holes (Dombrovsky 1971; Romankow-Zmudovska and Grala 1994). Because the relationship between population density and the density of holes is nonlinear (and rather noisy), this method is primarily useful for establishing if there are any periodicities in fluctuations.

The Issue of Multiple Species   Because most locations are inhabited by more than one rodent species, another conceptual issue for analysis, which has caused some controversy (Falck et al. 1995; Turchin 1995a), is to decide whether we should analyze the combined density of all species, or each species separately. This is a general issue that is of relevance not only in the phenomenological analysis but also in fitting mechanistic models to data (see section 12.3.3).

To make the discussion more concrete, let us suppose that we are dealing with a dynamical system consisting of two vole species and a predator, along the lines of the model investigated by Hanski and Henttonen (1996). If we analyze the pooled vole density, we have implicitly simplified the system from three-dimensional to two-dimensional. In other words, we have assumed that the vole **dynamical complex**, consisting of two species, interacts as a whole with the predator population. This simplification, of course, does some violence to the biology of the system because in reality the two vole species are likely to be characterized by different values of parameters, and their respective population trajectories will not be completely in synchrony with each other. On the other hand, we should keep in mind that having one state variable (population density) for each of the vole species also does violence to reality: each population is structured in a variety of ways, including age and spatial structure. So it is improper to reject the approach of pooling vole species on the grounds that in reality they are two separate populations. "In reality," all models are wrong. The properly posed question is which degree of simplification yields better results.

Analyzing each species separately, we implicitly deal with a higher-dimensional system. Thus, it is quite possible that dynamics of each vole species would be less regular than the dynamics of their pooled densities: in one peak one species may predominate, in the next peak another. Given long time series and low observation errors, we should be able to characterize these dynamics by fitting higher-order models. However, chances are that in an ecological setting, with short and noisy data, we would end up missing the high-dimensional signal, and assigning it to noise.

Furthermore, observation errors associated with pooled numbers are lower than errors for each species separately. This is especially important during the population troughs, when we see zero counts for many species if analyzed separately, while pooled numbers are much less affected by this problem. Again, lower measurement noise associated with pooled numbers gives us a better chance to correctly characterize the type of dynamics of the system.

In summary, if our goal is the characterization of dynamics, then my inclination would be to give more credence to results based on the analysis of pooled numbers. The arguments above suggest that this approach should perform better on short and noisy data sets characteristic of ecology. Indeed, empirical results suggest that the dynamics of pooled numbers are often characterized by higher signal/noise ratios than those for each species alone (Turchin 1996). On the other hand, it is always a good idea also to analyze each species separately, as a check on the results obtained with pooled numbers. Additionally, the analysis of nonpooled data may yield interesting patterns that could provide useful insights. An example is the systematic difference in the stability of fitted response surface models to the field vole versus bank vole data, discussed below. Thus, in the final analysis, I advocate a synthetic approach of analyzing the data in a variety of ways, and then interpreting the results in light of the considerations discussed above. An example of this omnivorous approach is the analysis of vole data below.

Multiple Observations per Year   The next issue to consider is what to do when there are multiple measurements of density per year. For example, many Fennoscandian data are routinely collected twice a year: once in spring–early summer, and once in late summer–early fall. Other data sets may have more than two or even a variable number of measurements per year. Analyzing data affected by seasonal influences as though it came from a completely endogenous system leads to severe problems. In particular, one may fit a more unstable model than the mechanism driving the data, because the phenomenological model will attempt to capture seasonal periodicity with (spurious) endogenous terms (Ellner and Turchin 1995). It is possible to add explicit seasonal terms to the model (see, e.g., discussion in Ellner and Turchin 1995), but in my experience it is better to reduce the data to one measurement per year. The main subject of this book is multiannual cycles, after all, not seasonal drivers. Thus, it is not a good idea to waste degrees of freedom on estimating seasonal terms that are peripheral to the main issue at hand. (Of course, there are situations where we are interested in characterizing seasonality, e.g., the fitting of the seasonally driven predator-prey model discussed below. In such cases, we shall want to utilize all the measurements within years that we have.)

Thus, the question is how we reduce multiple observations within a year to a single one. One possible approach is to add these measures together. The advantage of this approach is the reduced observation noise. However, if dynamics are too fast, then spring and fall measurements will not really be detecting approximately the same quantity, and by summing them we may lose an important part of signal. Because rodent dynamics are quite fast (oscillations of 3–5 years), I chose not to follow this approach (but it may be a reasonable one when dealing with, for example, 10-year cycles). The approach I did follow was to select either spring or fall data, depending on which season was characterized by higher average density (because higher density implies lower measurement noise and zero incidence). In vole data, it is the fall data that yield higher averages, while in lemming data the same is true for spring measurements.

Trends and Detrending   As with any collection of long-term data sets in ecology, a certain proportion of vole time series suffers from nonstationarity. As I argued in section 7.1.2, whether to detrend or not is an important issue in the analysis. Not removing a trend, where it is present, will result in not being able to characterize any complex dynamics underlying population fluctuations. On the other hand, unnecessarily detrending data may result in spuriously detecting regulation where there may not be any operating.

Population trends at some locations are fairly well documented. For example, during the 1980s at several northern Fennoscandian locations, one of the main vole species, *Microtus agrestis*, became rare. This change in the dynamical system apparently caused a decrease of least weasel density (who require *M. agrestis* for population persistence) and an increase in *C. glareolus* (see Hanski and Henttonen 1996 for a possible explanation of this pattern). Several other data sets in the collection were also clearly nonstationary, thus indicating that a flexible approach may be needed. Accordingly, I detrended those time series that had all of the following indications of nonstationarity: (1) visible trends in the mean on the time plot; (2) nonstationary type of ACF (i.e., ACF slopes down toward consistently negative values at high lags); and (3) higher signal/noise ratio in the detrended series, as indicated by fitting a series of response surface models of increasing order. The results I report below are all based on quadratic detrending for consistency.

12.2.2 Numerical Patterns

Northern European Voles: Pooled Data   The first collection of data I shall focus on is the database on population fluctuations of *Microtus* and *Clethrionomys* voles in northern Europe. This geographic area has the best collection of long-term time series on rodent population dynamics. Some of the data collection was begun before World War II by Russian researchers (this is the "Kola" data set; unfortunately the war interrupted data collection, so we shall analyze this series only from 1946, when trappings were resumed). Another remarkable data set, from Kilpisjärvi, Finnish Lapland, extends from 1949 to the present. Many more shorter data are available form the rest of Fennoscandia (Norway, Sweden, and Finland), Russia, Poland, and the British Isles. The summary of patterns detected by NLTSM analysis is given in table 12.1.

Let us look at the NLTSM results for the longest series, Kilpisjärvi (figure 12.1). Simply looking at the data on the time plot, we can see very clearly that there is a lot of structure and regularity in the dynamics of this system. However, the regularity is not of the familiar, periodic kind. While the data are undoubtedly characterized by statistical periodicity (ACF at the dominant peak of 5 years is significantly different from zero), this periodicity is not very strong (ACF[5] < 0.5). In fact, oscillations vary a lot in their lengths. Starting with the first complete oscillation, we count their lengths in years as follows: 6–4–4–6–5–3–5–5. First, we notice a slight nonstationarity in the period (longer periods during the first half). Second, cycle lengths vary by a factor of twofold (between 3 and 6 years). This is in contrast with, for example, larch budmoth dynamics, where each complete oscillation is exactly 9 years in length (figure 9.1).

Turning to the order of dynamics, all indications point in the same direction: PRCF shows clear spikes at the first two lags, the phase plot exhibits a characteristic cycling pattern, and cross-validation selects $p = 2$. Examining the response surface, we see that it slopes down in both the X and Y direction (unlike, e.g., the larch budmoth data, where the second lag clearly predominates). The importance of the first lag is emphasized by increasing the lag in the phase plot to 2, and observing no clear-cut pattern (not shown in figure 12.1).

The dominant Lyapunov exponent of the fitted response surface model is $\Lambda_\infty = +0.02$. It suggests that the dynamics of Kilpisjärvi voles are quasi-chaotic. We shall return to this issue with more sophisticated approaches later in this chapter. Checking on the result by analyzing the two halves of the data (since the series is long enough to allow this kind of test), we observe that $\Lambda_\infty = 0.11$ and 0.12, respectively. These values are somewhat larger than the one for the whole series, but still suggestive of quasi-chaotic dynamics.

Northern European Voles: Species Separated   The analysis above lumped together all vole species found at a locality, and now we need to check on its results by analyzing the three most common species separately. I shall particularly focus on the estimated Lyapunov exponent, as a good summary of dynamics (figure 12.2). In *M. agrestis*

TABLE 12.1. Summary of NLTSM results: voles in northern Europe; pooled densities at each location ("Lat." is the latitude of each site). Number of data points, n; measure of amplitude (SD of log-transformed densities), S; dominant period, T; the autocorrelation at the dominant period, ACF[T] ($**$ = ACF is significant at T; $*$ = ACF is significant at half-period); estimated process order, p; polynomial degree, q; the coefficient of prediction of the best model, R^2_{pred}; and the estimated dominant Lyapunov exponent, Λ_∞

Location	Lat.	n	S	T	ACF[T]	p	q	R^2_{pred}	Λ_∞
Finnmark	70	22	0.69	5	0.67**	3	1	0.57	0.18
Kilpisjärvi	69	44	0.63	5	0.46**	2	1	0.52	0.02
Pallasjärvi	68	23	0.73	4	0.50**	2	2	0.45	1.36
Kola	67	19	0.80	4	0.12*	2	2	0.30	0.70
Umeå	64	21	0.53	4	0.63**	2	2	0.68	0.16
Sotkamo	64	27	0.34	4	0.69**	3	2	0.55	0.52
Stromsund	64	19	0.53	4	0.41*	3	2	0.77	1.41
Ruotsala	63	20	0.62	3	0.77**	2	2	0.46	0.36
Alajoki	63	16	0.74	3	0.58**	2	1	0.35	−0.04
Loppi	61	21	0.27	6	0.57**	1	0	0	−∞
Karelia	61	21	0.35	3	0.03	1	0	0	−∞
Boda	61	28	0.54	4	0.63**	2	1	0.51	−0.16
Uppsala 1	60	19	0.46	5	0.23*	2	1	0.09	−0.24
Uppsala 2	60	17	0.46	4	0.37	3	1	0.05	0.07
Grimsö	59	22	0.45	3	0.33*	2	2	0.56	0.02
Zvenigorod	57	31	0.31	3	0.29	2	1	0.15	−0.73
Revinge	56	13	0.15	—		1	0	0	−∞
Tataria	56	23	0.34	4	0.24	1	0	0	−∞
Kielder	55	15	0.25	3	0.33*	2	1	0.39	−0.14

(*Continued*)

TABLE 12.1. (*Continued*)

Location	Lat.	n	S	T	ACF[T]	p	q	R^2_{pred}	Λ_∞
Serpukhov	55	28	0.32	3	0.26	1	0	0	$-\infty$
Tula	54	23	0.25	2	0.06*	1	0	0	$-\infty$
Bialowieza	52	21	0.42	—		1	1	0.08	-0.11
Wytham	51	22	0.27	—		1	2	0.12	-0.79

Data sources: Finnmark (Ekerholm et al. 2000); Kilpisjärvi and Pallasjärvi (Henttonen and Hanski 2000); Kola (Koshkina 1966); Umeå (Hörnfeldt 1994); Sotkamo and Loppi (Hanski et al. 1993, data collected by A. Kaikusalo); Stromsund and Uppsala (Hansson 1999); Ruotsala and Alajoki (Korpimäki 1994); Karelia, Tataria, Serpukhov, and Tula (Ivanter 1981); Boda (Marcström et al. 1990); Grimsö (Lindström et al. 1994); Zvenigorod (Ivankina 1987); Revinge (Lennart Hansson, personal communication); Kielder (Lambin et al. 2000); Bialowieza (Pucek et al. 1993); Wytham (Southern 1979).

Series that were detrended: Finnmark, Umeå, Sotkamo, Ruotsala, Alajoki, Uppsala 1, Grimsö, Zvenigorod, and Tula.

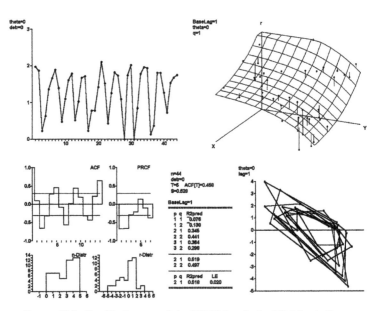

FIGURE 12.1. Graphical output of the NLTSM analysis: Kilpisjärvi. *Upper left:* time plot; *upper right:* response surface; *lower right:* phase plot.

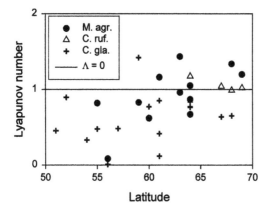

FIGURE 12.2. Relationship between trajectory stability and latitude, separated by vole species. Trajectory stability is quantified by the *Lyapunov number* (Lyapunov number is the antilog of the Lyapunov exponent). The advantage of using the Lyapunov number over the Lyapunov exponent is that it de-emphasizes variation in the negative range of Λ_∞ (e.g., $\Lambda_\infty = -1$ or -10 both mean approximately the same thing, strong stability, but look very different on the graph).

(which has been postulated as the key species driving the dynamical patterns in northern Fennoscandia; see Hanski et al. 1991; Hanski et al. 1993) we observe that the overall shape of the latitudinal pattern that we documented in the analysis of pooled species is preserved. Northern populations tend to be chaotic, they have much higher amplitude than southern populations, and the average period of oscillations tends to get shorter as we move south. However, this latitudinal pattern is not as clear-cut as it appears in the analysis of pooled species: the estimated Lyapunov exponent exhibits a lot of variation around the general trend.

C. rufocanus appears to exhibit somewhat milder oscillations than *M. agrestis*. Most interesting is the contrast in Lyapunov exponents: *rufocanus* estimates are clustered very tightly (mean $\Lambda_\infty \pm$ SE is 0.08 ± 0.04) in the quasi-chaotic range. Furthermore, the amplitude of fluctuations, as measured by S, at both sites where the two species co-occur (Pallasjärvi and Umeå), is less for *rufocanus* than for *agrestis*. However, the average period exhibits the same declining trend with decreased latitude (from 5 to 4 years; we do not see 3-year cycles, because *rufocanus* is a northern species, not found in lower latitudes where we could expect such short oscillations). Finally, *C. glareolus* exhibits the most stable dynamics of the three species: only one Λ_∞ estimate is in the chaotic range.

To summarize, the analysis of separate species clearly indicates that each species has its own pattern of fluctuations. *M. agrestis* exhibits the most violent oscillations, with estimated Lyapunov exponents ranging into the strongly chaotic region at some locations (with a caveat, however, that there appears to be a great degree of variation associated with this estimate). *C. rufocanus* is an intermediate species, whose dynamics are consistently characterized by quasi-chaotic oscillations. Finally, *C. glareolus* has the most stable (and noisy) dynamics of the three species. The latitudinal dynamical pattern is evident only in *M. agrestis*. Thus, it appears that the latitudinal shift in the dynamics of pooled species is partly due to shifting species composition (more oscillatory *M. agrestis* and *C. rufocanus* in the north, less oscillatory *C. glareolus* in the south), and partly due to *M. agrestis* becoming more stable toward the south.

Common Voles in Europe   Population dynamics of the common vole (*M. arvalis*) present a great contrast with the dynamical pattern of northern European voles discussed above. NLTSM analysis of data from France, Poland, and Russia suggested a preponderance of first-order dynamics (table 12.2). Few cases of periodicity were detected. With a single exception, all estimated Λ were negative. Yet, the dynamics of common voles are characterized by quite strong degrees of oscillation (e.g., see figure 12.3). In sum, analysis of *M. arvalis* data reveals no strong evidence for second-order oscillations; however, in some cases, such as the one depicted in figure 12.3, the evidence for *first-order* oscillation is quite strong.

North American Voles   A similar pattern was revealed by the analysis of six time series documenting fluctuations of North American voles in the genera *Microtus* and *Clethrionomys* (table 12.3). Again, the preponderant pattern is high amplitude (*S* ranged between 0.38 and 0.75), largely aperiodic (weak evidence of periodicity in only two cases), and first-order (in five out of six cases), with stable dynamics (all estimated Lyapunov exponents negative).

TABLE 12.2. Summary of NLTSM results: the common vole *M. arvalis*.
Same notation as in table 12.1

Location	n	S	T	ACF[T]	p	q	R^2_{pred}	Λ_∞
NW Tula	23	0.85	—	—	1	0	0	$-\infty$
SE Tula	23	0.84	3	0.22*	1	2	0.12	-0.54
Beauvoir	23	0.41	—	—	1	0	0	$-\infty$
Brioux	20	0.60	—	—	1	0	0	$-\infty$
Stupino	21	0.87	—	—	2	2	0.21	-0.32
Podole	17	0.63	—	—	1	0	0	$-\infty$
Zaraysk	21	0.61	—	—	1	0	0	$-\infty$
Odinzovsk	21	0.46	—	—	1	0	0	$-\infty$
Mozhaisk	21	0.61	—	—	1	2	0.02	-0.95
Potoschinski	21	0.66	—	—	1	0	0	$-\infty$
Verhnevolzhski	17	0.59	—	—	1	0	0	$-\infty$
Gorzow	16	0.42	—	—	1	1	0.14	-0.65
Pila	16	0.26	7	0.53**	3	1	0.36	-0.11
Poznan	16	0.41	4	0.24*	2	2	0.40	0.29
Leszno	16	0.30	—	—	1	0	0	$-\infty$
Opole	16	0.48	—	—	2	1	0.13	-0.46
Legnica	16	0.33	—	—	1	0	0	$-\infty$
Walbrzych	16	0.42	—	—	2	1	0.15	-0.73
Wroclaw	16	0.49	—	—	1	0	0	$-\infty$
Szczecin	16	0.33	—	—	1	0	0	$-\infty$

Data sources: Tula (Myasnikov 1976); Stupino through Verhnevolzhski (Dombrovsky 1971); Gorzow through Szczecin (Romankow-Zmudovska and Grala 1994).
Series that were detrended: Beauvoir and Pila.

Water Voles in Switzerland   The final vole species for which there are abundant time-series data is the water vole, *Arvicola terrestris* (table 12.4). Saucy (1988) presents data on fluctuations in seven locations. In many ways these trajectories resemble *M. agrestis* dynamics (second-order, periodic, with Lyapunov exponents ranging between

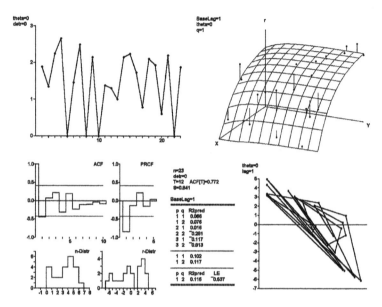

FIGURE 12.3. Graphical output of the NLTSM analysis: common vole, southeastern Tula. *Upper left:* time plot; *upper right:* response surface; *lower right:* phase plot.

TABLE 12.3. Summary of NLTSM results: *Microtus* and *Clethrionomys* voles in North America. Same notation as in table 12.1

Location	n	S	T	ACF[T]	p	q	R^2_{pred}	Λ_∞
Ontario	36	0.38	—	—	1	1	0.03	−1.55
Wyoming	19	0.53	—	—	1	0	0	−∞
California	19	0.44	—	—	1	0	0	−∞
Illinois	24	0.75	—	—	1	0	0	−∞
Yukon	13	0.57	4	0.24*	1	0	0	−∞
Vermont	16	0.47	4	0.30*	2	1	0.07	−0.60

Data sources: Ontario (Fryxell et al. 1998); Wyoming (Pinter 1988); California (Garsd and Howard 1981); Illinois (Getz et al. 1987); Yukon (Gilbert and Krebs 1991); Vermont (Brooks et al. 1998).
Series that were detrended: None.

TABLE 12.4. Summary of NLTSM results: the water vole *Arvicola terrestris*. Same notation as in table 12.1

Location	n	S	T	ACF[T]	p	q	R^2_{pred}	Λ_∞
Ste-Croix	47	0.81	6	0.53**	3	2	0.48	0.06
Chenit	26	0.37	6	0.29*	2	1	0.44	−0.03
La Brévine	27	0.37	6	0.16*	3	1	0.42	−0.03
Chateau d'Oex	34	0.53	6	0.29*	3	1	0.45	−0.32
Rougemont	41	0.39	6	0.46**	3	2	0.64	0.08
Bulle	20	0.34	5	0.40*	2	1	0.12	−0.61
Baraba	14	1.35	7	0.78**	2	1	0.74	0.23

Data sources: Ste-Croix through Bulle (Saucy 1988); Baraba (Evsikov and Mosjkin 1994).
Series that were detrended: Chenit, La Brévine, Rougemont, Bulle.

−0.61 and +0.23 (three out of seven are positive). However, one significant difference is that the average period of oscillations in the water vole is 6 years, compared to the typical 4-year cycle in smaller field voles.

Overview In summary, we can identify at least two groups of microtine rodent communities for which second-order oscillations have been established beyond reasonable doubt: (1) communities dominated by the field vole *Microtus agrestis* in many northern Fennoscandian localities—for instance, Pallasjärvi, Umeå, and Alajoki—and by an ecologically very similar species *Clethrionomys rufocanus* in Kilpisjärvi; (2) the water vole (*A. terrestris*) in Switzerland. To these two groups, we should add *Clethrionomys rufocanus* in Hokkaido (Saitoh 1987). Finally, there are lemmings, who will be discussed in section 12.5. Most likely there are other localities where small rodents exhibit second-order oscillations, but for which we at present lack definitive data or statistical analyses.

We can also identify several areas where microtine rodents do not oscillate: (1) forest communities dominated by *Clethrionomys glareolus* in Poland (Pucek et al. 1993) and in several central Russian localities—Tataria, Serpukhov, and Tula; (2) *Microtus agrestis* in southern England and southern Sweden; (3) all North American populations for which data are available (with one possible exception). These populations show neither significant periodicities in autocorrelation functions nor significant evidence for second-order dynamics (the possible exception is the Vermont data, which unfortunately are too short at 16 years to base strong conclusions on).

Finally, there is the interesting case of the common vole *M. arvalis*, in which evidence for first-order oscillations is quite strong at some localities.

The observation that population dynamics of small rodents can be so variable among species and geographic localities suggests to us that there is no such universal phenomenon as "the vole cycle" (Turchin et al. 2000), nor should we be looking for a single explanation of small rodent dynamics that applies to all species at all localities. Even if we are eventually successful in building a general theory for small rodent dynamics that encompasses all known empirical instances, the relative contributions of different factors postulated by such a theory would most certainly vary among different environments. The realization that small rodent dynamics are not a unitary phenomenon has one important corollary: empirically rejecting (or supporting) a particular hypothesis for one population does not imply that the hypothesis has been universally rejected/supported for all microtine populations.

12.3   HYPOTHESES AND MODELS

In this and next sections I focus on one system where a large body of information has been accumulated over the past fifty years—*Microtus* and *Clethrionomys* voles in Fennoscandia. Analyses discussed in the previous section have documented several repeatable patterns that call for explanation. First, population dynamics of voles in northern Fennoscandia are characterized by high-amplitude second-order periodic oscillations. Second, there is a striking change in the dynamical pattern with latitude: from chaotic or quasi-chaotic oscillations in the north to highly stable fluctuations in the south.

This north-south change is accompanied by reduced amplitude and a loss of periodicity.

Our task is to construct an explanation (formalized as the "best model") of population dynamics of Fennoscandian voles. The field of candidate hypotheses is narrowed to those that can in principle produce second-order oscillations. Thus, such a formerly popular explanation as the behavioral polymorphism hypothesis of Krebs (1978) is immediately eliminated from consideration, because the theoretical work by Warkowska-Dratnal and Stenseth (1985) and Stenseth and Lomnicki (1990) has conclusively demonstrated that this mechanism has stabilizing rather than oscillatory effect on dynamics. As I mentioned in section 12.1, the consensus at the 1996 North Dakota workshop was that three hypotheses for rodent cycles were most worthy of intensive investigation: (1) predation, (2) interactions with food, and (3) maternal effects (Batzli 1996). The task of translating the predation hypothesis into models specifically tailored to the Fennoscandian vole situation had already been accomplished by Hanski and coworkers, but the maternal effect and food hypotheses had not been developed to the same level. In this section I review the progress on this issue that has occurred since.

12.3.1 Maternal Effect Hypothesis

Two teams of investigators recently tackled the question of whether a model based on the maternal effect hypothesis is capable of generating small rodent cycles. The first model is a modification of the Ginzburg and Taneyhill (1994) equations (see section 3.4.1), proposed by Inchausti and Ginzburg (1998). The second model (Jorde et al. 2002) was developed specifically to address the question of vole cycles.

The Inchausti-Ginzburg Model   One conceptual problem in applying the Ginzburg and Taneyhill framework to vole population dynamics is that the model is framed in discrete time, while rodents breed continuously (even through the winter, when their population densities are low). Inchausti and Ginzburg tackled this problem by assuming that there are two nonoverlapping "breeding seasons" during each year. They acknowledged the artificiality of imposing discrete breeding seasons on a continuously reproducing population, but argued that the simplicity of the resulting model was actually an advantage over alternative "overparameterized models."

The equations proposed by Inchausti and Ginzburg are

$$N_{A,t} = N_{S,t} \frac{R_S X_{S,t}}{1 + X_{S,t}}$$

$$X_{A,t} = X_{S,t} \frac{M}{1 + N_{A,t}}$$

$$N_{S,t+1} = N_{A,t} \frac{R_A X_{A,t}}{1 + X_{A,t}} \tag{12.1}$$

$$X_{A,t+1} = X_{A,t} \frac{M}{1 + N_{S,t}}$$

where $N_{A,t}$ and $N_{S,t}$ are *scaled* vole densities in autumn and spring of year t, and $X_{A,t}$ and $X_{S,t}$ are the average quality of the mothers in autumn and spring, also scaled. Parameters R_S and R_A are the maximum reproductive rates during the spring and autumn reproductive seasons, respectively, and M is the maximum rate of increase of average maternal quality.

The Inchausti-Ginzburg model of the maternal effect hypothesis is capable of generating second-order population cycles for some parameter values (Inchausti and Ginzburg 1998: figure 5; see also Turchin and Hanski 2001: figure 1). However, in my opinion, the model suffers from two serious problems (see also Turchin and Hanski 2001). The first one is the model's main structural assumption, namely, that vole population dynamics can be represented as occurring at two discrete time steps per year. Leaving aside the issue of whether dynamics of a continuously breeding population should even be modeled with a discrete framework, I question the specific choice of two steps per year made by these authors (as opposed to three, four, or more). Their maternal effect model shares the general property of discrete second-order models: for most parameter values, cycle periods lie between 6 and 10 generations. A cycle period of six generations is an absolute minimum for this model, while cycles with periods of more than 10 generations require very small values of the intrinsic rate of population increase (Ginzburg and Taneyhill 1994). Thus, it is not surprising at all that assuming two time steps per year would result in the model predicting cycles of 3–5 years. The choice of the number of time steps per year largely determines the length of the predicted period (in years).

An alternative to assuming two time steps per year is to chose a time step corresponding to some biological property of the modeled population. Perhaps the most defensible choice is to equate the time step to the cohort generation time, T_c. We can estimate T_c using the simple model for vole population growth discussed by Turchin and Ostfeld (1997). T_c is primarily affected by survivorship. Assuming a maximum life span of 1 year and the monthly survival rate of 0.75 (as in Turchin and Ostfeld 1997) leads to an estimate of $T_c = 3.6$ months. This survival rate, however, was used to estimate the intrinsic rate of population increase, and did not include the effect of predation. A lower, and more realistic, monthly survival rate of 0.5 (see Norrdahl and Korpimäki 1995) would yield $T_c = 2$ months. In order to obtain $T_c = 6$ months, we have to push monthly survival beyond 0.95, which is an unrealistically high value for natural populations of small rodents. Thus, a more realistic assumption than the one proposed by Inchausti and Ginzburg is six generations per year, which would change the predicted cycle period to 1–2 years—much lower than the observed.

Inchausti and Ginzburg anticipated this objection, and they correctly state that it is possible to obtain a 4-year cycle with more than two generations per year. However, this can be accomplished only at the expense of assuming an unrealistically low r_0. The model of Inchausti and Ginzburg employs two growth rate parameters, R_S and R_A, that represent maximum discrete population growth rates during the spring and autumn breeding periods, respectively (see equations 12.1). These rates are related to the exponential intrinsic rate of population growth as follows: $r_0 = \ln(R_S R_A)$. In order for the Inchausti-Ginzburg model to generate cycles of 4–5 years, the parameter r_0 must be around 2–2.5 yr^{-1}. (This result does not depend on assuming two discrete breeding periods; e.g., a modification of the Inchausti-Ginzburg model with four breeding seasons yielded the same relationship between r_0 and the oscillation period.)

This value of r_0 is too low for fast-breeding rodents such as voles or lemmings. Estimates discussed in Turchin and Ostfeld (1997) place

r_0 around 5–6 yr^{-1} and certainly no less than 4 yr^{-1} (Hanski et al. 1993; Turchin and Ostfeld 1997; Turchin and Hanski 1997). For these more realistic values, the Inchausti-Ginzburg model produces 2-year cycles. In other words, while the maternal effect model is capable of qualitatively correct dynamics (second-order oscillations), it fails in making quantitatively correct predictions (oscillation period).

Turning to a different empirical pattern, the maternal effect model of Inchausti and Ginzburg (1998) explains the increase in the cycle period with increasing latitude by postulating an increased growth rate of voles with decreasing latitude. However, their own data (Inchausti and Ginzburg 1998: figure 3) give no support for such a latitudinal change in growth rate, and the model has a free parameter that also contributes to cycle period. Furthermore, the model of Inchausti and Ginzburg is unable to predict lack of cyclicity, such as observed in southern Fennoscandia, because the model generates cycles for all values of parameters.

The Jorde et al. Model   In contrast to the unstructured modeling approach of Ginzburg and coworkers, Jorde et al. (2002) employed the age-structured framework. More specifically, their models were based on a Leslie matrix with discrete time intervals of one month. This interval was chosen because the time both from conception to birth and from birth to sexual maturity is approximately one month. Jorde et al. assumed that density-dependent social inhibition, stress, or poor nutritional condition during high-density periods force young individuals to delay reproduction. As a result, individuals born at peak density do not mature immediately, and in seasonal environments they may even delay reproduction until the next spring. Furthermore, the reduced survival of offspring during the periods of population decline was hypothesized to be a result of maternal effects, mediated by (1) mothers being stressed or subject to poor nutritional conditions ("maternal stress"), or (2) mothers being old and senescent when reproducing ("maternal age"). Jorde et al. formulated two models (each based on one of these two mechanisms), as well as a third model that added seasonality to the maternal stress mechanism.

The theoretical investigation of these three models revealed that in order to obtain multiannual cycles, one needs to make rather extreme assumptions about the structure of delayed density dependence, and

about the parameter values. Thus, both nonseasonal models assume
a very strong effect of population density on delay in the maturation time. In these models, populations reach peak densities of about
200 voles/ha, yet the models assume that population density of only
50 voles/ha would be sufficient to force a maturation delay of 5.5
months (parameter F; see table 1 in Jorde et al. 2002). However, as
far as is known, intrinsic controls on female maturation have been
observed primarily after populations have approached peak density.
In the model with seasonality, when population density reaches 50
voles/ha, all juvenile females delay breeding until next year, which
adds an extra half-year to the average length of cycles. Furthermore, Jorde et al.'s assumption about senescence predicts that the
offspring of overwintered voles have dramatically reduced "intrinsic"
survival—less than 50% of the survival of females that reproduce in
their first year. Presently there are no data supporting such a prediction. Finally, the model is structurally fragile, because small changes
in parameter values may shift a large-amplitude cycle into practically
noncyclic dynamics. Even with these rather extreme assumptions, the
model predicts cycles of shorter periods than what is observed in the
data: model-predicted periods are 2–3 years (figure 4 in Jorde et al.
2002), or, when noise is added, at most about 3.5 years. Additionally,
the maternal effect model of Jorde et al. cannot be used to explain
the latitudinal changes in Fennoscandian vole dynamics.

Conclusions   The models considered by Inchausti and Ginzburg, and
by Jorde et al., suggest the following general conclusion. Although the
models based on maternal effect are certainly capable of generating
second-order oscillations, producing cycles that match the observed
vole dynamics even approximately requires rather strong assumptions
about model structure and parameter values. In particular, it turns out
to be quite difficult to generate longer period cycles (more than 3
years). Apparently, given the short generation times and fast reproductive rates of small rodents, were their dynamics driven by the
interaction between population density and average maternal quality,
the predicted population cycles would have periods of 2 years, or
even shorter.

12.3.2   Interaction with Food

The review of herbivore–food quantity theory in section 4.4.3 suggests that the food hypothesis encounters grave difficulties in attempting to explain the dynamical patterns of Fennoscandian voles. First, and most important, it seems highly likely that the food supply of the key vole species, *Microtus agrestis*, is better described with a regrowth model rather than with the logistic equation. Recollect that model (4.40) is very stable, especially if vegetation grows back rapidly. Thus, it is difficult to see how the interaction of small rodents with food could generate second-order oscillations. First-order cycles, in contrast, remain a distinct possibility, if vegetation regrowth is strongly affected by seasonality. Such dynamics would be characterized by winter crashes followed by 1–2 years of exponential increase until population density exceeds its winter food supply and crashes again, yielding 2-year cycles (3-year cycles if it takes 2 years of exponential growth to reach the peak). However, this type of dynamics is not what happens in Fennoscandia, even where 3-year cycles are observed. For example, in western Finland, it typically takes 1 year for vole densities to "bounce" back to high densities, where the population spends 2 years. Such a dynamical pattern is more consistent with the hypothesis that the eventual crash is caused by predators, because it takes two years for predators to increase to the point where they can greatly impact the prey population.

Second, the food hypothesis does not seem to help us with the explanation of the dynamical shift within Fennoscandia from stability in the south to oscillatory dynamics in the north. In fact, a food-vole model without predation would predict the opposite pattern to what is observed. As productivity of the plant community increases toward the south, vole dynamics should become less stable as a result of the "paradox of enrichment." But the opposite is true.

Although the food hypothesis does not appear capable by itself of explaining population oscillations of Fennoscandian voles, food may play an important role in stopping vole population increase and thus stabilizing the cycles of voles and their predators. Experimental results (Henttonen et al. 1987) support such a role for the food resources, since experimental addition of food did not prevent population crash but did result in higher peak densities. Additionally, the food hypothesis remains a viable contender for explaining population oscillations in some lemming populations (section 12.5).

12.3.3 Predation

The view of the importance of predation in vole dynamics, prevailing among ecologists, has changed dramatically and repeatedly over the three-quarters of the century during which voles have been intensively studied (see review in Hanski et al. 2001). An important conceptual breakthrough occurred during the 1970s and 1980s, when ecologists working on small rodent population dynamics came to realize that not all predators are alike and that not all rodent populations oscillate in the same manner. Thus, Andersson and Erlinge (1977) made an explicit distinction between three types of predators: resident specialists, nomadic specialists, and generalists. Andersson and Erlinge proposed that generalist predators can prevent sustained multiannual vole oscillations by switching to prey upon voles when their density is high. The idea was later specified to refer particularly to winter predation in temperate areas, such as southern Sweden, where voles are not protected by snow cover and where the impact of generalists is boosted by wintering avian predators, which stay in the area as long as vole density is sufficiently high to ensure survival (Erlinge et al. 1983). The bottom line was that predation by generalists can prevent cycles (Hanski et al. 2001).

If generalist predators have the capacity to inhibit rodent oscillations, resident specialists may have the opposite effect. Toward the end of the 1980s, several teams of ecologists working in northern Europe began to view predation by specialists as the sufficient cause for rodent declines (review in Hanski et al. 2001). These studies did not propose, however, that predation by specialist predators would be the only process needed to generate recurrent rodent oscillations. The rodent population growth rates are obviously so high that the predator populations would never catch up unless some other factors stop prey population growth at high densities. Possible mechanisms to do that include resource competition, social interactions (female territoriality and infanticide), and the action of generalist and nomadic specialist predators. The nomadic specialists represent a particular category of small rodent predators. These predators include many owls and diurnal raptors more or less specialized in tracking populations of small mammals in space over large regions, by moving to areas where the prey populations are currently most abundant (Korpimäki and Norrdahl 1991a). The dynamic outcome of such rapid spatial "switching" is similar to the sigmoid functional response of resident generalists (section 4.1.2). These insights led Hanski and coworkers (Hanski et al. 1991, 1993, 2001; Hanski and Korpimäki 1995; Turchin and Hanski 1997) to propose and develop a model of the interaction between voles and their specialist and generalist predators.

Developing Model Equations   The starting point for the investigation of vole-predator dynamics is provided by the Hanski model (equations 4.26 in section 4.2.4).

$$
\begin{aligned}
\frac{dN}{dt} &= r_0 N \left(1 - \frac{N}{k} \right) - \frac{cNP}{N+d} - \frac{gN^2}{N^2 + h^2} \\
\frac{dP}{dt} &= s_0 P \left(1 - q\frac{P}{N} \right)
\end{aligned}
\tag{12.2}
$$

As usual, N stands for population density of the prey (voles), and P is the population density of the specialist predator (least weasels).

The next step is to add exogenous drivers to the model. This force has a periodic component (seasonality) and an irregular, random component (environmental stochasticity, or process noise). We need seasonality because it has a profound effect on the rates of population growth in temperate and boreal population systems. Moreover, adding seasonality expands the range of the dynamical behaviors of which the model is capable, to include chaos. Process noise is also a key component of the model, because it modifies dynamical features of the system, such as the amplitude and period of oscillations, and can actually shift the dynamics from the stable into the chaotic region, and vice versa. Without explicitly including noise in the model, we cannot directly and quantitatively compare model output with data.

To keep things simple, we assume that seasonality affects only the growth rate of populations:

$$\frac{dN}{dt} = r(1 - e\sin 2\pi t)N - \frac{rN^2}{k} - \frac{cNP}{N + d} - \frac{gN^2}{N^2 + h^2}$$

$$\frac{dP}{dt} = s(1 - e\sin 2\pi t)P - \frac{qsP^2}{N} \tag{12.3}$$

This method of adding seasonality, in which the slope of density dependence does not vary with seasonality, was introduced and discussed in section 3.2.2. The consequence of this assumption is that the carrying capacity oscillates with season (figure 3.3), and can even become negative if $e > 1$, implying that during the worst period of the year the population always declines, even if the population density is very low. Increased population density would further accelerate the rate of decline.

Process noise is included by randomly perturbing the model parameters once a year (see the discussion in section 3.2.1). The strength of environmental stochasticity is measured by parameter σ^2.

Scaling the Model   We define new, scaled variables, $N' = N/K$ and $P' = KP/q$, and rewrite the model (12.3) as follows (dropping primes in order not to clutter the equations):

$$\frac{dN}{dt} = r(1 - e\sin 2\pi t)N - rN^2 - g\frac{N^2}{N^2 + h^2} - a\frac{NP}{N + d}$$

$$\frac{dP}{dt} = s(1 - e\sin 2\pi t)P - s\frac{P^2}{N} \tag{12.4}$$

The new parameter combinations are $g' = g/k$, $h' = h/k$, $d' = d/k$, and $a = c/q$. In the aseasonal model, we could reduce the number of parameters even further by appropriately scaling time, but in the seasonal model we cannot do this, because the units of t are fixed by our assumption that the period of seasonal forcing is exactly 1 year. We now have six biological parameters describing the endogenous interactions (r, s, g, h, a, and d), and two parameters for the exogenous forcing (e and σ). A detailed discussion of how parameters were estimated is in Turchin and Hanski (1997). These parameter estimates are as follows: $r = 5$, $e = 1$, $s = 1.25$, $d = 0.04$, $a = 15$, $h = 0.1$, $\sigma = 0.12$.

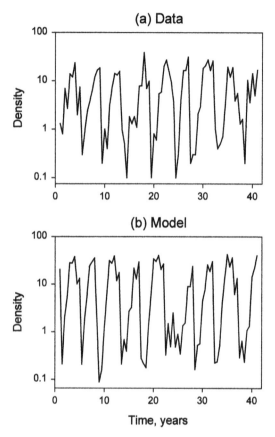

FIGURE 12.4. Comparison of (a) observed vole dynamics (Kilpisjärvi, Finnish Lapland, 1952–1992) and (b) dynamics predicted by the model with the median estimates of scaled parameters $r = 5$, $s = 1.25$, $d = 0.04$, $a = 15$, $e = 1$, and $\sigma = 0.12$ (generalist predation pressure g set to 0). Note that there are two density measurements per year (spring and fall).

Dynamics Predicted by the Model   Numerical solution of the model for the case of the far north, where the generalist predation pressure $g \approx 0$, shows that model-predicted dynamics are very similar to those observed at the northernmost location in Finnish Lapland, Kilpisjärvi (figure 12.4).

The model output matches correctly such probes as the periodicity, amplitude, estimated order, and Lyapunov stability. Numerical exploration of the parameter space (Turchin and Hanski 1997:857–859) suggests that these results are quite generic for model (12.4) within the range of parameters that we estimated. In particular, all combinations of parameters are characterized by periodicities in the range of 3–5 years, with mild statistical periodicity strengths (ACF[T] between 0.2 and 0.6). Furthermore, most parameter combinations yield chaotic dynamics, as measured by the dominant Lyapunov exponent. However, the system is not strongly chaotic, with typical positive Λ_∞ found in the range 0–0.2. (This observation suggests that the model mostly predicts quasi-chaotic behavior.)

When we increase the generalist predation pressure from zero to values characterizing southern Fennoscandia, the model predicts a rather abrupt shift in dynamics: from high-amplitude quasi-chaotic oscillations in the north to low-amplitude stable fluctuations in the south. This shift occurs for g hypothesized for latitudes of around 61–63°N. In this region, dynamics are still perceptibly periodic, but with a shorter period of around 3 years. In more southern locations, dynamics cease to be periodic, and also lose any indications of second-order regulation. These predictions match very closely the observed patterns in vole dynamics (section 12.2.2).

12.4 FITTING THE PREDATION MODEL BY NLF

As reviewed earlier in this chapter, there are currently two distinct lines of evidence suggesting that dynamics of northern vole populations undergo a shift from stable to chaotic (or quasi-chaotic) dynamics: flexible phenomenological models fitted to time-series data, and mechanistic models based on life history and experimental data. However, each of the two approaches has weaknesses as well as strengths. For instance, choosing model complexity, in particular the process order, p, is a critical step in the nonlinear time-series analysis, because incorrect estimation of p can greatly affect the resulting characterization of the dynamics (section 7.2). In contrast, the order of a mechanistic model is determined by biological assumptions, as are all other aspects of model complexity. However, parameter estimates for the mechanistic model proposed to explain Fennoscandian vole oscillations were derived from heterogeneous life history and experimental information, which has an unavoidable element of subjectivity and does not yield very precise estimates. Can we avoid these problems by fitting the mechanistic model directly to the time-series data? The discussion of nonlinear forecasting fitting methods in section 8.5 suggests that we indeed can do it. The specific approach dealing with the Fennoscandian vole data was implemented by Stephen Ellner and myself in a recent article (Turchin and Ellner 2000a).

The overall question that we sought to answer was, what are the stability properties of the dynamic process underlying vole oscillations in northern Fennoscandia? Instead of attempting to answer this question directly by analyzing time-series data (for the reasons explained above), we approached it in two steps. The first, and most critical step, was to ascertain that we had the best possible model for this system within the limits of our present knowledge. Conclusions about the dynamics are only as good as the model on which they are based. We defined the "best" model as a model that is based on defensible ecological mechanisms with biologically plausible parameters, and that does the best possible job of capturing the dynamical patterns observed in time-series data. The second, and much easier, step was to use this model to calculate various measures of stability, such as the global Lyapunov exponent.

Our primary focus was on the analysis of the best data set from Fennoscandia, the 41-year-long record of vole dynamics at Kilpisjärvi, and then we supplemented the insights from this data set with analyses of shorter series from other locations (Pallasjärvi, Umeå, Sotkamo, Loppi, and Grimsö). The first step of the NLF analysis was to determine the baseline prediction accuracy for Kilpisjärvi, that is, R^2_{data} (see section 8.5). Using kernel regression methods, we estimated R^2_{data} as 0.51 (Turchin and Ellner 2000a: table 1), suggesting that at least one-half of variance in vole population numbers is explained by endogenous, density-dependent factors.

In the second step, we used the vole-weasel predation model (equations 12.4) with parameters previously estimated by Turchin and Hanski (1997) to *predict* the Kilpisjärvi data. We found that the predation model predicted the data much better than the data-based algorithm, since R^2_{atlas} was 0.59. Applying the same procedure to the other data sets, we found that, in all cases but one (Sotkamo), $R^2_{atlas} > R^2_{data}$. Because the time-series data were not used in selecting the parameter values on which model predictions were based, this result provides a strong endorsement of model (12.4), as well as the specialist/generalist predation hypothesis itself.

In the final step, we inverted the problem, and asked what parameter values maximize the prediction accuracy, R^2_{atlas}? When addressing this question, we did not attempt to fit all model parameters. For example, we fixed the strength of exogenous forcing at $\sigma = 0.12$, as previously estimated by Turchin and Hanski (1997). Second, we fixed $h = 0.1$, because a reasonably good estimate of this parameter is available. Thus, our analysis attempted to derive estimates of the following parameters: r, s, d, a, e, and g (for explanation of parameters, see section 12.3.3). Applying the NLF method to Kilpisjärvi data, we found that best parameter values, for which R^2_{atlas} increased to 0.66, were not dissimilar to the biologically derived estimates. In particular, all parameter values were within the biologically reasonable ranges, with parameter a deviating the most (NLF-estimated $a = 8$–9, while the previous estimate was $a = 15$).

The other five data sets were much shorter than Kilpisjärvi, so we reduced the parameter set to be fitted even further, by focusing on those four parameters that exert the strongest effect on model dynamics (Turchin and Hanski 1997): d, a, e, and g. We fixed r and s at values intermediate between those postulated by Turchin and Hanski and those estimated for Kilpisjärvi data (this involved less than 10% change). Our fitting results suggest that as latitude is decreased, we obtain increasingly noisier fits and less reliable parameter estimates, paralleling the decrease in the strength of the endogenous signal (as measured by R^2_{data}). At the same time, estimated parameter g, quantifying the strength of generalist predation, grew from small values at northern locations to greater values at southern locations, just as the generalist/specialist theory predicts (for details, see Turchin and Ellner 2000a: table 1).

Stability Properties of the Fitted Model What do these fitting results tell us concerning the qualitative nature of vole dynamics? One way to answer this question is to calculate the global and local Lyapunov exponents of the predation model with the parameters estimated on the Kilpisjärvi data. We also quantified uncertainty associated with the estimate by using a nonparametric bootstrap to calculate 90% confidence intervals associated with each estimate. Our results suggest that the estimated Λ_∞ is statistically indistinguishable from zero (the 90% confidence interval is [–0.07, 0.08]). Thus, according to this particular measure of sensitive dependence on initial conditions, the system is poised right on the boundary between stability and chaos. (We also calculated the *local* Lyapunov exponents characterizing the model with best parameter estimates, and found that the system spends more than half of the time in the chaotic regime characterized by sensitive dependence on initial conditions. For details, see Turchin and Ellner 2000a.)

The fact that we obtained reasonably small confidence intervals indicates that the dynamics are not overly sensitive to changes in parameter values near to those estimated for Kilpisjärvi. This outcome might be surprising given that deterministic predator-prey models with strong periodic forcing have a complicated bifurcation diagram that includes limit cycles, the period-doubling route to chaos, and the quasiperiodic route to chaos. However, it is important to remember that our model includes a component representing exogenous perturbations of the system (process noise). This leads to the phenomenon known as the "noise-induced stabilization of chaotic transients" (Rand and Wilson 1991), which is explained in section 5.3.3. The smoothing effect of noise allows us to estimate the stability properties of the dynamics despite our inability to estimate parameters with great precision. It also provides further evidence for a point we have argued elsewhere (Ellner and Turchin 1995): that properties of the noise-free "skeleton" model (Tong 1995) may be very misleading for the ecological context where exogenous noise cannot be ignored.

In summary, our analysis suggested a good fit between model predictions and data in the longest data set, Kilpisjärvi. Probably the most remarkable result was that a priori estimates of model parameters, derived without considering time-series data, predicted Kilpisjärvi data better than the phenomenological, data-based algorithm. Other data sets were also well predicted by the model (with the exception of Sotkamo), but results of fitting these data with the model met with variable degrees of success. However, the estimate of generalist predation intensity, g, tended to increase as latitude decreased. This result is in agreement with the prediction of the generalist/specialist predation hypothesis (Turchin and Hanski 1997).

The two measures that we used to quantify sensitive dependence on initial conditions in model (12.4), with parameters estimated from Kilpisjärvi data, provide somewhat different (but complementary) viewpoints on the dynamical nature of vole oscillations. The estimated global Lyapunov exponent (Λ_∞) suggests that in the long run there is neither strong divergence nor strong convergence of trajectories (since the estimate is within ± 0.08). What happens in the long run (on the order of hundreds and thousands of years), however, may not be a dynamical measure of high relevance for ecological systems. We expect to see few natural populations that would oscillate unchanged for such long periods of time, unaffected by either long-term environmental trends or evolutionary processes. Thus, when characterizing population dynamics Λ_∞, at the very least, should be supplemented by shorter-term measures of sensitive dependence on initial conditions. It is therefore both significant and exciting to find that dynamics of vole populations in northern Fennoscandia are characterized by alternating periods of rapid trajectory convergence or divergence.

12.5 LEMMINGS

Population dynamics of lemmings are of considerable interest, not least because it was the observation of regularity in lemming outbreaks that lead Charles Elton to become involved in the study of population cycles (section 1.1.1). The study of lemming dynamics, however, lagged behind that of their microtine relatives, simply because lemmings inhabit such distant and difficult-to-work-with locations. As we shall see below, key events for lemming population dynamics take place during winter (which may last up to 10 months in some locations lemmings inhabit). It is easy to imagine difficulties accompanying empirical research on organisms who are active under a meter or more of snow, while temperatures above go down to $-50°C$. Nevertheless, ecologists managed to gather quite a large collection of data, including some experimental work in North America (Alaska and Canada), Russia (Wrangel Island), and Fennoscandia. In particular, we now have several time series on lemming population fluctuations. In this review of lemming population dynamics, I begin, as usual, by considering empirical patterns in long-term records of their fluctuations. Next, I describe a test that we designed to distinguish between two possible explanations of lemming cycles—a predator-prey versus a herbivore-vegetation interaction (Turchin et al. 2000). Finally, I review our efforts to model lemming cycles at one particular location, Barrow, Alaska (Turchin and Batzli 2001). My focus throughout this section is on the lemmings in the genus *Lemmus*: the Norwegian lemming (*L. lemmus*) and the brown lemming (*L. sibericus*). The ecology and population dynamics of the collared lemming appear to be quite a different story (e.g., see Reid et al. 1995).

12.5.1 Numerical Patterns

Several data sets on lemming fluctuations are available from Fennoscandian, North American, and Russian locations (table 12.5). Most data were obtained by snap-trapping, except for the category "indirect indices," which includes two data sets. The provenance of the first one, published by Schultz (1969), is unclear, and probably includes some guesstimates. Still, I include it because this data set has been widely cited in the secondary literature. The last data set (Taimyr) is a very interesting one, because it does not refer to a direct index of lemming numbers, but is actually the proportion of juveniles among dark-bellied brant geese. These geese breed on the Taimyr Peninsula, Russia, and overwinter in western Europe, particularly in the Wadden Zee area of the Netherlands, Germany, and Denmark, where the birds are counted every year. Several studies have shown that fluctuations in the abundance of lemmings in Taimyr have a huge effect on the breeding success of brant geese (Underhill et al. 1993; Summers et al. 1998). Breeding success is highest during years of high lemming densities, because arctic foxes, a major source of egg mortality, have an abundant alternative food supply. Breeding success is lowest during the year after lemming peak, when arctic foxes are abundant and very hungry. Thus, percent of brant goose juveniles provides an indirect index of lemming dynamics, mediated by a two-link connection through the trophic web: first, the lemming–arctic fox link and, second, the arctic fox–brant goose connection.

TABLE 12.5. Summary of NLTSM results: lemmings. Same notation as in table 12.1

Location	n	S	T	ACF[T]	p	q	R^2_{pred}	Λ
L. lemmus								
Finse	25	1.25	3	0.37*	2	2	0.43	2.01
Kilpisjärvi	25	0.92	—	—	3	2	0.31	1.01
Finnmark	20	1.06	—	—	3	1	0.21	0.08
L. sibericus								
Barrow	18	0.77	—	—	2	2	0.35	0.53
Kolyma	12	0.79	4	0.85**	2	1	0.49	0.13
Indirect indices								
Pt. Barrow	21	0.73	3	0.43*	2	1	0.50	0.05
Taimyr	39	0.97	3	0.56**	3	1	0.32	0.34

Data sources: Finse (Framstad et al. 1997); Kilpisjärvi (Turchin et al. 2000, data collected by H. Henttonen); Finnmark (Ekerholm et al. 2000); Barrow (Pitelka 1976); Kolyma (Potapov 1997); Pt. Barrow (Schultz 1969); Taimyr (*Wadden Sea Newsletter*, 1997).
Series that were detrended: none.

The results of the analysis of lemming data are very consistent. All data sets are classified as second-order or higher. By contrast, evidence for periodicity is not very strong: several data sets are characterized by 3–4 year periods, but others appear to be aperiodic. The amplitude of oscillations is huge: of the data analyzed in this book, only the larch budmoth cycle is capable of matching such huge swings in abundance. Note, also, that lemming densities periodically "drop off the radar screen" (see figure 12.5). In order to analyze these data, we must substitute a small number instead of 0 observations, which has a side effect of reducing the amplitude. Thus, S-values appearing in table 12.5 are an underestimate. But the most striking pattern is observed in the Lyapunov exponents, whose estimates are uniformly positive, and some (e.g., Finse, which is incidentally the best data set in terms of length and accuracy of measurements) are truly huge.

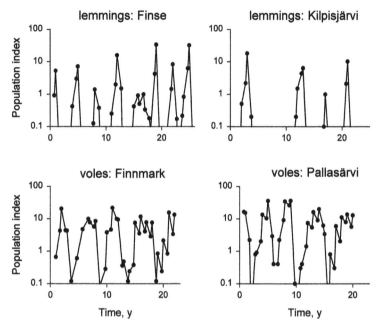

FIGURE 12.5. Population trajectories of lemmings in Finse and Kilpisjärvi; and voles in Finnmark and Pallasjärvi.

Taken together with weak presence of statistical periodicity and huge amplitude, these results strongly suggest that we are dealing here with a case of strong chaos.

12.5.2 Testing Alternative Trophic Hypotheses

As I discussed in section 4.2.3, many models of resource-consumer interactions predict that population cycles of the two species will be characterized by distinct shapes: blunt rounded peaks for resources, and sharp angular peaks in consumer density (see, e.g., figure 4.3). This general pattern arises in predator-prey, host-parasitoid, and herbivore-vegetation models, but is particularly prominent in models developed specifically for small rodents (Turchin et al. 2000). Consider the interaction between rodents and their specialized predators, weasels (e.g., section 12.3.3). A model of this system must include a self-limitation term in the prey equation, because without rodent self-limitation no cycle can occur—the rodent population will escape predator control by virtue of its much higher rate of growth. In a typical rodent-predator cycle, prey density rapidly builds up to the point where its further growth is prevented by self-limitation mechanisms, while predators can start increasing only after prey density grows beyond the threshold where predators can maintain positive energy balance. By the time predators have increased to the point where they cause prey to decline, prey population has spent a prolonged period of time at peak densities. As a result, the shape of prey peaks will be blunt.

By contrast, if rodents are "predators" (i.e., the mechanism driving cycles is the interaction between rodents and their food supply), then their population will grow exponentially while food abundance is above the ZPG threshold. When food is depleted, rodent density will immediately collapse as a result of starvation (or emigration). Thus, rodent population trajectories should be characterized by "saw-shaped" dynamics with sharp peaks if rodents are consumers, rather than the resource.

Distinctive topological features of prey versus predator cycles are a generic feature of two-species trophic models. Predator peaks must be sharp: we cannot "flatten" them by adding predator self-limitation, because this leads to a loss of oscillation (stabilization of the system). The bluntness of prey peaks can be more variable, since the length that prey spend at the upper density threshold will depend on the relative growth rates of prey and predators. If prey growth rate is similar to (or slower than) that of predators, it may be difficult in practice to detect the plateau phase in prey dynamics. However, because rodents are characterized by much faster reproductive rates than their predators, peak topology should be a particularly useful diagnostic for their dynamics.

These theoretical observations allow us to design an empirical test to distinguish between the two rival hypotheses for lemming cycles, one invoking the interaction with the food supply, and the other with predators. To pursue this idea, we located all time-series data on the population dynamics of Norwegian lemmings (*Lemmus lemmus*) that were at least 20 years long, and had at least two observations per year (Turchin et al. 2000). We also analyzed a comparable set of three vole series from the northernmost part of Fennoscandia (latitude > 68°N), where vole populations exhibit high-amplitude oscillations. Two of the lemming data, contrasted with two vole data sets, are plotted in figure 12.5 (the third vole series, Kilpisjärvi, is shown in figure 12.4a). Visual examination of the data plotted on logarithmic scale immediately suggests a striking difference between the lemming and vole time series. Lemmings have very sharp peaks, with rarely more than one observation period at the peak, while vole populations spend at least 2 years in the vicinity of maximum densities before their populations collapse. We applied several statistical tests to these data, and they confirmed the results of the visual inspection (for details, see Turchin et al. 2000). Our results, therefore, are consistent with the hypothesis that lemmings are functional predators, while voles are functional prey.

An additional feature of numerical dynamics, the variability of peak densities, provides a further clue about dynamical mechanisms that may be responsible for lemming and vole cycles. Resources should have rather stereotypical dynamics at the population peak, because they hit the population ceiling imposed by density-dependent regulation. By contrast, consumer dynamics do not have a comparable "hard ceiling," because their density will start to decline when consumers run out of food. Precisely when this occurs depends very much on the timing of consumer increase with respect to seasonality. Consider a threshold consumer density, $N_{threshold}$, above which consumers are expected to run out of winter food supply and therefore experience a winter crash. If this threshold is reached in the fall, then peak density would be equal to $N_{threshold}$. However, if $N_{threshold}$ is achieved in the spring, then consumer density will increase much beyond it during the favorable summer conditions, and the resulting winter crash will be much deeper. As a result, peak density of consumers (as well as the depth of collapse) should be highly variable. In fact, this mechanism, based on the interaction between seasonality and predator-prey dynamics, is at the root of mathematical chaos that may arise in rodent population models. Again, numerical results are consistent with our main hypothesis (figure 12.5; statistics in Turchin et al. 2000: table 1).

12.5.3   Lemming-Vegetation Dynamics at Barrow

The survey of general models in section 4.4 suggested that it is rather easy for the interaction between plants and herbivores to produce population cycles if vegetation dynamics are appropriately described by a logistic growth equation. Coupled with the results in the previous section, this suggests that we should attempt to develop an empirically based model for the lemmings-vegetation system, to check whether this system could exhibit population oscillations for biologically reasonable functional forms and parameter values. This is the question that was addressed by George Batzli and myself in a recent paper (Turchin and Batzli 2001). Our focus was on the interaction between the brown lemming, *Lemmus sibiricus* (formerly known as *trimucronatus*), and its food supply.

Model Equations   We began by making the important distinction between the two kinds of vegetation that together comprise the food supply of brown lemmings: mosses and shoots of vascular plants (primarily graminoids, grasses and sedges). Graminoids have higher nutritional value and provide the bulk of summer food for lemmings (Batzli 1993). Graminoid biomass, however, is greatly reduced by seasonal die-off in September–October, before it is preserved by being frozen. During winter, therefore, lemmings switch to a much higher utilization of mosses (see table 1 in Batzli 1993). The seasonal dietary shift from graminoids to green mosses appears to be paralleled by increased ability of brown lemmings to utilize mosses. Experimental feeding of brown lemmings with green mosses showed that during winter months animals can survive for long periods on this kind of food, while during summer they usually die within 2–3 days (Chernyavsky et al. 1981).

Let V and M be the edible biomass (in kg of dry weight per ha) of vascular plants and mosses, respectively. We modeled dynamics of graminoid shoot biomass with a modification of the regrowth equation that takes into account the seasonal dynamics: growth in summer (two months from melt-off in mid-June to first heavy frosts in mid-August), rapid die-off of 90% of biomass during the transition period between summer and winter, and no change under snow during the winter months, except for consumption by lemmings. Moss dynamics were modeled as a seasonally modified logistic equation with growth in summer and neither growth nor decline in winter. We assumed no direct competition between mosses and graminoids, so that in the absence of herbivory, both resources would increase to their respective "carrying capacities" (maximum standing crops), k_V and k_M.

We assumed that lemming populations are regulated solely by food availability. This is probably not a bad assumption, because there is no evidence that lemming density is regulated by direct intraspecific competition, including social interactions. However, the mortality imposed by predators on lemmings can be very high, especially in summer (Chernyavsky and Dorogoy 1988). This consideration suggests that our model should eventually be modified to include such predation effects.

Let N be the density of lemmings (individuals per ha). Lemming consumption of vegetation is modeled as the hyperbolic functional response, and the lemming equation is in the Rosenzweig-MacArthur form (i.e., the amount of food consumed directly affects the growth/decline rate of N). Because winter conditions impose much greater energetic demands on lemmings, the ZPG parameter η is assumed to change with season, in turn affecting the maximum rate of lemming population growth when food is abundant, r_0. The maximum rate of increase is, therefore, high in summer (assumed to be about 6 yr^{-1}), while during the rest of the year it is lower at about 4 yr^{-1} (see Turchin and Ostfeld 1997 for the discussion of population growth rates characterizing arvicoline rodents).

The equations resulting from these assumptions are

$$\frac{dV}{dt} = u(\tau)\left(1 - \frac{V}{k_V}\right) - \frac{aVN}{V + M + b}$$

$$\frac{dM}{dt} = v(\tau)M\left(1 - \frac{M}{k_M}\right) - \frac{aMN}{V + M + b} \qquad (12.5)$$

$$\frac{dN}{dt} = \xi N\left[\frac{a(V + M)}{V + M + b} - \eta(\tau)\right]$$

The variable τ indicates season ($0 < \tau < 1$), with $\tau = 0$ corresponding to the fall (transition between summer and winter). Seasonal dynamics are included in growth rates $u(\tau)$ and $v(t)$ and the ZPG consumption rate, $\eta(\tau)$, which are the following functions of time:

- Winter ($5/6 \leq \tau < 1$): $u(\tau) = 0$, $v(\tau) = 0$, and $\eta(\tau) = \eta_w$.
- Summer ($5/6 \leq \tau < 1$): $u(\tau) = u_s$, $v(\tau) = v_s$, and $\eta(\tau) = \eta_s$.
- In addition, at $\tau = 0$ (transition between summer and winter), $V(t)$ is reduced by 90%.

Estimates (and ranges) of model parameters are derived in Turchin and Batzli (2001). The only deviation from the account in that paper is that I assumed that lemmings have no preference for vasculars. That is, I set the discounting factor to 1; this assumption does not strongly affect the model dynamics. As discussed in Turchin and Batzli (2001), introducing preference for vascular plants tends to reduce the period of oscillations somewhat.

Dynamics of the Model For the median values of parameters ($u_s = 10$ Mg ha^{-1} yr^{-1}, $v_s = 12$ yr^{-1}, $k_V = 1$ Mg ha^{-1}, $k_M = 2$ Mg ha^{-1}, $a = 15$ kg yr^{-1} ind^{-1}, $b = 70$ kg ha^{-1}, $\chi = 10.7a$, $\eta_s = 0.44a$, $\eta_w = 0.63a$), the model exhibits sustained oscillations with a period of about 7 years and densities varying by more than two orders of magnitude. Model-predicted dynamics do not appear to be too dissimilar to the observed lemming oscillations at Barrow (figure 12.6). Although this is not a quantitative, rigorous comparison (for one thing, we have not added any stochasticity to the model), some points of qualitative similarities are immediately apparent. First, both trajectories exhibit sharp angular peaks, consistent with the hypothesis that oscillations are generated by a herbivore–logistic vegetation system. Second, both data and model trajectories tend to spend some time at low densities. In the model we know precisely why this happens: moss biomass is extremely low after a peak, and takes several years to increase to the point where lemmings can utilize it. Vascular biomass bounces right back up, but is not sufficient to sustain the lemming population. As a result, lemmings continue to decline for several years after the peak, but at a slower rate than would happen in the absence of vascular vegetation.

Model-predicted dynamics do not mimic the data in all particulars. For example, the predicted period is a bit longer than what is

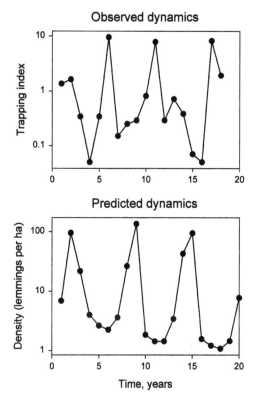

FIGURE 12.6. Population trajectory of the brown lemming in Barrow, Alaska, compared with predictions by the Barrow model.

observed, and the observed dynamics are less regular than the predicted (especially at low lemming densities). This is as it should be, because the functional forms and parameter values of the Barrow model are not as well grounded in empirical data as one would wish. In fact, it is surprising how well this "first cut" approach did in replicating certain features of the data. Furthermore, it is possible that rather straightforward additions to the model would bring it more in line with data. For example, adding stochasticity, particularly in the growth rates of vegetation, may have a disproportionate effect on the regularity of dynamics, especially in the low phase, where a better growing season for vascular vegetation may actually allow lemmings to temporarily increase, even before their main winter supply, mosses, grows back. Another feature of lemming biology that should eventually be investigated is the influence of generalist predators (jaegers, snowy owls, and arctic foxes).

12.6   SYNTHESIS

In this section I begin by summarizing the modeling and empirical results on vole and lemming dynamics, and then attempt some generalizations about population dynamics of rodents in general.

12.6.1   Summary of Findings

The Field Vole in Northern Fennoscandia   The question of why Fennoscandian voles exhibit population oscillations appears to have the clearest answer. Of the three most likely hypotheses deemed worthy of intensive study at the 1996 vole workshops, two appear to meet with insuperable difficulties. If the dynamics of food supply of Fennoscandian voles are indeed better modeled with a regrowth rather than a logistic equation, then theory is quite emphatic that the rodent-vegetation interaction cannot exhibit the observed second-order cycles. Experimental evidence supports this theoretical conclusion. First, the experiment of Klemola et al. (2001) clearly demonstrated a lack of delayed effect of depleting food on next year's population growth rate in the field vole at Alajoki (Finland). A very similar result was previously obtained in an analogous experiment on an ecologically similar species, *M. pennsylvanicus*, in New York (Ostfeld et al. 1993). Second, adding food to an experimental population of voles in Pallasjärvi (northern Fennoscandia) throughout a population cycle resulted in higher peak densities, but failed to prevent or even delay the population decline (Henttonen et al. 1987).

The maternal effect hypothesis similarly encounters grave theoretical difficulties. However, the lack of match between theory and data is not qualitative, as with the food hypothesis, but quantitative. It is clear that models based on the maternal effect hypothesis are capable of producing second-order oscillations. However, they can do that only for rather extreme assumptions about the delayed effect of population density on the average quality. Furthermore, maternal effect models encounter a difficulty in attempting to match certain quantitative features of vole cycles, most notably the oscillation period. Put simply, these models predict too short cycle periods. On the empirical side, the maternal effect hypothesis has not yet been properly challenged by an experiment. Yet, now that we have explicit models capable of making quantitative predictions, conducting experimental tests should be a priority. The fact that the maternal effect explanation has not yet survived such a test is another reason to downgrade the *current* degree of belief associated with this hypothesis.

The remaining explanation of vole cycles in northern Fennoscandia, based on the interaction between voles and their specialist predators, is by contrast richly supported both theoretically and empirically. Models of vole-weasel interaction, parameterized with independent data, successfully predict all the major quantitative features of the observed dynamical pattern. What is even more striking is that the model outperforms the data-based one-step-ahead prediction algorithm. Since this was achieved without any circularity (model parameters were *not* estimated with time-series data), it is a strong empirical endorsement of the predation hypothesis (and, incidentally, provides a nice example of strong inference in population ecology!). Furthermore, the predation hypothesis has survived a number of experimental tests. Most notable, in this respect, is the work by Korpimäki, Norrdahl, and Klemola at Alajoki in western Finland. These investigators showed that predator removal can prevent summer declines in vole density (Korpimäki and Norrdahl 1998; Klemola et al. 2000). Although this is a strong endorsement of the predation hypothesis, I am more impressed by the experiment in which these investigators tracked vole mortality throughout one increase-peak-decrease period (Norrdahl and Korpimäki 1995). They showed that avian predation was directly density dependent (without a lag), so that predation rate due to owls and kestrels was similar in the increase and decline phase of the cycle. By contrast, predation rate due to mustelids was characterized by marked delay: it was practically nonexistent during the increase, and very heavy during the decline phase (see also Korpimäki and Norrdahl 1991b; Korpimäki and Norrdahl 1991a; Korpimäki et al. 1991; Korpimäki 1993; and see Steen 1995) for very similar results obtained for a different species,

M. oeconomus, in a Norwegian location. These observations are, of course, precisely in line with the hypothesized roles of avian (treated by theory as generalist) and weasel (treated as specialist) predators.

My general conclusion that the predation-based explanation is currently the best-supported hypothesis when considered against the alternatives does not mean that it is necessarily the correct one, and that the relative degrees of belief assigned to other hypothesis will not change with time. What I would like to claim, though, is that the question of microtine cycles can certainly be considered as an example of "mature science" in ecology. All the hallmarks of mature science are present: hypotheses translated into models, models have empirically based functional forms and parameters, multiple hypotheses/models contrasted with each other using data, explicit theoretical predictions tested with experiments, and, finally, a strong differentiation (in terms of degrees of belief) among various proposed explanations (to put it simply, there is one clear winner).

Latitudinal Shift in Vole Dynamics within Northern Europe   While the question of cycles in northern Fennoscandia appears to have a clear-cut answer, the explanation of the latitudinal gradient in dynamics, based on variation in generalist predator pressure, needs to be hedged with some caveats. First, the analysis of the expanded data set focusing specifically on the field vole confirms the presence of a dynamical gradient, as quantified by changes in the amplitude, period, and Lyapunov stability (for the latter, see figure 12.2). Thus, the original conjecture of Hansson, Henttonen, and Hanski (Hansson and Henttonen 1985, 1988; Hanski et al. 1991) is vindicated. However, there is one aspect of this dynamical gradient that remains unclear: is there a shift from second- to first-order dynamics in the southern part of the field vole range? Note that the only indication of such a shift comes from a single location, Revinge (which also happens to be the shortest series). Dynamics at another southern location, Kielder, are classified as second-order and periodic, although of quite small amplitude. Thus, the question of whether population dynamics of *M. agrestis* in the southern part of its range are stabilized to the extreme of first-order aperiodic fluctuations must wait for further data.

What about the role of generalist predators in explaining the dynamical gradient in *M. agrestis*? This explanation is well supported both theoretically (the modeling results when a generalist predation term is added) and empirically (for review, see Hanski et al. 2001). However, its logical status is not quite as strong as the explanation of northern Fennoscandian cycles based on vole interaction with specialist predators, because there are several alternative hypotheses. One is that increased generalist predation on weasels also is capable of stabilizing the oscillations (this is based on my numerical exploration of the appropriate modification of model 12.4). Second, Hansson (1999) proposed a hypothesis based on habitat fragmentation changes with latitude. These alternatives have not been formally contrasted with the explanation based on generalist predation. In other words, the generalist predation model has not yet survived such stringent tests as formal contrasting against explicit alternatives, or as an experimental challenge.

Lemming Dynamics   The second case study investigated in depth in this chapter, lemmings, offers a very interesting contrast to the vole story, because the predation hypothesis appears to be rejected by the preponderance of evidence. First, as we discussed in section 12.5.2, the predation model makes wrong predictions about the peak shapes characterizing lemming trajectories. Second, mustelids—potential specialist predators of lemmings—are completely absent at some locations (as on Wrangel Island; see Chernyavsky and Tkachev 1982), while at other sites they only infrequently invade lemming habitat (typically, after vole peak years). Other predators, even mammals such as arctic foxes, are unlikely to be able to play the role of specialist predators, because they do not prey on lemmings in winter—the period taking up most of the year, during which key events determining dynamics happen.

Intrinsic hypotheses also appear to be an unlikely source of explanations for the lemming cycle (Stenseth and Ims 1993b), although it must be admitted that we need to know more about intrinsic factors that may influence lemming population dynamics. This leaves interactions with food as the most plausible hypothesis for lemming oscillations. So far, the food hypothesis has survived the first challenge: it turns out that a model based on it can generate oscillations similar to those observed for biologically reasonable values of parameters. The next step should be an experimental challenge of the food hypothesis, although it will be a difficult task to set up, since any food manipulation would have to be conducted during winter. Still, this definitely would be a worthwhile attempt, given the intrinsic interest of lemming dynamics. To conclude, I believe that the food quantity hypothesis is best supported among the alternatives, given the current state of affairs. However, its degree of support is nowhere near as solid as that for the specialist predation explanation of the field vole cycle.

12.6.2 Toward a General Trophic Theory of Rodent Dynamics

The contrasting "best models" for voles versus lemmings have an important methodological consequence. Historically, population ecologists tended to look for "universal" explanations of small rodent population dynamics (Chitty 1996). Given the weight of evidence reviewed in this chapter, such a view no longer appears to be tenable. Is this a pessimistic conclusion, then—is each species in each location driven by a unique and idiosyncratic combination of mechanisms? I do not think so, because I believe that we can seek generality at the next level of explanation. In other words, I propose that instead of looking for a general explanation based on the specific ecological mechanism, we should instead base it on a general *explanatory framework*, which in my opinion should be based on the theory of trophic interactions. What I have in mind is a general model of the following kind:

$$\frac{dV}{dt} = f(\cdot) - \frac{aVN}{V+b}$$

$$\frac{dN}{dt} = \xi N\left(\frac{aV}{b+V} - \eta\right) - \frac{r_0}{k}N^2 - \frac{cNP}{d+N} - \frac{gN^2}{h^2+N^2} \quad (12.6)$$

$$\frac{dP}{dt} = \chi P\left(\frac{cN}{d+N} - \mu\right) - \frac{s_0 P^2}{\kappa N}$$

Here V, N, and P are densities of vegetation biomass, rodents, and specialist predators, and all parameters have their usual interpretation (see the list of mathematical symbols at the front of this book). Model (12.7) is not meant to be applied to any specific case study; it is

too unwieldy for that. In fact, it is not even complete—we need seasonality and stochastic environmental noise (I decided not to encum-
ber the equations with extra notation). It may be further complicated
by adding other rodent species to it, along the lines investigated by
Hanski and Henttonen (1996). This model, instead, is a *theoretical
framework*, in the sense that for different species, and for different
habitats, we would drop some terms and retain others, depending on
the ecological characteristics of the studied system. Thus, for *Lemmus lemmus* in a location like Finnmark, we would probably drop the
specialist predators, perhaps the rodent self-limitation term ($r_0 N^2 / k$),
and substitute a logistic growth term in place of $f(\cdot)$. With added
exogenous drivers (seasonality and noise) this could be a reasonable
model for the system. A model for the field vole in Pallasjärvi, on the
other hand, would certainly retain the specialist predators, although
we might decide to drop the generalist predation term (since generalist predation is quite weak at this northern location). We may decide
to model vegetation explicitly, substituting a regrowth term in place
of $f(\cdot)$, or we might decide to model vegetation implicitly, by dropping V as the state variable, and using a simple logistic term in the
herbivore equation.

I would further argue that model (12.7) provides a good framework
for most other studied rodent populations. It is certainly possible to
construct explanations of their dynamics, based on model (12.7), with
different qualitative structural assumptions, and different quantitative
choices of parameters. These explanations, discussed below, are in
my opinion plausible, because they do not violate known features of
empirical systems, but also speculative, because most are not supported by explicit modeling and empirical tests. Nevertheless, they
may provide useful working hypotheses for further work.

The Bank Vole   One important characteristic of rodents is the
degree to which their diet depends on energy-poor (mosses, grasses)
versus energy-rich foods (seeds, insects). The more a rodent relies
on difficult-to-digest foods, the larger digestive machinery—gut—it
needs to "drag around." In consequence, rodents characterized by
more herbivorous habits tend to be less agile and easier for predators
to catch than rodents with more omnivorous habits (Hansson 1987).
The rodent species commonly found in northern Europe, thus, can

be ranged in a herbivory gradient from moss eaters to omnivores as follows: Norwegian lemmings, field voles, gray-sided voles, bank voles, yellow-necked mice. Of particular interest is the contrast between the herbivorous field voles and the more omnivorous bank voles, who also consume fungi, berries, and seeds. We can write structurally the same model for both species, for example,

$$\frac{dN}{dt} = r_0 N \left(1 - \frac{N}{k}\right) - \frac{cNP}{d+N} - \frac{gN^2}{h^2+N^2}$$

$$\frac{dP}{dt} = \chi P \left(\frac{cN}{d+N} - \mu\right) - \frac{s_0 P^2}{\kappa N}$$

(12.7)

This model is practically the same as the "standard" vole model (equations 12.2), but the predation equation has the variable-territory form, instead of the form developed by May (recollect that the latter can be thought of as an approximation for the former; see section 4.2.2). However, although model equations may be the same for the two vole species, parameter values can be quite different. The first difference is in the search efficiency of predators, a. If weasels are much more likely to capture a field vole compared to a bank vole, then a in the bank vole model will be much lower than in the corresponding model for field voles. Since the half-saturation constant is related to the search efficiency inversely (section 4.1.1), bank vole's d should be much greater than field vole's d.

A further difference is that the carrying capacity of bank voles is much lower than that for field voles, partly because the bank vole food supply is more limited, partly because they have more stringent social controls on density. Now we recollect the fact that one of the strongest influences on the stability of predator-prey models such as equations (12.7) is the parameter ratio d/k (the "paradox of enrichment"). We see that due to higher d and lower k, this ratio will be much larger for bank voles than for field voles. Thus, we reach an inescapable conclusion that bank vole dynamics should be much more stable than dynamics of field voles. The contrast between the estimates of Λ_∞ in figure 12.2 provides a strong empirical confirmation of this conjecture.

Yet another possible characteristic of bank vole food, distinguishing it from that of field voles, is that it may exhibit a great degree of fluctuation in certain habitats. For example, seed production in broadleaf forests in central Europe varies enormously from year to year (this is known as masting). Accordingly, we would expect that bank voles would rapidly increase in years of high mast and crash in years following mast. The degree of crash should be accentuated by the action of predators. This theoretical scenario appears to match closely the situation observed in the Bialowieza Forest of eastern Poland (Pucek et al. 1993).

The Gray-Sided Vole   The gray-sided vole, *Clethrionomys rufocanus*, is in its ecological characteristics intermediate between the field and bank voles. However, like the field vole it is largely herbivorous. Thus, the argument above applies, except that we would expect that gray-sided vole parameters would be more similar to those for the field vole than for the bank vole. Thus, our expectation is that gray-sided vole dynamics should be quite similar to those of field voles, but perhaps a shade more stable. This is precisely what we see in the time-series data (figure 12.2). The Lyapunov exponents for four gray-sided vole series sit right in the quasi-chaotic region, while Λ for the northern populations of the field vole extend into the more positive range.

The Common Vole   Time-series analysis of the common vole (*M. arvalis*) data suggests that its dynamics are dominated by first-order mechanisms. However, first-order cycles with periods of 2–3 years appear to be not unusual. This observation suggests that specialist predators do not play an important role in the dynamics of *M. arvalis* (see also Jedrzejewski and Jedrzejewska 1996). One possible explanation may be the strong presence of avian predators that may keep weasel densities well below the threshold where they can impact vole populations. First-order cycles may, therefore, result from the interaction between voles and their food supply (modeled with a regrowth equation) in a seasonal environment. It is clear that this scenario is again a special case of the framework model (equations 12.7).

Other Rodents   Another example of second-order cycles examined in section 12.2.2, in addition to field voles, is water voles in Switzerland. There are some data suggesting that these cycles are driven by interaction between water voles and their specialist predators, stoats (Debrot 1981). Note that the average period of 6 years characterizing water vole oscillations is consistent with general theory predicting that larger body sizes, via lower intrinsic rates of population change, should cause longer-period oscillations.

Turning to wood mice, we might consider their dynamics as an even more extreme example of omnivory than the contrast between the field and bank voles, discussed above. In fact, the ratio d/k characterizing mouse-weasel interaction may be so high as to prevent a weasel population from subsisting solely on mice. Thus, we should expect first-order aperiodic dynamics in forest mouse populations, characterized by a great degree of fluctuation where their food supply is variable due to masting. Examples are mice in Bialowieza (Pucek et al. 1993) and southern England (Southern 1979).

In the scenarios I considered above, the focal species was treated as the "main player." Yet rodent populations are embedded within multispecies communities, and dynamics of numerically subordinate species may be more affected by the ecological properties of the numerically dominant species than of their own. It seems now generally accepted that the dynamics of bank voles in northern Fennoscandia are oscillatory because they are affected by weasel predation "spilling over" after population collapses of the dominant species, the field vole (Hanski et al. 2001). A similar mechanism may explain the irregular dynamics of voles in North America, especially those living in small clear-cuts within forests. There, mouse outbreaks resulting from high-mast years may be followed by years of high predation activity that spills over onto vole populations (Ostfeld and Keesing 2000).

As a final thought, I should add that when discussing vole dynamics in this chapter, I have largely followed my rule of ignoring spatial complications. Yet, a synthesis of this issue cannot be complete without an explicit consideration of space. Lacking space to address spatial aspects here, I therefore refer the reader to two papers: one on the landscape perspective by Hansson (1999), and the second on the role of optimal to marginal patch areas by Lidicker (2000).

CHAPTER 13

Snowshoe Hare

13.1 INTRODUCTION

The snowshoe hare–lynx population cycles, like cycles in rodents, lie at the very beginnings of the systematic study of complex population dynamics (Finerty 1980). Although rodent cycles chronologically were first to be noticed by Charles Elton (see section 1.1.1), at the time there were no quantitative time-series data for rodents to be analyzed. But there were long-term records of fur returns at the Hudson's Bay Company, so for several decades the main focus of research shifted from the shorter rodent cycles in Fennoscandia to the 10-year hare-lynx cycles in boreal North America.

Early theories attempting to explain the biological mechanisms of the 10-year cycle focused on exogenous factors, starting with Elton himself, who suggested that lynx dynamics may be driven by the sunspot cycle (Elton 1924; for other exogenous hypotheses, see Royama 1992: table 5.1). Although recently the sunspot hypothesis was partially resurrected (Sinclair et al. 1993), it was not suggested to be the primary cause of the cycle, but at best a mechanism explaining some secondary features of data (such as geographic synchrony and variation in amplitude). Although the argument advanced by Sinclair et al. (1993) does not appear compelling to me, here is not the place to discuss it, since my emphasis is on the main mechanism, or mechanisms, that drive 10-year cycles (and Sinclair et al. are the first to stress that they do not doubt the endogenous nature of this main mechanism).

The current consensus in the field appears to be that two main factors are responsible for snowshoe hare cycles: availability of food and predators. Exactly how these two factors interact in driving the cycle, however, remains somewhat controversial. Keith (1990) argued that winter food shortages stop hare population increase and initiate the decline. This is followed by increased death rates due to predation, which drives hare density to its cyclic low. On the basis of food addition experiments that failed to prevent declines, Krebs et al. (1986) initially argued that predation alone is sufficient to account for the population crash. However, this pure predation hypothesis was later abandoned, because a manipulative experiment suggested that simultaneous removal of predators and addition of food resulted in much higher hare densities than either treatment by itself (Krebs et al. 1995). Thus, quantitative details of the interaction between predation and food shortage in explaining hare cycles are currently not completely understood. An interesting idea relevant to this issue is the predation-sensitive foraging hypothesis of Hik (1995). Additionally, other authors argued that factors such as sublethal nematode parasitism (Murray et al. 1997) and predator-induced stress (Boonstra et al. 1998) are important modifying factors of the cycle, in addition to predation and food. Finally, some still argue that the case for predation is far from proven (Chitty 1996).

In this section I attempt to synthesize the insights from time-series analyses, mathematical modeling, and field experiments. Because all the major approaches to hare dynamics have been described in primary literature, in this chapter I use a somewhat different exposition style than in previous chapters. Instead of delving into the nitty-gritty details (for which I refer the reader to the relevant primary literature), my focus is on connections between the empirical and theoretical approaches. Furthermore, I attempt to place the dynamics of hares within the general context of other taxa discussed in the book.

13.2 NUMERICAL PATTERNS

To start placing the 10-year cycles in context, let us first consult a recent survey of 700 long time series data on population fluctuations (Kendall et al. 1998). These authors concluded that cycle incidence is particularly common in mammal populations. Fully 70% of mammal species had at least one cyclic population, and out of the total of 328 mammal data series one-third were cyclic. A typical pattern within species that exhibit both cycling and noncycling populations is a latitudinal gradient in cyclicity. Thus, the hare-lynx cycle is not an unusual occurrence, and understanding its causes may help to reveal some general mechanisms of mammalian population dynamics (especially when taken in conjunction with insights from rodent and cervid case studies).

The time series of lynx fur return statistics (Elton and Nicholson 1942) is the most famous example of oscillatory dynamics in ecology, discussed in every ecological textbook I know of, and is probably the most analyzed data set in ecology (Moran 1953; Leigh 1968; Bulmer 1974; Tong 1977; Finerty 1979; Schaffer 1985; Turchin and Taylor 1992; Sinclair et al. 1993; Stenseth et al. 1999 is only a partial list). The best overview of these data and analyses is still Royama (1992: chapter 5), whose conclusions I largely follow in this section.

There are at least three difficulties with interpreting lynx data, having to do with the properties of trappers and traders as a "measuring apparatus" of lynx dynamics. First, fur returns have a nonlinear relationship to lynx density, since the amplitude of oscillations in fur returns probably exaggerates the cyclic changes in lynx population (Royama 1992:171–175). Because company traders paid trappers a fixed price per pelt, trappers must have been discouraged to catch animals when scarce. This property of the "measuring apparatus" would tend to stretch the lows. To this observation, I would like to add the following one. It is possible that at high lynx densities the ability of the trappers to catch and process animals was saturated (both the number of trappers that could be recruited to deploy traps, and the number of traps that they had must have been limited). If this supposition is correct, then during the cyclic peaks the fur returns data must have compressed the actual degree of change. In other words, I suggest that the trappers behaved as generalist predators characterized by a sigmoid total response. The consequence of this behavior was not only an inflation of amplitude but also some distortion of the cycle topology when reflected in fur returns data.

The second imperfection of trapping as the measuring apparatus of density is the possibility that lynx are more readily trapped when the

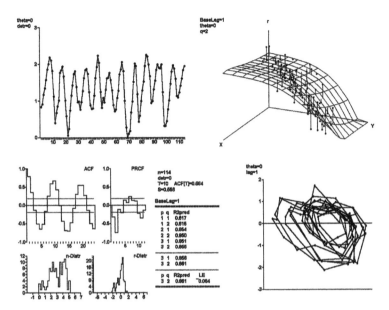

FIGURE 13.1. Graphical output of the NLTSM analysis of lynx data: the Mackenzie River District. *Upper left:* time plot; *upper right:* response surface; *lower right:* phase plot.

population of their chief prey, the snowshoe hare, declines (Ward and Krebs 1985). The third source of measurement error is establishment of new collecting stations, changes in the boundaries of some districts, missed years, and so on. The Mackenzie River District underwent the minimum degree of changes among the ten districts, and, probably as a result, shows the most stable oscillation pattern (Royama 1992:175).

The NLTSM analysis of the Mackenzie River data is shown in figure 13.1. As we see, the analysis suggests that the dynamical process driving lynx oscillations is characterized by order 2–3 (although the selected order is 3, OCV values associated with $d = 2$ are almost as good: see the analysis panel in figure 13.1). This result agrees with most other analyses (e.g., Royama 1992; Stenseth et al. 1997). The only exception is Tong (1977), who used an eleventh-order autoregressive process to model the lynx data. From the ecological point of view, however, it is very doubtful that the order 11 reflects any real ecological processes. What seems more likely is that this estimate is an artifact of using a linear autoregressive (AR) approach: what appears to be happening is that a linear model is incapable of adequately capturing the nonlinear nature of the oscillation (remember that a linear AR process cannot produce a stable limit cycle; it can only generate oscillatory approach to a stable point). As a result, a very high-order AR term is needed to bring the output of the statistical model closer in line with the pattern shown by data. In the words of Royama (1992:184), the linear eleventh-order AR model is "like a polynomial curve fitting and provides no insight into the ecological structure of the lynx population process."

Pelt harvest statistics for snowshoe hare are not as extensive as those for lynx. The longest series is from the Hudson's Bay Company, tabulated by MacLulich (1957). The hare data are characterized by a lower signal/noise ratio than the lynx series, but the periodic nature of the series is very apparent.

Having data for both lynx and hare might tempt us to probe the topology of cyclic peaks, in order to determine whether our ideas about which species is prey and which predator coincide with the analysis results. Unfortunately, this is a case where peak topology analysis cannot be used, because it critically depends on the assumption that the analyzed population index is linearly related to the underlying population density. As we know from the discussion of the nature of fur return data, this is not the case. Thus, these data should provide us with good indications of periodicity, signal/noise ratios, and order, but are not useful for determining the finer probes (or even the amplitude).

There is another use to which we can put these data, however: to obtain an estimate of the phase lag—the time by which predator density lags behind prey density. One potential problem, of course, is the higher trappability of lynx during the periods of their prey decline. This feature of the data should result in recorded lynx declines occurring somewhat later than the actual ones. However, in my judgment this effect should be numerically not very strong, resulting in a postponement of lynx decline by perhaps a year. This caveat should be kept in mind. But there is an additional, historical reason for the analysis of phase lag between prey and predator densities. Gilpin (1973), using the data tabulated by Leigh (1968), noticed that the peaks of lynx numbers tended to occur slightly ahead of hare peaks, prompting the title of his paper "Do Hares Eat Lynx?" This observation has become part of ecological folklore (and even the subject of serious attempts to explain it), and not all ecologists know that it is based on faulty data.

As Royama (1992:233) noted, there are two sources of errors in the Leigh-Gilpin analysis. First, there was a confusion about assigning years to data points, having to do with when the furs were collected, and when they were marketed (one year later). This resulted in Leigh's shifting lynx data one year back in time. Second, Leigh-Gilpin data on lynx fluctuations were pooled from different districts across Canada, while the hare data came from the shores of Hudson Bay. Royama suggested that, in order to check on the presence of phase lag, we need to compare the hare series with lynx data collected in the two districts adjacent to Hudson Bay (North Central and James Bay). Thus, the previous claims that hare and lynx dynamics do not fit the pattern predicted by predator-prey theory are unwarranted. Furthermore, the Hudson's Bay Company is not the only source of data where we can check whether the phase lag between prey and predators conforms to theoretical predictions. Thus, Bulmer (1974) located two data sets from northern Russia. Both data sets clearly show that lynx peaks lag behind hare by 2–3 years.

13.3 MODELS

Given the amount of effort lavished on empirical studies and statistical analysis of time-series data from the hare-lynx system, it is surprising that so little attention has been devoted to developing and parameterizing mechanistic models. This attitude is, most likely, a result of a strong antitheoretical bent of some of the most influential empirical ecologists working in the field (e.g., see Krebs 1995). In any case, all the mechanism-based models of the snowshoe hare–lynx cycle known to me date from the 1990s (Akcakaya 1992; Ives and Murray 1997; King and Schaffer 2001; a pioneering attempt by Leigh 1968 must be judged as failure, partly because he did not go beyond the overly simplistic Lotka-Volterra model, and partly because his data set was flawed; see above). In this section, I primarily focus on the King and Schaffer paper, as the most developed and credible approach, but first a few words about the earlier modeling efforts.

The approach of Akcakaya (1992), in my opinion, suffers from his adoption of the ratio-dependent framework for modeling predator-prey interactions. It may be expedient to use the ratio-dependent functional response in cases where it provides a better approximation of data than the hyperbolic response (see section 4.1.1). However, Akcakaya (1992) did not demonstrate this, and thus I see no reason to prefer the less mechanistic version to the theoretically sound and empirically highly supported hyperbolic response. Furthermore, a predator-prey model based on the ratio-dependent response, as Akcakaya shows, cannot exhibit stable limit cycles: when the non-trivial equilibrium is unstable, all trajectories approach the origin (i.e., both prey and predator go extinct). Since, obviously, the hare-lynx system exhibits bounded oscillations, this feature of the model presents a conceptual difficulty. Akcakaya solves it by positing the existence of an absolute refuge. When such a refuge is added to the model, it becomes capable of generating stable limit cycles. Apart from the biological implausibility of an absolute refuge for the hare-lynx system, its introduction into the model requires Akcakaya to estimate yet another parameter without any independent empirical basis. To summarize, the decision to use the ratio-dependent functional response imposes at least two costs: lack of firm mechanistic basis and the need to unnecessarily complicate the model (to remind the reader, simple models based on the hyperbolic functional response, such as the Rosenzweig-MacArthur, naturally generate either stable points or stable limit cycles, depending on parameter values, without any need to introduce additional features, such as refuges). Because of these criticisms, I think that the models discussed next are better approaches to the theoretical investigation of the hare-lynx cycle. However, I do not want my critique of Akcakaya (1992) to appear too harsh, because, after all, it was the first comprehensive attempt at a theoretical/empirical synthesis of the hare-lynx oscillations, which showed that a simple predator-prey model with biologically reasonable parameters is capable of producing limit cycles of appropriate period and amplitude.

The goal of the series of models developed by Ives and Murray (1997) was to investigate the effect of sublethal parasitism on predator-prey dynamics. They showed that even a relatively small increase in the vulnerability of nematode-infested hare to predation may substantially destabilize the interaction between hare and their predators (lynx, coyotes, hawks, and owls). Thus, a hare-predator-parasite model may exhibit stable limit cycles, even though the reduced hare-predator system is characterized by a stable point equilibrium. One problematic feature of this very interesting paper is the use by the authors of the discrete modeling framework (mammalian interactive systems operate in continuous time, unlike forest insect systems that are naturally modeled with difference equations, and the time step of one year may be too large in relation to the length of the cycle if we wish to employ discrete equations as approximations of the underlying continuous process). However, it is likely that the main message of Ives and Murray would stay unchanged were they to use the continuous time approach, because it is a general feature of dynamical systems that increasing their dimensionality usually decreases their stability (May 1974b). This is all I will say about the paper by Ives and Murray (1997), because the main question that I wish to pursue here is what is the identity of the primary mechanism driving the snowshoe hare population cycles. By contrast, Ives and Murray postulate an answer to this question (predation), and address a secondary issue, how another ecological mechanism (parasitism) may affect the workings of the primary one.

The King and Schaffer (2001) Model   As was discussed in the beginning of section 13, the current consensus in the field appears to be that the snowshoe hare cycle is driven by a trophic mechanism, or a combination of trophic mechanisms. Thus, three hypotheses can be explicitly formulated: (1) hare-vegetation interaction (Lack 1954), (2) hare-predator interaction (Trostel et al. 1987), and (3) the three trophic levels hypothesis (Keith 1990). To theoretically investigate the plausibility of these three hypotheses, King and Schaffer (2001) proposed a model of vegetation-hare-predators interaction, estimated its parameters, and compared the model's predictions to time-series data and the experimental data obtained by Krebs et al. (1995). Although the King-Schaffer model is a translation of the tritrophic hypothesis into mathematical language, they could also have used two simplified versions of the model to assess the dynamical effects of food and predation acting alone (more on this below).

The model of King and Schaffer falls squarely within the "standard theory" of trophic interactions, as outlined in chapter 4. Somewhat simplifying, we can write its equations as follows:

$$\frac{dV}{dt} = u_0 \left(1 - \frac{V}{m}\right) - \frac{aVN}{b + V}$$

$$\frac{dN}{dt} = \xi N \left(\frac{aV}{b + V} - \eta\right) - \frac{cNP}{d + N} \qquad (13.1)$$

$$\frac{dP}{dt} = \chi P \left(\frac{cN}{d + N} - \mu\right)$$

This is the minimal model of King and Schaffer (2001: equation 9), which I further simplified by suppressing dependence of some parameters on season, in order to see the model structure more clearly. The state variables of the model are V, the biomass of hare browse; N, hare population density; and P, density of hare predators. Note that both V and P do not refer to single-species populations, but to "dynamical complexes" of browse species (aspen, willow, etc.) and predators (lynx, coyotes, hawks, and owls), respectively. This is clearly an oversimplification, but perhaps not a fatal one.

In the vegetation equation, the first term reflects vegetation dynamics in the absence of herbivores, modeled with the regrowth equation (section 4.4.2). The second term is the standard hyperbolic functional response of herbivores to fluctuating forage biomass. In the herbivore equation, the first term represents herbivores' numerical response, written in the form of equation (4.15) (see section 4.2.1). The second term represents predators' functional response, again using the hyperbolic form. Finally, the predator equation is the same as in the Rosenzweig-MacArthur model (section 4.2.1). We see that this is a rather generic model, and, in fact, fits well within the general framework for modeling vegetation-rodent-predator systems advanced in section 12.6.

The actual model that King and Schaffer analyze is a more complex version of equation (13.2), in which several parameters vary seasonally. This modification both increases the realism of the model (since seasonal effects are quite strong in boreal ecosystems), and leads the model to exhibit an array of complex dynamical behaviors. The authors use bifurcation methods to map out the fascinating mathematical structures known as "resonance horns" or "Arnold's tongues."

For the results of this analysis, I refer the interested reader to the papers by King and Schaffer (1999, 2001). For the purposes of this book, we are primarily interested in determining how King and Schaffer's model fits within the general picture of mammalian cycles (and, more generally, complex population dynamics). Because seasonality has a rather mild effect on the main probes of interest, amplitude and period, I will go ahead and use the simplified nonseasonal version of the King-Schaffer model (but we should keep in mind the possibility of more complex behaviors, such as multiple coexisting attractors, in the model with seasonality).

Using a variety of available ecological information (but not time-series data), King and Schaffer estimated biologically plausible ranges for all the parameter values of their model. They found that for these parameters the model generically predicts oscillations of between 8 and 12 years in period, and somewhat greater variation in the amplitude (with most peak/trough ratios lying between 10 and 200). This result is in basic agreement with the empirically observed period of 10 years, and peak/trough ratios of 13–141 (see table 2 in King and Schaffer 2001). A "typical" trajectory is shown in figure 13.2a. To generate this output, I solved equations (13.2) for 100 years to allow the transients to die out, and then for the further 50 years, which are plotted in the figure. The parameter values are those which King and Schaffer used to compare their model output with the experimental data of Krebs et al. (1995; this comparison will be discussed below): the regrowth rate $u_0 = 84$ Mg km^{-2} yr^{-1}, carrying capacity of vegetation $m = 120$ Mg km^{-2}, herbivore saturation rate $a = 0.66$ Mg hare^{-1} yr^{-1}, herbivore half-saturation constant $b = 40$ Mg km^{-2}, herbivore conversion constant $\xi = 5.55$ hare Mg^{-1}, herbivore ZPG consumption rate $\eta = 0.324$ Mg hare^{-1} yr^{-1}, predator saturation rate $c = 600$ hare pred^{-1} yr^{-1}, predator half-saturation constant $d = 50$ hare km^{-2}, predator conversion constant $\chi = 0.0044$ pred hare^{-1}, and predator ZPG consumption rate $\mu = 374$ hare pred^{-1} yr^{-1}. (All these parameters are explained in chapter 4). As we see in figure 13.2a, for these values of parameters the model predicts oscillations of about one order of magnitude (trough and peak densities are, respectively, 14 and 180 hare km^{-2}) and a period of 10 years.

Assuming that the King and Schaffer model with these parameter values is a reasonable description of the actual population dynamics of

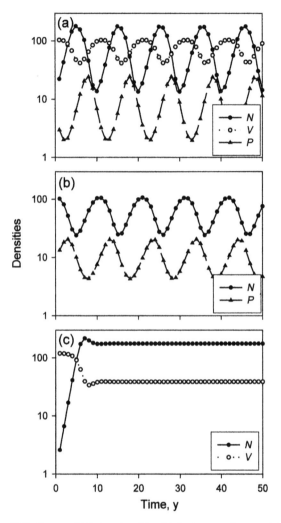

FIGURE 13.2. Output of (a) the tritrophic, (b) the hare-predator, and (c) the vegetation-hare models.

snowshoe hare (we shall return to this question below when we compare the model output to the results of the manipulative field experiment), we can use it to really dissect the dynamical roles of food
and predators in generating the hare population oscillations. The main
question that motivates me is to determine whether the oscillation is

a truly tritrophic phenomenon, or whether it is primarily due to one or another trophic link (food versus predators). We can answer this question by the following thought experiment: let us "turn off" food or predators in the model, and observe whether the basic oscillatory dynamics remain after either manipulation. Turning predators off is easiest. We can simply set $P = 0$ in model (13.2), which leads us to model (4.40), discussed in section 4.4.3. This model is globally stable for all values of its parameters. Numerically solving it for the parameter values given in the previous paragraph, we observe that, after a very mild overshoot, both hare and vegetation trajectories settle down to a very stable equilibrium (figure 13.2c). One might criticize this result by pointing out that the appropriate method for turning off the dynamical effects of predators is not to set them to zero, but to stop the variation in the predation intensity with hare density. One way of addressing this criticism would be to increase the parameter η appropriately. Numerical investigation shows that increasing η does not cause the model to cycle. When η is set too high, the hare population simply goes extinct. To summarize the insights from the vegetation-hare submodel, it appears that predation is necessary for cycles. However, it is still possible that cycles are a true tritrophic phenomenon, and to check on this we need to remove the dynamical effects of food from the tritrophic model.

Unlike predation, we cannot simply set food to zero, because in that case the hare population would simply go extinct due to starvation. The opposite approach, fixing food biomass at some constant level (similarly to setting predation mortality at a constant, as we did in the previous paragraph), does not work either. Because food limitation is what keeps hare numbers in check in the King-Schaffer model, removing the ability of hares to depress food level leads to hare population simply growing forever. Thus, what we need to do is to model the effect of food as a strictly first-order factor that acts without any lags. This is accomplished by substituting a logistic term into the hare equation:

$$\frac{dN}{dt} = r_0 N \left(1 - \frac{N}{k}\right) - \frac{cPN}{d + N}$$

$$\frac{dP}{dt} = \chi P \left(\frac{cN}{d + N} - \mu\right)$$

We immediately recognize these equations as the Rosenzweig-MacArthur model. The logistic parameters can be obtained from the hare set as follows. The intrinsic rate of population growth $r_0 = \xi(am/(b+m) - \eta)$, which is the hare growth rate obtained when vegetation is at the maximum, $V = m$, and predators are at the minimum, $P = 0$. To obtain the carrying capacity, we simply solve for the equilibrium density that hares achieve in a pure vegetation-herbivore model. Solving the resulting model, we obtain the dynamics depicted in figure 13.2b. The remarkable result is that the basic pattern of oscillations is little changed by shifting from the King-Schaffer model to the Rosenzweig-MacArthur one. The period remains at the same value, 10 years, although the amplitude is somewhat decreased.

The results of this thought experiment, therefore, suggest very strongly that the primary factor responsible for cycles *in the model* is predation. Food acting on its own is unable to cause sustained oscillations, and when acting together with predation, its only role is to increase the amplitude of the cycles. Inasmuch as the model is a faithful description of the real situation, therefore, we can transfer this insight on the functioning of the snowshoe hare–based ecosystem.

13.4 EXPERIMENTS

There are two intensively studied field sites in boreal North America: one near Rochester, Alberta, and the other in the Kluane Lake area of the Yukon Territory (there was also a shorter study in Minnesota during the 1930s; see Green and Evans 1940). The research articles stemming from these two long-term studies are too numerous to list here, but the Rochester study was reviewed by Keith (Keith and Windberg 1978; Keith 1990), and the Kluane study is described in Krebs et al. (2001). Both studies yielded a wealth of data, including time-series data on actual population densities (as opposed to fur return–based indices) of snowshoe hare and its predators, estimates of predation mortality, food depletion, parasite loads, movement rates of radio-collared hares, and much more. These data have been extremely valuable both for model parameterization (section 13.3) and for testing various theories of why hare populations cycle.

The Rochester Study   In their studies of the Rochester hare population, Keith and coworkers employed a mensurative approach (in this review of the Rochester results I follow Keith 1990). They chose several variables thought to be important in driving the hare cycle and then carefully measured them throughout the complete oscillation, with a particular focus on the peak and decline phases. The initial emphasis of the research group was on food, because it was hypothesized as the primary cycle driver, with predation thought to play a secondary role of deepening and extending low-density phases (Keith 1974; Keith and Windberg 1978). Accordingly, Pease et al. (1979) measured the food available to hares (woody browse) during six winters (1970–1975) of population decline following the 1970 peak. Their results indicated that hares experienced food shortages during 1970 and 1971, that is, during the peak and the first post-peak years. After that, the amount of browse was more than sufficient to support the existing hare density, yet the population continued to decline for several more years (the first year of increase was in 1977).

Functional and numerical responses of the main hare predators were monitored at Rochester during the complete 1965–1975 cycle. From this information and a knowledge of hare densities, Keith and coworkers calculated rates of overwinter predation on hares (Keith et al. 1977). Predators killed an estimated 10–15% of the hare population during the increase and the peak years, 1966–1970 (hereafter "winter of 1970" is short for winter of 1970–1971). After that, predation rates increased sharply, reaching 43% in 1972, with predators accounting for about 70% of total deaths. By the fifth winter after peak (last year of decline), predators were still responsible for 28% of mortality (twice the increase phase level). Thus, unlike the impact of food shortage, predation acts in a markedly delayed density-dependent manner.

Furthermore, subsequent research showed that the indirect calculations of predation impact on hare populations, which were based on estimates of predator numbers and their kill rates, underestimated the actual hare mortality due to predators during the decline years (possibly because the researchers did not properly take into account surplus killing). Radiotelemetry studies of hare mortality at Rochester during the subsequent 1975–1985 cycle (peak in 1980) suggested that predators were the proximate cause of most deaths much earlier in declines than first envisioned, and the predation impact during the decline years was much more severe. For example, 80–90% of hare deaths during the second winter (1981) of decline were due to predation (Keith et al. 1984). During the same year, food was apparently not short in three out of eight study areas, yet hare density continued to collapse everywhere (Keith et al. 1984).

Taking into account that calculations of Keith et al. (1977) miss some predation events (e.g., those due to surplus killing), we have a rough estimate that overwinter mortality of hares due to predators should peak around 50%. Is this mortality severe enough to drive the cycle in hare density? Clearly, we have here a benchmark that we can use to test the King and Schaffer model. The model predicts that instantaneous predation mortality should peak at $\delta_{pred} = 1.6 \text{ yr}^{-1}$ (King and Schaffer 2001: figure 10a). To translate this number into the overwinter mortality rate, we note that Keith et al. assumed winter length at 151 days. Thus, overwinter survival is equal to $\exp[-\delta_{pred}t] = \exp[-1.6 \text{ yr}^{-1}151/365 \text{ yr}] = 0.52$, and proportion killed is 1 minus that, or 0.48. As we see, this theoretical prediction is in good agreement with the empirical morality rate.

Summarizing this admittedly sketchy review of the Rochester study (for a more detailed description, I refer the reader to Keith 1990), we see that these findings are in substantial agreement with the predictions of the King and Schaffer model (section 13.3). In particular, the availability of food acts as an essentially first-order variable, whose effect on hare dynamics quickly dissipates after the first year of decline. As the model suggests, food appears to prevent indefinite growth of the hare population, and gives it the first nudge toward decline. It is predation, however, that causes a prolonged (4-yr) decline phase, and is therefore the primary factor causing the oscillation. Furthermore, the quantitative predictions of the King and Schaffer model about the maximum magnitude of predator-caused mortality during the collapse phase are also in excellent agreement with the Keith et al. (1984) data.

The Kluane Study   In their study of the Kluane Lake hare population, the University of British Columbia (UBC) group has consistently employed a manipulative experimental approach (see, e.g., Krebs 1996). Two of their long-term experiments deserve particular mention. In the first, researchers provided rabbit chow as winter food to three experimental areas, and compared their hare densities to those in six control areas where no food supplementation took place. The experiment took place during 1977–1984, covering the increase, peak (in 1980–1981), and decline phases of the cycle. The major findings from this experiment were as follows. Generally, hare populations in areas with extra winter food reached peak densities three times those of the controls (although in one unusual replicate, for obscure reasons, the density was little affected by food supplementation). However, in all food addition areas hare density decreased essentially synchronously with the control areas (the beginning of decline was delayed by 6 months in one area, but not in others). Based on these results, the authors concluded that food shortage is not a necessary cause of the cycle.

In the second experiment, which took place during the next hare cycle (1986–1994), Krebs and coworkers (1995) extended the spectrum of experimental treatments. Each experimental unit consisted of a huge 1 km × 1 km block of undisturbed boreal forest. Three areas were used as unmanipulated controls. Two other areas were provided with supplemental food. In one further area, mammalian predators were excluded with an electrified fence (the investigators also covered 10 ha in the predator-exclusion treatment with monofilament, but this approach was ineffective in reducing avian predation). Finally, in one area the food addition and predator exclusion treatments were combined. (In addition, two areas were used for a fertilizer addition treatment, but because adding nutrients had virtually no effect on hare densities, I mention it only briefly on p. 361.)

Summarizing their results and conclusions from this ambitious experiment, Krebs et al. (1995) wrote: "Predator exclosure doubled, and food addition tripled hare density during the cyclic peak and decline. Predator exclosure combined with food addition increased density 11-fold.... Food and predation together had a more than additive effect, which suggests that a three-trophic-level interaction generates hare cycles." This conclusion seems to imply (and, certainly, was so interpreted by most ecologists) that food and predation play qualitatively similar roles. It reflects an essentially linear thinking, epitomized by the analysis of variance: predation explains two units of difference, food another three units, and interaction between the two factors the remaining six units, for a total of eleven. This interpretation of the experimental results is radically different from the message of the King and Schaffer model. As King and Schaffer repeatedly emphasize, the fundamental cause of multiannual cycles is that winter browse sets the peak density, while predation causes the crash. In the terminology of this book, food is a first-order mechanism, while predation is a second-order mechanism.

Do the Krebs et al. results contradict the insights of the King and Schaffer model? To answer this question, King and Schaffer employed their model to simulate the results of experimental manipulations. They modeled supplemental feeding by adding a constant term V_{suppl} to the right-hand side of the V equation, representing the rate at which food was added by experimenters. Partial predator exclosure was modeled by multiplying the predator density in the N equation by a factor Q ($Q = 0$ corresponds to all predators excluded, $Q = 1$ means no predators excluded; King and Schaffer assumed $Q = 0.2$). The results of the simulation are shown in figure 13.3). Even though there are certain differences in detail, we see that the broad patterns in the experimental data are well captured by the model. Thus, the model predicts a much greater peak density in food addition treatments, and a delayed crash in predator-exclosure treatments. Note, in particular, how hare density in the combined treatment collapses to a low value, even though only 20% of predators have been assumed to be able to get at hares in exclosures (in reality, this is an underestimate, because avian predators constitute about 40% of the predation community).

Further support for the model comes from comparing predation mortality predicted by the model with that observed during the Krebs et al. experiment. Krebs et al. report that annual survival during the decline phase in the controls was 0.7%, while in the predator exclosure it was 9.5%. If we assume that mortality in the exclosure was the base rate δ_0, while in the controls mortality rate was a sum of the base rate and predation rate, $\delta_0 + \delta_{pred}$, then a quick calculation shows that

$$\delta_{pred} = \ln S_{excl} - \ln S_{contr} = \ln 0.095 - \ln 0.007 = 2.6$$

In other words, the measured death rate due to predators is even greater than that assumed by the model! This difference is probably

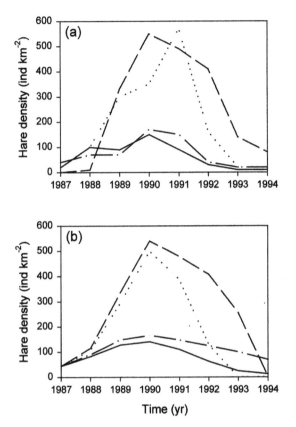

FIGURE 13.3. Comparison of (a) the experimental results of Krebs et al. with (b) simulations using the King and Schaffer model, with modifications allowing for food addition and partial predator exclusion. Treatments: controls (solid line), supplemental feeding (dotted line), partial predator exclosure (dash-dot), and combined treatment (dashed line). (After King and Schaffer 2001: figure 9.)

not statistically significant, as the next calculation suggests. In the fertilizer addition treatments, the average survival rate was 1.9%, which was not statistically different from the 0.7% rate in the controls. Substituting this survival rate in place of 0.9% in the above equation, we obtain an estimate of the instantaneous predation rate $\delta_{pred} = 1.6 \text{ yr}^{-1}$, which is the same as predicted by the model.

13.5   SYNTHESIS

Concluding my quick review of the Kluane Lake research program, I would like to express my personal opinion of the approach taken by the UBC group, with a specific focus on the 1986–1994 experiment. On the positive side, I am awed by the sheer magnitude and daring of this experiment. Its use of huge 1 km^2 blocks of forest, its duration (8 years), the excellent group of researchers working together toward the same goal, and their struggle to implement experimental treatments, particularly the predator exclosure, evoke admiration. Some ecologists have criticized this experiment for employing only single replicates for the key predator exclosure and combined treatments. However, such criticisms are unfair. If we want to understand what causes hare cycles, we have to use large experimental units, and our ability to replicate them has to be limited. I personally think that using hare-permeable fences was a mistake, because it allowed hares to immigrate to areas of high food, and emigrate from areas protected from predators. Thus, hare movements confounded the key processes that were studied (reproduction and mortality). In fact, one of the key findings of the experiment, that food addition substantially elevated hare densities, was due more to hare immigration than to their elevated reproduction or survival (Boutin 1984). Still, having conducted my own long-term field manipulations, I know very well that no design is perfect—one always has to balance conflicting demands. In the case of the Kluane Lake experiment, the investigators clearly wished to avoid the "Krebs" effect, in which enclosed populations may exhibit unnaturally high population densities. In any case, trying to second-guess decisions made by a group of excellent field biologists with decades of experience with the system is futile. Furthermore, the study yielded an enormous treasure trove of information about the functioning of the Kluane Lake system (Krebs et al. 2001), and it is unlikely that, given limited resources, more could have been obtained.

On the negative side, the data from the Kluane Lake study, in my opinion, were poorly analyzed. One central idea of this book is that we simply cannot test hypotheses about complex nonlinear dynamics without explicit mathematical models. Yet, neither the construction of predictions to be tested, nor the analysis of results of the Kluane Lake experiment, were theoretically informed. For a specific example, building the case for the tritrophic nature of the hare cycle on the degree by which various experimental treatments increase population density during the peak and decline is based on faulty reasoning, as I tried to show above. These are strong words, but in this case, I believe, my criticism is fair. Charles Krebs, the leader of the UBC group, is on record with antitheory statements such as "avoid mathematical models" (Krebs 1995). In his opinion, mathematical models should be employed only after everything about the system has been learned using the experimental approach. In my opinion, our general approach should be exactly the reverse: first model the system (preferably with several alternative models), then test model predictions with experiments. Thus, it is important to show how the methodological stance of the UBC group has impaired their long-term research efficiency.

It is clear that the initial intent of the Kluane Lake experiment was to determine which of the treatments would "stop the cycle." The experiment failed to do that, yet the investigators still concluded that the combination of two factors manipulated in the treatments is the cause of the cycle. This logic is, of course, vulnerable to criticism, as was gleefully pointed out by Dennis Chitty (1996). Unlike Chitty, I am persuaded by arguments of Krebs et al. (1995) that hare density declined in the predator exclosures because not all predators were excluded, and because hares emigrated through the fences. The main observation that persuades me that Krebs et al. are correct in assigning predation the main role in driving the cycle is the sheer magnitude of predation mortality during the decline phase of the hare cycle, when practically all deaths were due to predators. Furthermore, as I discussed above, the magnitude of predation mortality is consistent with that assumed by the model. In other words, the observed predation rate is numerically "strong enough" to drive the cycle. Of course, this observation was already made by the group studying the Rochester population. Thus, in my opinion, the key contribution of the Kluane Lake experiment was not the effect of treatments on hare density, but the demonstration that partially excluding predators results in a huge decrease in mortality rate. This result addresses a potential criticism of the Rochester study that if predators would not eat the hares, they would die of some other cause anyway. For example, it is conceivable that the hare cycle is driven solely by factor X (e.g., stress), and predators can eat only the hares about to die as a result of high X. In such a situation, we would observe high predation mortality during the decline, and yet would be wrong to conclude that predation is the cause of the cycle. The Kluane Lake experiment excluded this possibility.

CHAPTER 14

Ungulates

14.1 INTRODUCTION

Mechanisms underlying population dynamics of North American cervids (such as white-tailed deer, reindeer, elk, and moose) are a subject of some controversy. The current debate centers on the importance of predation versus interactions with food, and the dynamical role of exogenous factors (as far as is known, social population controls are lacking in deer). For example, Mech et al. (1987) argued that the main determinant of moose dynamics is their interaction with food as modified by weather (specifically, the cumulative effect of snowy winters), and that predation by wolves is secondary to winter weather in influencing moose populations. By contrast, Messier and Crête (1985) and Messier (1991) advocated predation as the most important factor shaping moose population dynamics (for a further exchange, see McRoberts et al. 1995; Messier 1995; Post and Stenseth 1998). Subsequently, Boutin (1992) concluded that evidence for predation as an important factor in moose population change was not convincing, but Van Ballenberghe and Ballard (1994) and Messier (1994) presented more data and models suggesting that predation by wolves, especially when supplemented by predation by bears, can substantially reduce moose population density. Another controversy focused on factors determining moose dynamics in predator-free environments (Saether et al. 1996; Saether 1997; Crête 1998; Saether et al. 1998).

Such debates are a healthy sign, especially because many data are available, and each new cycle of controversy tends to engage an increasingly wider spectrum of empirical information. However, I believe that the discussion could profit from greater theoretical rigor. Accordingly, my main goal in this chapter is to place the ideas and data employed by ungulate ecologists within the conceptual framework developed in the book. Note that this objective is much less ambitious than in other case studies, where I generally attempt to arrive at some conclusion about ecological mechanisms that are responsible for the observed fluctuation patterns. The organization of the chapter reflects this shift in objectives. Instead of following my usual procedure of starting with an overview of numerical patterns, I jump right away into model building. My focus is on trophic interactions: food and predators. Model development is then followed by comparisons between models dynamics and whatever long-term data are available.

A Terminological Note The current debate on deer population dynamics appears to center on the question of what are the limiting and what are the regulating factors in deer population dynamics. Probably the clearest definition of these concepts is given by Messier (1991). *Limiting* factors are those responsible for year-to-year changes in the rate of population growth, while *regulating* factors are those that act in a density-dependent manner, and serve to keep population density within a certain range. Thus, regulating factors are a subset of limiting factors characterized by negative-feedback mechanisms.

I prefer to use an alternative classification (section 5.1). Like Messier and other ungulate ecologists, I focus on factors that explain variation in the population rate of change, quantified as the realized per capita rate of change $r(t)$. **Null factors** are those that do not explain any variation in $r(t)$. **Exogenous factors** are those that affect $r(t)$, and therefore population density, but are not themselves affected by density. They correspond to nonregulating limiting factors in the classification of Messier. **Endogenous factors** are those that reflect feedbacks from population density. Endogenous factors correspond to the regulating factors in Messier's classification. I further subdivide endogenous factors into first- and second-order ones. **First-order** endogenous mechanisms directly translate the changes of population density into an effect on $r(t)$, acting without an appreciable lag.

Second-order endogenous mechanisms are those in which population density affects $r(t)$ indirectly, by first acting on some other state variable, $X(t)$, and then by $X(t)$ affecting $r(t)$. Second-order factors, thus, act with a time lag. Recollect also that in practice the distinction between first- and second-order factors is not absolute, so we usually treat factors acting with a lag less than generation time as first-order, and those acting with a lag of greater than generation time as second-order (section 2.5).

Although the definitions used by Messier are internally self-consistent, my preference for the exogenous/endogenous classification is due to the following reasons. First, *limiting* has all the wrong mnemonics, because it evokes a mechanism that would keep population density "within limits," which is of course the opposite of the sense in which it is used. I do not think it is a good idea to use terminology in which the technical sense is the opposite of the "common" sense. By contrast, endogenous and exogenous factors evoke mechanisms that are either part of the dynamical feedbacks or not. Second, limiting and regulating categories are nested within each other, leading to awkward usage; it is much better to use nonoverlapping concepts. Additionally, "regulating" factors do not have an associated classification comparable to first- and second-order endogenous factors. Finally, different people use different definitions of what are limiting and what are regulating factors, even within the ungulate ecological community. For example, Boutin (1992:117) defines a regulatory factor as the one whose magnitude is a *direct* function of population density. This definition, of course, leaves second-order factors beyond the pale. In fact, Boutin (1992: figures 1 and 2) finds no or an inverse relationship between predation rate on moose and moose population density, which he uses to argue against an important dynamical role of predation. His argument, however, ignores the possibility (indeed, probability) that predation may cause delayed density dependence in moose dynamics. Definitions of "population limitation" are even more diverse. Thus, Kunkel and Pletscher (1999) defined a limiting factor as "one that far outweighs others in impeding the rate of increase." Sinclair (1991), by contrast, called the limiting factors those that set the position of the equilibrium. In other words, there are at least three wildly different definitions of the term "limiting factors" commonly used within the same scientific community!

14.2   INTERACTION WITH FOOD

The first decision to make when building a herbivore-vegetation model is whether to base it on the logistic or the regrowth functions (section 4.4.2). Models of deer-vegetation interaction typically use the logistic function (Schmitz and Sinclair 1997; Boyce and Anderson 1999). However, this assumption does not make sense, because typically only a minor proportion of vegetation is accessible to mammalian browsers (I focus on winter food supply as the most important population bottleneck). Browsers such as deer consume only twigs less than a certain diameter (e.g., Vivas and Saether 1987). Additionally, most twigs are inaccessible because they are too high to reach. Thus, the dynamics of food supply for deer should be better approximated with a regrowth equation. The only possible exception to this rule would be situations where deer cause high plant mortality (even in such cases, the pure logistic model would not be appropriate, and we would want to use a more complex equation combing elements of both logistic and regrowth).

Thus, we can write the following simple model for the deer-vegetation interaction (see model 4.40 in section 4.4.3):

$$
\begin{aligned}
\frac{dV}{dt} &= u_0\left(1 - \frac{V}{m}\right) - \frac{aVN}{b + V} \\
\frac{dN}{dt} &= \xi N\left(\frac{aV}{b + V} - \eta\right)
\end{aligned}
\tag{14.1}
$$

As usual, V denotes the vegetation biomass density, N is the population density of grazers, u_0 is the regrowth rate of vegetation at $V = 0$, m is the maximum standing crop, a and b are the parameters of the herbivore's functional response, ξ is the conversion efficiency, and η is the herbivore ZPG consumption. Note that I assumed a hyperbolic functional response for herbivores as the simplest yet realistic form (section 4.1.1). Schmitz and Sinclair (1997: figure 13.4) present some data suggesting that white-tailed deer may be characterized by a sigmoid functional response. However, their conclusion was based on observations of deer foraging in plots of variable forage density that were available to deer *simultaneously*, and thus reflect **aggregative response**, rather than a functional response (see section 4.1.2 for this key point).

We can attempt a crude parameterization of model (14.1) for the moose system, because published information gives us at least an order-of-magnitude indication of parameter values (I stress that none of these estimates should be taken too seriously; my main purpose is to generate a reference set that provides a starting point for a numerical investigation of model dynamics). Thus, Vivas and Saether (1987) studied moose foraging behavior in relation to winter browse availability (I interpret V as the biomass of winter food, because summer food appears to be always present in abundance). They state that the highest density of food biomass they used in experiments (72.5–80.5 g m^{-2}) approximated the densest naturally occurring areas in their experimental area. Thus, I set m equal to 100 g m^{-2}, which translates into $m = 100$ Mg km^{-2} (I standardized all parameters to the following units: forage in Mg of dry weight, area in km^2, time in yr, and moose density in ind km^{-2}). Fitting the hyperbolic functional response curve to the data on moose food intake as a function of forage availability (Vivas and Saether 1987: figure 3a), I obtain an estimate of b varying between 20 and 60 g m^{-2} (depending on a different method; the estimate is not very precise because there are only five data points). Thus, I set $b = 40$ Mg km^{-2}.

For estimates of parameters a and η I turn to Crête and Bédard (1975). These authors estimated daily consumption by moose as 2.5 kg ind^{-1} day^{-1}, and stated that it is comparable to the theoretical estimate, based on metabolic rate, of Gasaway and Coady (1974). Since the interpretation of parameter η is the food intake rate at which moose population just breaks even (deaths equal births), I set $\eta = 2.5$ kg ind^{-1} day$^{-1} = 1$ Mg ind^{-1} yr^{-1}. Maximum intake rates observed by various authors are about twice the estimate of η (Crête and Bédard 1975:373), so I set $a = 2$ Mg ind^{-1} yr^{-1}.

Parameter ξ is related to the intrinsic rate of moose population growth, r_0, and other parameters as follows:

$$\xi = \frac{r_0}{am/(b+m) - \eta}$$

Using $r_0 = 0.2$ yr^{-1} (Van Ballenberghe 1983; Fryxell 1988; Eberhardt 1998:382), I calculate $\xi = 0.467$. Finally, we need an estimate of u_0, or annual regrowth rate of moose forage. To estimate this parameter, I capitalized on the observation of Messier (1994:484) that the equilibrium density of moose at Isle Royale in the absence of predators should be $N^* = 2$ ind km^{-2}. Because the equilibrium density in model 14.1 is

$$N^* = \frac{u_0}{a - \eta} \left(\frac{a}{\eta} - 1 - \frac{b}{m} \right)$$

we can solve it for u_0, and obtain an estimate of $u_0 = 3.33$ Mg km^{-2} yr^{-1}.

Numerical solution of model (14.1) with these parameters produces a trajectory of moose density that increases to a peak of 6.5 moose per km^2, and then collapses to about 1 moose per km^2 (figure 14.3a, below). We can compare this trajectory to the observed fluctuations of moose at Isle Royale (Mech 1966:21–22). The first moose probably arrived on Isle Royale in the early 1900s. By 1915, there were around 200 individuals. The peak of around 3,000 moose (5.5 ind km^{-2}) was reached in 1934, after which the population collapsed to 400–500 (just under 1 ind km^{-2}). A very similar sequence of events occurred during the 1990s, when moose population reached a peak of 2,500 in 1996, followed by an 80% decline over the next two years. We see that the model-predicted trajectory mimics these density fluctuations reasonably well. However, the model predicts a somewhat higher moose density at the peak than observed. An even more problematic feature of model output is that it predicts a slow decline of the moose population, taking more than 10 years to reach the low density after the overshoot, although the actual population at Isle Royale collapsed much faster. Still, for a very simple model and very crude estimates of parameters, the match between the predictions and the observations is not bad. In any case, this problem can be "fixed" by a judicious choice of parameters (e.g., decreasing b generally causes a faster collapse after peak). The point here is not to "massage" the model to make it fit the observation, but to answer the question of whether the model is capable of correct dynamics for parameter values within the reasonable ranges.

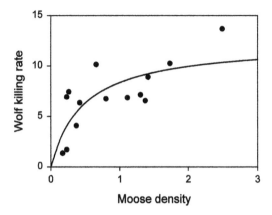

FIGURE 14.1. Wolf functional response to moose density. Units of moose density: moose per square km; killing rate is in moose per wolf per year. The fitted curve is $f(N) = 12.3N/(0.47 + N)$. Data from Messier (1994: table 2).

14.3   INTERACTION WITH PREDATORS

The role of wolf predation in the dynamics of moose populations has generated a certain amount of controversy (section 14.1). One interesting idea posits that vegetation-moose-wolf systems are characterized by multiple stable state dynamics (Messier and Crête 1985). This proposal was elaborated and put on an empirical basis in an important article by Messier (1994).

Messier reviewed a number of empirical studies of moose-wolf systems at different locations in Alaska, Canada, and the northern United States (1994: table 2). In particular, Messier collected together data on the killing rate by wolves as a function of moose density (figure 14.1). Fitting the hyperbolic functional response to these data, we obtain parameter estimates $c = 12.3$ moose wolf^{-1} yr^{-1} and $d = 0.47$ moose km^{-2}. The hyperbolic function explains 54% of variance in the killing rate. I also fitted the Beddington predator interference functional response to these data. This three-parameter function explained 60% of variance in the data, which improvement I deemed not sufficient to warrant adding an extra parameter. Finally, I investigated the applicability of the ratio-dependent response.

This two-parameter functional form fitted data very poorly (17% of variance explained), so I do not consider it further.

If we assume that moose population dynamics in the absence of predators are governed by the logistic, and that predator functional response is of the hyperbolic kind, as suggested by data in figure 14.1, then we can write the following equation for moose dynamics:

$$\frac{dN}{dt} = r_0 N \left(1 - \frac{N}{k} \right) - \frac{cPN}{d + N} \qquad (14.2)$$

We have already estimated parameter r_0 as 0.2 yr^{-1} (see section 14.2), and moose carrying capacity in the absence of wolf predation was estimated by Messier as 2 moose km^{-2} (1994:484).

So far our analysis proceeds in parallel with that of Messier. The predator equation, however, is where we must part ways. Messier modeled wolf numerical response by fitting a function to the data shown in figure 14.2. The data were constructed by recording moose and wolf densities at different study locations and during different time periods (e.g., Messier separated the Isle Royale time series into 5-year chunks, and used average densities of moose and wolves during each time period as a data point). This approach was correctly criticized by Eberhardt (2000) because it assumes that wolf density immediately mirrors any change in prey density. In other words, Messier assumed that the predator equation has the following form:

$$\log_{10} P = \frac{58.7(N - 0.03)}{0.76 + N}$$

(using the function fitted by Messier to the data in figure 14.2), instead of the normal procedure of modeling the predator's rate of change (section 4.1.3). Because wolf density is a slow variable, taking several years to respond to any changes in moose density, we should follow the standard approach, rather than that proposed by Messier.

The simplest equation we can write for wolf density would be based on the assumption that wolf numerical response is a linear function of their functional response:

$$\frac{dP}{dt} = \chi \left(\frac{cN}{d + N} - \mu \right) P \qquad (14.3)$$

which, of course, leads to the Rosenzweig-MacArthur model (section 4.2). Note that no self-limitation terms have been added to this

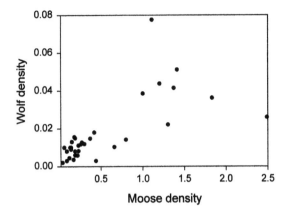

FIGURE 14.2. Relationship between moose and wolf densities, $ind\,km^{-2}$. Data from Messier (1994: table 2).

equation. Since wolf packs are fiercely territorial, we should consider adding such terms at later stages of investigation.

We now need to estimate two additional parameters: χ and μ. Parameter χ we estimate using the same logic as in section 14.2: it is related to the intrinsic rate of predator increase, s_0, as follows: $\chi = s_0[ck/(d+k) - \mu]^{-1}$. Parameter μ we estimate using the information about wolf energetics (Fuller and Keith 1980:594–595). In particular, the minimum consumption rate that allows a wolf to maintain its body weight is 0.06 kg of meat per kg of wolf per day. The consumption rate needed for maximum rate of population increase is 0.013 (same units) (Mech 1977, cited in Fuller and Keith 1980). This suggests that $\mu \approx 0.5c$. Together these relationships imply $\mu = 6$ and $\chi = 0.1$. Dynamics of the model for these parameters are shown in figure 14.3b. Given the notorious propensity of the Rosenzweig-MacArthur model to oscillations, perhaps we should not be surprised by its prediction of long-term high-amplitude cycles. As we know (section 4.2.1), the Rosenzweig-MacArthur model has only one stabilizing mechanism, logistic growth of the resource (vegetation). Yet, as I noted above, the assumption that wolf population density is regulated only indirectly, via its food supply, is highly unrealistic. I explored the effect of adding self-limitation to the wolf equation by switching from the Rosenzweig-MacArthur to the Bazykin model (for equations, see section 4.2.1). This model needs an extra parameter, κ,

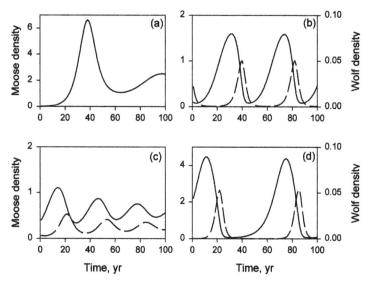

FIGURE 14.3. Dynamics predicted by various ungulate models. In all graphs, solid curves represent moose density, and dashed curves represent wolf density (both densities in ind km^{-2}). (a) Vegetation-moose model (equations 14.1) with parameters $u_0 = 3.33$, $m = 100$, $a = 2$, $\eta = 1$, $\xi = 0.467$ and $b = 40$. (b) Moose-wolf model (equations 14.2 and 14.3 with parameters $r_0 = 0.2$, $k = 2$, $c = 12$, $d = 0.5$, $\mu = 6$, $\chi = 0.1$.) (c) Moose-wolf model with wolf self-limitation term, same parameters as in (b) plus $\kappa = 0.1$. (d) Vegetation-moose-wolf model (equations 14.4), same parameters.

the maximum density reached by predators not limited by their food supplies. The data on maximum wolf densities suggests that such a limit should be around 0.1 wolf km^{-2} (figure 14.2). Dynamics of the Bazykin model with this value of κ are shown in figure 14.3c. We note that the addition of the self-limitation term adds substantially to the stability of the moose-wolf interaction. In fact, the Bazykin model for reference values of parameters is characterized by a stable equilibrium (which is, however, approached very slowly; thus, in the presence of even a little environmental stochasticity, the model would exhibit sustained second-order oscillations). Interestingly, the equilibrium wolf density is around 0.01 wolf km^{-2}, which is an order of magnitude less than the socially imposed limit, κ. A numerical investigation of the dynamical effect of κ showed that it primarily affects stability of the system, and has a very weak effect on the equilibrium wolf density, P^*, which is primarily set by interaction with the prey. One must reduce κ substantially (pushing it down to below 0.01) in order to reduce P^*.

Finally, we combine the vegetation-moose and moose-predators interactions together to investigate the dynamics of the tritrophic system:

$$\frac{dV}{dt} = u_0\left(1 - \frac{V}{m}\right) - \frac{aVN}{b+V}$$

$$\frac{dN}{dt} = \xi N\left(\frac{aV}{b+V} - \eta\right) - \frac{cPN}{d+N} \qquad (14.4)$$

$$\frac{dP}{dt} = \chi\left(\frac{cN}{d+N} - \mu\right)P - \frac{s_0}{\kappa}P^2$$

The dynamics of this model for the reference set of parameters are shown in figure 14.3d. Note that by putting together two components, each characterized by a mild degree of oscillations (figure 14.3a and c), we obtain a much less stable total system (figure 14.3d).

Before summarizing the results of the theoretical investigation into ungulate trophic dynamics, I wish to stress very much that none of the specific parameter estimates I used above should be taken at all seriously. My purpose was not to try to even approximate an empirically based model for ungulate dynamics, but rather to derive a parameter reference set to provide a starting point for numerical investigations of dynamics characterizing various specific models. Therefore, none of the patterns depicted in figure 14.3 are meant as specific predictions for any particular empirical system. However, numerical investigations of the models suggest that the following *qualitative patterns* are quite robust. First, the degree by which ungulate populations should overshoot their carrying capacity, set by food supply, and then collapse depends primarily on two parameters: m and u_0. The greater is m, the higher the densities that will be achieved by ungulates at the peak. The smaller is u_0, the lower the density to which ungulates will collapse. This observation suggests that overshoot-collapse dynamics are most likely where vegetation is characterized by lowest productivity but high standing crops in the absence of herbivores, such as arctic habitats. Second, the food-chain-length effect (i.e., replacing the herbivore logistic term with the vegetation-herbivore interaction leads to increased oscillation) also depends on the regrowth rate of vegetation, u_0. Making vegetation regrow faster, while keeping herbivore carrying capacity fixed, brings the dynamics of the tritrophic model closer in line with the simplification that ignores vegetation dynamics. Third, one initially unanticipated aspect of modeled dynamics is the propensity for oscillations, characterizing quite a wide range of parameters around the reference set. Finally, all models predict a rather long period of oscillations: perhaps two or three cycles per century. The cycle period primarily depends on intrinsic growth rates of ungulates, as well as those of predators and the regrowth rate of vegetation. Because intrinsic rates of increase of moose and wolves are known with a reasonably high degree of precision, average periods implied by models appear to be one robust *quantitative* prediction.

14.4   NUMERICAL DYNAMICS

I now turn to the review of empirically observed dynamics of ungulate populations. My purpose is not a rigorous test of the models presented in sections 14.2 and 14.3 (quite simply, the theory has not yet been developed to the point where it can be so tested). Instead I aim to survey what long-term data on cervid dynamics are available and check whether there is sufficient correspondence between the insights from models and empirical patterns to encourage further development of hypotheses and models.

The first conjecture worth checking out is the qualitative result from model (14.1) that overshoot-collapse dynamics are more likely to be found in systems with low vegetation productivity, such as the reindeer habitat. One classic data set, which has made it into most ecology textbooks, is the dynamics of reindeer introduced on two Pribilof Islands off the coast of Alaska (figure 14.4a and b). As is well known, on one island, St. George, the reindeer population went though a mild peak, collapsed, and then fluctuated at a low density. By contrast, the St. Paul Island population reached a peak density an order of magnitude higher, and then collapsed to extinction. I do not know whether these different outcomes are a result of stochastic events, or of a difference between the ecological characteristics of

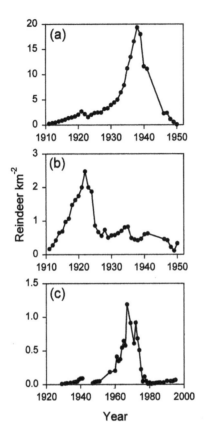

FIGURE 14.4. Population dynamics of reindeer: (a) St. Paul Island, Alaska (Scheffer 1951); (b) St. George Island, Alaska (Scheffer 1951); (c) Lapland Wildlife Refuge, Russia (Lopatin and Abaturov 2000).

the two islands. The vegetation-herbivore model suggests one possible scenario: if St. Paul had a greater standing crop, then reindeer should have reached a higher peak density, and also collapsed lower, than at St. George. Only future research can tell whether this hypothesis has a basis in reality. However, boom-bust dynamics appear to be a typical feature of unmanaged (and free of predation) reindeer populations. Another example (figure 14.4c) comes from the Lapland Wildlife Refuge, located on the Kola Peninsula (Russia). It is significant that Lopatin and Abaturov (2000) also consider the vegetation-herbivore interaction as the most likely explanation. They advance and

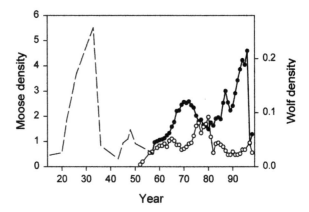

FIGURE 14.5. Moose and wolf population densities at Isle Royale (in
ind km^{-2}). Broken line: moose dynamics prior to systematic census (from
Mech 1966). Filled and hollow circles: censused densities of moose and
wolves, respectively (Peterson and Page 1988; Post et al. 1999).

parameterize a model that is very close in spirit to the one I advanced
above (although their model is somewhat more complex including,
e.g., age structure in the deer population).

Another much analyzed data set concerns the dynamics of a moose
population on Isle Royale (figure 14.5). The story is well known (it
also is featured in most ecology texts), and need not be repeated here.
There are two points important to the focus of this chapter. The first
one is that moose population numbers exhibit quite violent oscillations, which is in line with the predictions of the theory developed in
the previous section. The second observation is that ecological mechanisms apparently responsible for oscillatory dynamics have repeatedly
changed in nature. The first oscillation, with the peak in the early
1930s, was undoubtedly a result of moose overeating their food supply. Note that the pointed shape of the peak (as best as we can say,
given the fragmentary nature of data) and rapidity of collapse is consistent with this explanation (see section 4.2.3 for a general idea and
section 12.5.2 for an application). The second oscillation, with the
peak in the early 1970s, has a blunt rounded peak and slower decline,
which is consistent with the hypothesis of moose-predator oscillation.
Finally, for reasons that are still obscure (see Peterson 1999 for a
discussion), the wolf population never recovered after its collapse in

the early 1980s. Thus, the third moose oscillation, ending with an abrupt collapse in late winter of 1996, apparently was driven again by the vegetation-moose interaction. The much higher peak densities achieved by moose during the first and third oscillations, compared with the second, are entirely consistent with this repeated shift in mechanisms driving the cycles.

How much are violent dynamics of moose on Isle Royale due to the insular nature of this population? In other words, are sustained population oscillations the rule or exception for deer dynamics? We can answer this question only with more time-series data. However, the difficulty lies in the very long periods—theory predicts 30 to 50 years (figure 14.3)—characterizing the hypothesized ungulate oscillations. Nevertheless, I searched the literature for any long-term data on either deer or their predator dynamics. The best data in terms of length and resolution I found were records of fur returns. The data from the Hudson's Bay Company (figure 14.6) give us an indication of deer dynamics in the eighteenth and nineteenth centuries, prior to massive hunting and habitat destruction, leading to large-scale extinctions of deer and wolf populations. These data have several problems, however. The first set (figure 14.6, a top) lumps together wolves and coyotes (unfortunately, the Hudson's Bay Company did not keep separate records). This is actually a very fascinating data set. Superimposed on longer oscillations, we clearly see a 10-year cycle. Bulmer (1974) documented this 10-year periodicity statistically in the second half of the series (Bulmer suggests that the proportion of coyotes increased toward the end of the covered period). If we "subtract" (by eye) the 10-year cycles, than we see three longer oscillations with peaks in roughly 1770, 1810, and 1850. If, as is likely, these oscillations are due to wolf pelts, then their period is consistent with the theoretical prediction of 30–50 years.

The data on elk and deer pelts are more fragmentary, but are not contaminated with the 10-year cycle (figure 14.6, middle). Four troughs are visible or can be inferred in this data: mid-1750s, 1784, 1815, and soon after 1850. The visible peaks in the early 1770s and 1806, and the inferred peak around 1840, coincide with, or precede, the inferred wolf peaks, which is again consistent with the hypothesized predator-prey relationship. It is interesting that previous

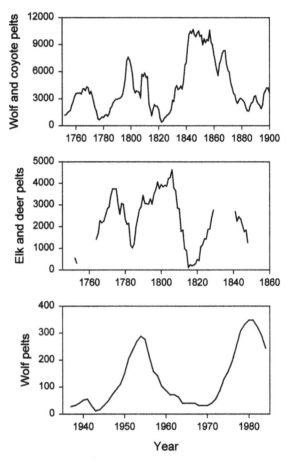

FIGURE 14.6. Fur return statistics: *top*, wolf and coyote pelts from Hudson's Bay Company (Jones 1914); *middle*, elk and deer pelts from Hudson's Bay Company (Jones 1914); *bottom*, wolf pelts in Leningrad region (Nazarov 1988).

workers apparently have completely missed the possibility of oscillatory dynamics in deer and their predators. For example, in one of the best reviews of the Hudson's Bay data, Finerty (1980:81) suggests that wolves are not cycling, and does not even present the deer data.

As to Eurasia, I could not locate any fur return data on deer dynamics, but wolf pelt series are available. One example is shown in figure 14.6 (bottom). Furthermore, there are a number of shorter or fragmentary data sets coming from other localities. For example, population numbers of deer (white-tailed and moose) apparently fluctuated quasi-cyclically in Algonquin Park, eastern Ontario, Canada, with periods of high density during the 1920s, 1950s, and 1980s (Forbes and Theberge 1996). At another location in Ontario, white-tailed deer numbers went through two consecutive troughs in the mid-1950s and around 1980 (Fryxell et al. 1991). The moose population in Finland went through three oscillations since 1930 (Lehtonen 1998).

14.5 SYNTHESIS

Whereas in the previous case studies my main objective was to identify the ecological mechanisms driving population oscillations, the goal of this chapter is much less ambitious. I believe that the current controversy about cervid dynamics can profit from the conceptual and theoretical framework described in the book. In particular, I think that the language of the dynamical systems theory (e.g., exogenous versus endogenous, and first- versus second-order dynamics and mechanisms) provides a better conceptual framework than the one based on limiting versus regulating factors. Furthermore, I propose the "generic mammalian herbivore model" (section 4.6) as an integrative framework for investigating cervid population dynamics.

My initial expectations in developing trophic models for cervids were to find generally stable dynamics, and therefore the original goal of the investigation was to attempt to predict the statics (e.g., mean population densities) characterizing deer in various circumstances—with or without predators, in vegetation communities characterized by high versus low productivity, etc. It was a considerable surprise, therefore, to see rather violent dynamics predicted by models in sections 14.2 and 14.3, and particularly by the tritrophic model. The tritrophic model oscillates for a wide variety of trophic interaction parameters. It appears that the main source of stability in the model is intraspecific competition among the predators (wolf territoriality), because by far the strongest effect on stability is exerted by parameter κ. This qualitative prediction from the models, that second-order oscillations should not be uncommon in cervid populations, motivated my survey of the long-term data (section 14.4). My general impression after this search is that wherever deer populations are not heavily affected by humans, oscillations with a period of roughly 30–50 years appear to be the rule, rather than an exception.

If cervid dynamics indeed turn out to be more prone to oscillation than they are given credit for, then there are major implications for deer population ecology. This would mean that the current discourse about the limiting and regulating factors largely misses an important point. If deer populations exhibit sustained second-order cycles, the main question becomes what factors are responsible for the oscillatory nature of dynamics (which are the second-order endogenous mechanisms, in my terminology), which factors ensure that the oscillation does not get out of hand (first-order mechanisms), and which factors are responsible for stochastic fluctuations in the realized per capita rate of change (exogenous mechanisms).

The very preliminary results discussed in this chapter suggest that a further investigation melding mechanistically based theory with empirical parameter estimates and tests employing time-series data is highly warranted. In this book I have focused on trophic interactions (and, clearly, more fine-tuning of functional forms and parameter estimates are needed). However, another important mechanism that needs to be explicitly added to the model is the effect of exogenous factors on ungulate population change. This task can be accomplished within the generic mammalian herbivore framework by making model parameters either stochastic variables, or functions of measured exogenous variables (see section 3.2). A particularly promising direction is the current investigation of population effects of variability in winter weather (Post and Stenseth 1998, 1999; Post et al. 1999). Another direction is an investigation of the effects of adding age structure to trophic models (this is particularly interesting because vulnerability to predation may be greatly affected by age; see, e.g., Durant 1998). Finally, it has been proposed that predation by black bears may cause an appearance of a low-density stable equilibrium in the moose-wolf system (Crête 1987; Messier 1994). This proposal also needs to be evaluated with empirically based models.

CHAPTER 15

General Conclusions

15.1 WHAT MECHANISMS DRIVE OSCILLATIONS
IN NATURE?

Now that we have done so much work trying to understand the specific mechanisms responsible for complex population dynamics in
each of the case studies (chapters 9–14), it is time to step back and
see if any patterns emerge. Table 15.1 brings together the conclusions for these case studies, together with some other studies that I
did not have space to review in this book, but for which sufficient
information exists for informed judgment. One pattern is immediately
obvious: all cases for which we can reach a reasonably well-supported
conclusion belong to one general category of ecological processes:
trophic interactions. The majority of cases are either specialist predators or parasitoids (which, essentially, amounts to the same thing). In
addition, we see one case each of food quantity, food quality (as a
contributing factor), microparasite, and macroparasite. There is not a
single case where an intrinsic hypothesis has provided a theoretically
sound and empirically supported explanation of complex dynamics in
nature. In one case, the red grouse in Scotland, an intrinsic mechanism (kin favoritism hypothesis) remains a viable contender, but this
explanation has not yet reached the point where it has been translated into an empirically based model, nor has it yet been subjected
to an experimental test (chapter 11). Thus, given the present state of
knowledge, I conclude that the overwhelming majority of examples
of population oscillations in nature are explained by the mechanism
of specialist predation (including parasitoids), with a few additional

TABLE 15.1. Summary of case studies with their "best-supported models" (an empirically based model exists for all studies, although the degree of knowledge about parameter estimates is variable). The columns "1st order" and "2nd order" indicate the mechanisms on which models are based. The "Expt." column indicates whether a manipulative experiment was performed to test the hypotheses on which models are built

| System | Mechanisms | | Expt. |
	1st order	2nd order	
Larch budmoth	food quant.	(1) parasitoids (2) food qual.	—
So. pine beetle	intrasp. comp.	predators	Yes
Grouse	food	macroparasite	Yes
Vole	(1) food quant. (2) social interact.	predators	Yes
Lemming	none[1]	food quant.	—
Hare	food quant.	predators	Yes
Moose	food quant.	predators	—[2]
Measles	supply of susceptibles	microparasite	—[3]

[1] The best model is stabilized by the logistic term in the vegetation equation.

[2] Although no formal manipulative experiments have been performed in this system, the invasion of Isle Royale by wolves constitutes a natural experiment.

[3] The immunization programs may be considered as a manipulative experiment.

cases involving other kinds of trophic interaction (food and parasites). We do not yet have examples of complex dynamics driven by intrinsic mechanisms, interspecific competition, mutualisms, commensalisms, etc.

I should add that there are laboratory examples of intrinsic cycles, for example, flour beetles in the genus *Tribolium*; although even there one might argue that cannibalism is a kind of a trophic interaction (Costantino et al. 1995; Costantino et al. 1997; Dennis et al. 1995). There is also a recent report claiming to demonstrate stable population cycles in lizards driven by natural selection—the Chitty hypothesis (Sinervo et al. 2000; Bjornstad 2001). This is an extremely interesting result, provided that it withstands the test of time. However, I should note that the lizard example differs from the case studies in table 15.1 in one important regard: it is a two-year (and two-generation) cycle. In other words, this is an example of a first-order oscillation, rather than second-order dynamics, which characterize all case studies examined in this book. And as I argued in chapter 3, intrinsic population mechanisms are much more likely to lead to stable dynamics or first-order cycles, rather than longer-period second-order oscillations. Thus, taking all the caveats into account, I still must conclude that the current state of evidence overwhelmingly supports trophic mechanisms as drivers of *second-order oscillations*.

Perhaps in retrospect this conclusion might appear self-evident. After all, the very first models of ecological cycles advanced by Lotka and Volterra were written specifically for predator-prey interactions. Yet, there was a long period starting in the 1960s and extending up to the late 1980s when the predation hypothesis was held in extremely low esteem (e.g., Krebs and Myers 1974)—a whole generation of population ecologists swayed by ideas of the "doomed surplus" (Errington 1963) and genetic polymorphism hypothesis (Chitty 1967).

On the other hand, we also do not find a unique solution to the puzzle of population cycles—the same specific mechanism that would explain population oscillations in every single case. Different species, and even the same species in different locations, can exhibit very different dynamical patterns. This observation suggests that a specific mix of mechanisms underlying these fluctuations is quite variable. Perhaps the best example is the variety of dynamical patterns observed among rodents: predator-prey cycles in voles *M. agrestis* and *C. rufocanus* in northern Fennoscandia and, most likely, water voles in Switzerland; herbivore-food oscillations in lemmings; stable dynamics with occasional outbreaks due to massive inputs of high-quality food in bank voles inhabiting forests of central Europe; first-order oscillations of the common vole *M. arvalis;* and irregular high-amplitude fluctuations of North American voles. From what we know about the connection between these time-series patterns and ecological mechanisms, such a great variety of patterns almost certainly is matched by a great variety of ways in which specific mechanisms are mixed in each system (and in some case we have empirical evidence for this, as in the contrast between voles and lemmings). Thus, the hope for a universal mechanism of population cycles (Chitty 1996) must be abandoned once and for all. Furthermore, the fact that we do not yet have a single instance of a second-order oscillation driven by an intrinsic mechanism is no guarantee that such population systems are completely absent in nature (although it does seem likely that they are relatively rare; otherwise we should have encountered at least one by now).

To summarize, there appears to be one general category of ecological mechanisms that is responsible for population oscillations: trophic interactions. Yet, within this general class, specific mechanisms combine in a variety of intricate ways leading to a complex tapestry of different kinds of complex ecological dynamics.

15.2   STRUCTURE OF DENSITY DEPENDENCE

One of the general themes recurrent throughout the book is the value of time-series analysis as a tool for characterizing patterns in observed population trajectories, and connecting them with possible mechanisms that produce these patterns. It is true that the relationship between time-series patterns and mechanisms is one-to-many, so that we cannot infer what specific mechanism (or set of mechanisms) is responsible for each observed pattern. However, there is sufficient congruence between time-series patterns and mechanisms that allows us in many cases to reduce the list of plausible mechanisms for a specific empirical system. The quantitative tools we use to connect time-series patterns to ecological mechanisms are **probes**. Practical experience with using various probes in the case studies, reviewed in this book, suggests that not all probes are equal; some are more useful than others in helping us to reduce the list of potential mechanisms: the average period, the estimated process order, and the topological characteristics of cycles.

The average period is one of the easiest-calculated probes, yet it is extremely useful in guiding further modeling efforts and experimental tests. The average period can provide some indications about the process order (see below), since the periods of generation, first-order, and second-order cycles tend to fall into classes of one-two-many generations (section 3.5). Even within one category, for example, second-order oscillations, the average period can provide useful diagnostic indications about potential mechanisms driving dynamics.

Second-order cycles typically result from an interaction between two populations, and the period is determined by characteristics of both species (most important, by their respective r_0). If species interacting with the focal population are very different in their biological characteristics, then different hypotheses imply very different periods. Application of this logic can yield strong inference, as is illustrated by the case of southern pine beetle cycles (chapter 10), where parasitoid, specialist predator, and host hypotheses yielded very different and practically nonoverlapping frequency distributions of predicted periods. Thus, the average period can be an extremely useful diagnostic tool in the analysis of population oscillations. It is interesting that another easily calculated index, the amplitude, is of much less value in narrowing down the field of possible mechanisms. This is probably a result of the ability of all mechanisms that in principle can induce oscillations to generate a wide spectrum of amplitudes by rather minor variation in parameters.

The second very important concept that helps us connect patterns with mechanisms is the **process order**. Order is more controversial than period, and not all ecologists agree that it is a robust and useful concept. In fact, it appears that the distinction between first- and second-order dynamics particularly appeals to ecologists who have worked on time-series analysis. By contrast, mathematical ecologists may prefer to use different classifications. I acknowledge that there is some validity to their criticisms of the concept of order. Real life is messy, and we cannot neatly separate all empirical cases of fluctuations into nonoverlapping discrete classes. As can be seen in table 15.1, best models for well-studied case studies always have both first- and second-order components (in the only apparent exception, lemmings, the first-order component is in the vegetation equation). This is not surprising, because second-order mechanisms are needed to generate oscillations, while first-order mechanisms are required to prevent such oscillations from diverging. Thus, even though actual dynamics are not pure first- or second-order, the mechanisms can be classified by their order.

There is another interesting observation related to the concept of process order. Historically, discrete first-order models, such as the Ricker, played a very important role in motivating ecologists to study complex population dynamics. Such models produce characteristic

first-order oscillations: limit cycles with periods of 2–3 generations, or chaotic oscillations of similarly short, but variable, durations. Yet, such dynamics are rare in nature; in fact, I know of no good example of a classic first-order oscillation outside the lab (with the possible exception of Sinervo et al. 2000). Certainly, we have examples where oscillations appear to be first-order *on the scale of years:* the tree-dwelling aphid *Drepanosiphum platanoidis* (Dixon 1990), the common vole *Microtus arvalis* (see figure 12.3), and the wood mouse *Apodemus sylvaticus* (Southern 1979). Description as first-order cycles, however, is good only on the phenomenological level, as all these populations have multiple generations per year. Thus, a more mechanistic description would not be first-order. For example, if we were to model common vole dynamics, we would probably add an explicit equation for seasonal regrowth of its food supply. What I am trying to say is that a first-order discrete-time equation would not seem (at least, to me) to be a reasonable *mechanistic* model for this system. And this is the kind of system where we would be most tempted to use it, because of results from the phenomenological time-series analysis. I will discuss further implications of the apparent scarcity of first-order oscillations in nature in the next section (15.3).

The discussion in the preceding paragraph highlights one of the difficulties of applying the concept of order in practical applications: the specific choice of base lag may (and usually does) affect the estimate of order. In the common vole example, the estimate of order as 1 critically depends on choosing the timescale of 1 year. If we had more finely sampled data, and analyzed them with base lag equal to average generation time, we would certainly select a more complex model than the first-order equation (most likely, we would also have to include seasonality in the model explicitly). This observation suggests several remarks. First, we should always treat the estimated order as a quantity that is conditioned on the base lag choice. In some cases, different choices of base lag may yield the same maximum order measured in natural units (e.g., the analysis of *Plodia interpunctella* dynamics in Turchin and Ellner 2000b: table 2). In most cases, however, we will probably get diverging results, and must live with it. Second, not all potential choices of base lag are equal, and some make more biological sense than others (practical advice on selecting the base lag can be found in section 7.2.2). Third, the fact that order is conditioned on the base lag choice does not matter when we use it as a means for quantitative contrast between model predictions and data. We simply choose whatever base lag seems most appropriate, and then estimate order from both data and model output, using the same τ.

Peak shape is another probe that I found to be of great use in applications, although it is perhaps the most controversial, simply because it was proposed very recently (Turchin et al. 2000), and has not yet been tested by time. Case studies in part III highlight both the potential and the limitations of this probe. The lemming case, of course, provides the best example where this idea was used to most effect. However, the change in peak shapes during the history of the moose population at Isle Royale is at least consistent with the hypothesized mechanisms responsible for moose density collapse in all three oscillations (chapter 14). By contrast, the peak shape diagnostic was not useful in the analysis of hare-lynx interaction. First, long-term lynx data—fur returns—are based on an index that is nonlinearly related to population density. Second, trajectories predicted by the best model for this system predict a very slight difference in peak shapes between prey and predators (figure 13.2), which would be difficult to detect in noisy and short data sets.

Peak shape is not the only shape probe that we might gainfully employ in analysis. Another potential probe is the asymmetry of oscillations: crudely, the ratio of increase to decrease periods. Ecological oscillations are typically asymmetric (right-skewed), because the rate of increase is limited by r_0, while the rate of collapse is potentially unlimited (Ginzburg and Inchausti 1997). However, the degree of asymmetry may vary between different populations, and this provides an important clue to underlying mechanisms. One example among the case studies in part III is the contrast between dynamics of English and Scottish populations of the red grouse. In addition to being characterized by shorter periods, English populations are more asymmetric, exhibiting a much more abrupt collapse (compare figures 11.2 and 11.4). On the basis of this and other observations it seems likely that red grouse populations in these two areas may be driven by a different mix of ecological mechanisms. Only future work will show whether this suggestion has a basis in reality, but it appears that an index of asymmetry has a potential as another useful diagnostic probe.

15.3   WHAT ABOUT CHAOS?

Original impetus to the study of chaotic dynamics in ecology came from very simple first-order discrete models, such as the Ricker or the discrete logistic (May 1974a, 1976). The empirical observation of paucity of first-order oscillations in nature, remarked on in the previous section, suggests that direct analogues of *strong chaos*, as envisioned by pioneers, are rare, or even absent from nature. By "strong chaos" I mean very irregular-looking dynamics (lacking even statistical periodicities) with very positive Λ_∞ (1 or greater). Such dynamics can be obtained from, for example, the Ricker model with $r_0 = 4$.

In fact, I am inclined to the opinion that first-order discrete equations cannot really be considered as even remotely mechanistic models for population dynamics. As I discussed before, density dependence is not an ecological mechanism but a phenomenological summary of ecological mechanisms that act as density feedbacks (with various lags). And the assumption of discrete generations fits well only a few real-life systems (of the case studies examined in this book, only the larch budmoth). Thus, in my opinion, Ricker-style models are closer to the phenomenological end of spectrum; they belong to the class of models for the structure of density dependence.

Chaotic dynamics that we see in nature arise from a different class of models. Let us consider two examples for which we have well-developed models that show chaotic dynamics. (Remember that, properly speaking, chaos is a characteristic not of an empirical system, but of a model we might have for this system. Thus, saying that a certain natural population is chaotic is a shorthand way of saying that "the best-supported model we have for this system is chaotic.") The two systems, measles and Fennoscandian voles, are similar in that ecological mechanisms driving their oscillations are well known, that we have reasonably good parameter estimates, and that models for both systems have been extensively and quantitatively tested against time-series data. These two systems are also similar in the dynamical mechanism that brings about chaotic dynamics: the interplay between a consumer-resource cycle (microparasite for measles and specialist predator for voles) and seasonality. Finally, both systems are not strongly chaotic in the sense defined above. In fact, long-term dominant Lyapunov exponents are very near zero in both cases; thus, these systems are **quasi-chaotic:** roughly half of local Lyapunov exponents are negative, half are positive, and trajectories are characterized by recurrent short-term episodes of sensitive dependence on initial conditions. Judging by the empirical observation that most "chaotic" systems are actually characterized by Λ_∞ very close to zero (Ellner and Turchin 1995; Turchin and Ellner 2000a), Ellner and Turchin have hypothesized that quasi-chaos is actually the most common case of complex population dynamics. Analyses of case studies in this book appear to further confirm this conjecture, since the majority of them are characterized by not very positive Lyapunov exponents (see tables 9.1, 11.1, 12.1, and 12.4).

The only possible exception to this general observation is the case of Norwegian and brown lemmings. Time-series analyses of available data consistently indicate very positive Λ_∞ (table 12.5). Furthermore, if inferences made in section 12.5 are based in reality, then the mix of ecological mechanisms that we identified there, a vegetation-herbivore interaction of the Rosenzweig-MacArthur type combined with very strong seasonality, is extremely prone to chaotic behaviors. Finally, lemming outbreaks *look* chaotic (Oksanen and Oksanen 1992)! On the other hand, we need also remember that the number of data sets is not as extensive as we would wish, and many lemming series suffer from some methodological problems (or we even use a different species—brant geese—as an indirect index of lemming fluctuations). While the mechanistic explanation based on the vegetation-herbivore interaction seems plausible, and has been empirically tested, no formal manipulative experiments have yet been done. Thus, the case for strong chaos in lemming dynamics must be classified as tentative, given the current state of knowledge.

To summarize, the intensive "search for holy chaos" (thanks to Alfredo Ascioti for coining this phrase), inspired by the insights of early workers, in one sense was a failure, because so far we have not found any direct analogues of chaotic Ricker-style dynamics in nature. This became clear very soon after the initial excitement (Hassell et al. 1976). In another sense, however, the search for chaos was extremely productive. It resulted in multifarious indirect benefits to population ecology, and played an extremely important role, in my opinion, in bringing about the very synthesis that is occurring in our field. The critical insight that allowed us to move beyond Ricker-like models was Bill Schaffer's importation of phase-space reconstruction from physical literature, melded with the older tradition of using linear autoregressive models in the analysis of population data (Moran, Bulmer, Royama, Berryman). When these two approaches were synthesized, it emerged that first-order approaches substantially underestimate the frequency of complex dynamical behaviors in nature (Turchin and Taylor 1992). Subsequent research showed that a most frequently found kind of complex dynamics in nature is quasi-chaotic second-order oscillations, characterized by various degrees of statistical periodicities and typical Λ_∞ in the $[-0.1, 0.1]$ range (Ellner and Turchin 1995; Turchin and Ellner 2000a). The overwhelming majority of case studies reviewed in this book are of this type (again, with the possible exception of lemmings).

15.4   POPULATION ECOLOGY: A MATURE SCIENCE

In the preface I posed the question, is population ecology on the brink of maturity? Is it becoming a predictive science? Now that we have worked through the intervening hundreds of pages (and remember, I could review only a fraction of what is currently done in the area of population dynamics), I believe that the answer is a resounding yes. In fact, my whole book can be taken as an extended answer to this question. In part I, I reviewed the ecological theory relevant to modeling complex dynamics, and it seems to me to be beyond doubt that population ecology has a mature theory. We now have an excellent grasp of which features of model-predicted dynamics reflect the deep underlying ecological principles, and which are particular to the specific mathematical framework used (e.g., difference vs. differential vs. delayed differential equations). We have good understanding, buttressed by an extensive empirical base, of certain functional forms we use as model components (the best example is the typology of functional responses). Finally, we are essentially done with characterizing the dynamics of pairwise species interactions, we have made a lot of progress with three-species systems, and we are now building a general theory of multispecies communities (Holt 1977; Holt and

Lawton 1993; Hastings and Powell 1991; Abrams and Roth 1994; Abrams 1998; Gurney and Nisbet 1998; and many other papers).

It also seems incontrovertible that there are a number of general themes emerging from the rich theory base we have for population dynamics. One such theme is the argument I construct in chapter 2 that population dynamics are underlaid by a set of foundational principles, which are analogous to laws in physics. One may argue with me about the precise logical status of these principles; also, principles (postulates, theorems) may be added or subtracted from the list I propose. Yet, it seems clear to me that population dynamics theory is not a random set of models; there are some general threads running through it. Another theme, perhaps also controversial, is the process order (see section 15.2). Finally, a third theme, perhaps the least controversial, is the connection between mechanisms and parameters, on one hand, and such obvious properties of oscillations as the average period, on the other hand.

The maturity of mathematical theory for population dynamics is matched by equally impressive advances along the empirical direction, as well as the theoretical/empirical synthesis. Not the smallest achievement is the accumulation of an enormous database on population fluctuations in nature (thousands of time-series stored at the Centre for Population Biology, Silwood Park, UK). But there are also several empirical systems that have been intensively studied by multiple teams of investigators over a period of decades. The combination of empirical and theoretical work is beginning to bear fruit. Thus, one of the hallmarks of mature science is its ability to make predictions (which sometimes are even confirmed by experiment). With all the caveats about not equating prediction with forecasting, I should point out that under certain conditions dynamical models can serve as forecasting tools in ecology. One example is the larch budmoth, where a three-parameter model based on host-parasitoid interaction predicted future log-transformed density with 95% accuracy. The other example is the vole *Clethrionomys rufocanus* at Kilpisjärvi, where a mechanistic model with previously estimated parameters beat the data-based forecasting scheme.

But the connection between the maturity of a field and the ability to forecast is not strong. Sometimes forecasting is possible without

any understanding of the underlying mechanisms (many economet-
ric applications in which the Box and Jenkins approach is used). In
other cases, we may perfectly well understand the mechanisms but
lack forecasting ability (weather prediction). The hallmark of mature
science is rather the ability to make predictions: in the strongest case,
when two or more models make conflicting predictions about some
aspect of the system that we can empirically observe or measure.
Population dynamics has certainly matured to this point, although it
reached this stage literally within the last decade. It is very interesting
from the point of view of history of science that the first manipulative experiments of population oscillations suddenly began to be
published very recently, since 1995 (Krebs et al. 1995; Korpimäki and
Norrdahl 1998; Hudson et al. 1998; Turchin et al. 1999). The last two
experiments tested explicit and quantitative predictions of the theory
(and, in another sign of maturity, the predictions were published prior
to conducting the experiments). Additionally, manipulations of nature
are not the only way to conduct an experiment. An experiment is any
empirically based choice between predictions of two or more hypothe-
ses/models. The empirical test of predation vs. herbivory hypotheses
for lemming cycles, using the peak shape as the telltale characteristic,
certainly qualifies as an experiment.

The final observation is that it is striking how simple the models that we used for modeling specific case studies are. We typically
use just two or three mechanistic ingredients in constructing these
models (table 15.1). Certainly, none of the best models that we set-
tled on are simple to the point of being monofactorial. Monofactorial
(single-mechanism) explanations do not work in the explanation of
complex population dynamics, probably because we typically need
at least two mechanisms, one to cause oscillations, and the other to
prevent them from diverging (second- and first-order mechanisms, in
my classification). On the other hand, we did not need to build large
models with many ecological mechanisms to achieve success. In fact,
large models such as the conceptual scheme proposed by Lidicker
(1988, 1991) or the gypsy moth life system simulation (Shehan 1988)
do not lead to breakthroughs in our understanding of population
cycles (although when complex models are boiled down to their main
dynamic constituents, a significant insight can result; see Sharov and
Colbert 1996). Thus, an *empirical* conclusion (in the sense that this

approach seems to produce the best progress) is that neither mono-factorial nor highly polyfactorial models work; best models seem to be *oligofactorial*.

If we think about it, we should be very surprised by the success of these models of intermediate complexity. Take the Kilpisjärvi case study, where a rather simple model produced excellent results (here I am referring to the model without generalist predators). This model has only two state variables (although perhaps we should count sea-sonality as another), five scaled "biological" parameters, plus a noise parameter. This is a very simple model (although capable of quite complex dynamics). Now think about real voles, characterized by age, sex, and size structure, interacting socially with other voles, compet-ing against both conspecific and heterospecific rodents. These voles construct burrows and nests, feed on a wide variety of plants, are hunted by a wide variety of predators. Their spatial distribution is highly heterogeneous, and is affected by distributions of food, pre-dation risk, fixed features of the environment, and results of human activity. This is an incredibly complex, messy real-life system. There is no reason why a relatively simple oligofactorial model should be able to capture its dynamics. Yet it does. And models of intermediate complexity do well in several other oscillatory systems. There is a pattern here, but I do not think we fully understand, yet, why this hap-pens. (After writing this paragraph, I discovered that physicists have also been puzzled by the "unreasonable effectiveness of mathematics in natural sciences"; see Wigner 1970.)

Of course ecological models are not as precise as models in some physical applications, such as planetary motions. To give a quantita-tive measure to what I mean when I talk about "success," I define it as getting R^2_{pred} in the range of 50–90%. This is modest prediction ability, compared with the best physical applications, but ecologists should not feel bad about it. The complex systems with which we deal are unlikely to allow better prediction ability (and as I discuss in the above paragraph, it is amazing that we can get any predictability at all). And do not forget that in science it is not the absolute prediction that counts, but the ability to construct alternative theories/models, and distinguish between them on the basis of their relative prediction ability. Thus, if one theory yields R^2_{pred} of only 10%, but the alterna-tive is even worse, at $R^2_{pred} = 1\%$, then we have made progress by rejecting the worse alternative. We are doing good science, although clearly our best current model will not be a useful forecasting tool in applications.

Finally, ecological systems affected by oscillations are typically more predictable than stable noisy systems (unless we are able to understand and model "noise" as exogenous signal, e.g., by including fluctuations of food supply—masting—or climatic changes explicitly in the model). There is an important message here, too. Presumably, the stable noisy systems are affected by ecological mechanisms similar to those that drive oscillations. In fact, there are several examples in which oscillatory populations smoothly grade into stable ones (e.g., Fennoscandian voles). Thus, it is unlikely that oscillatory systems are *qualitatively* different from the stable ones; the difference is rather a matter of quantitative parameter values. This means that by building and testing a theory for complex population dynamics we are simultaneously perfecting a theory for all kinds of population dynamics, including simple ones. In a sense, population oscillations *are* the planetary motions of ecology, because they provide us with systems where we can hone and test our approaches, before applying them to population fluctuations in general.

Glossary

aggregative response — Local increase in predator numbers (density) resulting from predator movements in response to spatial variation in prey density.

Allee effect — A nonlinear relationship between the per capita rate of population change, $r(t)$, and population density, N, in which $r(t)$ increases for small N, reaches a peak at some intermediate N, and declines thereafter. Usually, it is also assumed that $r(t)$ is negative for N near zero (the strong Allee effect).

ancillary data — Quantitative information in addition to primary data that may be useful in modeling population dynamics of the focal species. May include data on various aspects of biology of the focal species (e.g., age structure, or changes in average body size), as well as time-series data on population fluctuations of interacting species (e.g., resources or predators).

attractor — A geometric object in the phase space that attracts all trajectories starting within its domain of attraction. Examples of attractors include stable equilibria, stable limit cycles, and strange attractors.

chaos — Bounded fluctuations with sensitive dependence on initial conditions.

chaotic oscillations — Dynamics of mixed deterministic/stochastic systems characterized by positive Lyapunov exponents (trajectory divergence). In practice must be characterized by a fairly large signal/noise ratio (otherwise, we shall not be able to detect the chaotic component). May or may not have statistical periodicity (in most ecological applications some periodicity is present).

coefficient of prediction — The proportion of variance in log-transformed density predicted by the model R^2_{pred} (see section 7.2.4 for formula).

DDE models — Delayed differential equation models are similar to ODE models, but their right-hand sides include lagged state variables, for example, $N(t - \tau)$, where τ is the time delay.

density dependence — Some (nonconstant) functional relationship between the per capita rate of population change and population density, perhaps involving time lags.

domain of attraction — A part of the phase space including the attractor itself, as well as all initial conditions from which a trajectory will converge to the attractor.

dominant period — The lag at which the autocorrelation function has its first maximum.

dynamical complex — Two or more state variables affected in a similar fashion by the processes causing fluctuations in a dynamical system. As a result, dynamics of these variables become synchronized, allowing us to replace them with a single averaged variable without a great loss of predictive power. An example is two specialist enemies preying on the same species, and having similar characteristics (functional responses, population growth rates).

dynamical dimension — The number of state variables (and equations) in a dynamical model.

endogenous factors — The density-dependent component of population dynamics, or population feedbacks. May act without an appreciable lag (direct or undelayed density dependence; first-order feedbacks), or involve a lag (delayed density dependence, or second-order feedbacks).

exogenous factors — Density-independent mechanisms that affect population density without being affected by it. In other words, there is no dynamic feedback between these processes and population density.

experiment — A planned comparison between data and a novel, nontrivial prediction derived from a hypothesis.

extrinsic factors — All processes affecting the focal population that are not intrinsic factors. Not to be confused with exogenous (a specialist predator is an extrinsic but endogenous factor).

first-order oscillations — Oscillations arising in first-order dynamical systems, that is, systems in which population feedbacks operate with time lags of one generation or less. Typical periods are in the range of 2–4 generation times. One diagnostic feature of first-order cycles is rapid (one time step in duration) population crashes.

fluctuations — A generic term for any kind of population dynamics (i.e., temporal changes in population density or numbers).

focal species — The species under investigation whose dynamics we are attempting to understand.

fully specified model — A mathematical model that is specified at all three levels, from most general to most specific: (1) all state variables identified, and interpreted in biological terms; (2) all functional forms given explicitly; and (3) all parameter values given. In other words, the model is so completely described that it can be implemented as a computer simulation without any additional information.

functional forms — Mathematical functions that relate state variables (and their rates of change) to each other. There often are discrete choices for alternative functional forms, for example, Type II versus Type III functional response. However, in certain cases one can write a more general parametric form that includes the alternative functional froms as special cases. For example, substituting quadratic exponents in Type III functional response with a parameter θ gives us a general functional response (the "phenomenological form"), which includes Type II and Type III responses as special cases ($\theta = 1$ and $\theta = 2$, respectively).

functional response — The rate at which an individual predator kills prey (thus, its units are prey individuals per predator per unit of time). The three general kinds of simple functional responses are linear (or Type I), hyperbolic (or Type II), and sigmoid (or Type III).

generation cycles — Population oscillations with period approximately equal to one generation. These dynamics arise in age- and stage-structured models, and in DDE models. Sometimes I refer to these dynamics as zero-order oscillations.

intrinsic factor — A process or mechanism pertaining to the focal population. Examples include age and stage structure, dynamics of individual quality and maternal effects, intraspecific competition, and cooperative processes leading to an Allee effect.

logistic-like models — ODE models in which the per capita rate of change, $r(t) = dN/(N \, dt)$, is characterized by a maximum at $N = 0$, is positive for all $N < k$, and negative for all $N > k$ (where k is the carrying capacity).

Lyapunov exponents — A global Lyapunov exponent is the long-term ($t \to \infty$) exponential rate of trajectory divergence. Local Lyapunov exponents measure trajectory divergence over a finite interval of time (e.g., a year). Local exponents depend on the current state of the system, thus characterizing variation in sensitivity to initial conditions in different parts of the phase space.

measurement noise — Errors affecting only the observations. Measurement noise does not affect the actual population dynamics.

mechanism in population ecology — A description of what individuals do that serves as a buildling block in a population model. Mechanisms relevant to the theory of population dynamics come in two flavors. They can be explicitly based on individuals (e.g., a detailed description of an individual predator searching, capturing, and consuming prey). Alternatively, a mechanism can average over some ensemble of individuals, abstracting some important feature of individual behavior or performance (e.g., predators' functional response, or density dependence in fecundity).

metastable dynamics — A special case of multiple coexisting attractors, in which attractors are stable equilibria. Depending on initial conditions, the trajectory will approach either one or the other equilibrium, and stay there indefinitely (unless perturbed).

methodological individualism — The principle that population dynamics can be ultimately understood only with individually based mechanisms and theories. This is an example of a reductionist approach to doing science.

monotonically damped stable point — A stable point attractor approached exponentially (without overshooting). This is dynamics characterizing purely deterministic models.

multiple coexisting attractors — A possible configuration of the phase space in nonlinear models, in which more than one attractor is present. The phase space is divided into multiple domains of attraction, each associated with its attractor.

neutral oscillations — Oscillations, poised on the boundary separating stable oscillations (which converge to a stable point or stable limit cycle) from unstable oscillations (which diverge without bound). The amplitude of neutral oscillations is determined by the initial conditions: the further away from the equilibrium the trajectory starts, the greater will be the amplitude of oscillations.

noise-induced chaos—Population dynamics resulting when a deterministically stable endogenous component (Lyapunov exponent is negative) is combined with a stochastic exogenous component, yielding a chaotic system (Lyapunov exponent is positive).

nonstationarity—Absence of stationarity that occurs because the mechanisms generating population fluctuations change with time.

null factors—Processes that have no effect on the per capita rate of population change.

numerical response—The rate of change of predator population as a function of prey and predator densities.

ODE models—Models employing ordinary differential equations, that is, equations with a first temporal derivative (e.g., dN/dt) on the left-hand side. Right-hand sides are functions of undelayed state variables.

oscillations—Population dynamics that have some element of regularity, allowing some degree of prediction of future population density. Usually oscillations are characterized by statistical periodicities, but it is also possible to have aperiodic oscillations with a strong deterministic component (chaos).

oscillatory damped stable point—A stable point attractor approached in an oscillatory manner (with overshooting). This is dynamics characterizing purely deterministic models.

paradox of enrichment—Increasing prey's carrying capacity in the Rosenzweig-MacArthur (and related) models leads to a destabilization of the point equilibrium and emergence of limit cycles.

parameters—In a specific realization of a dynamical model (e.g., predicting trajectory from certain initial conditions), parameters are constants, while state variables change with time. However, parameter values can change between different realizations. The most common reason is when we apply a model to a different situation (different habitat, or even different species).

PDE models—Partial differential equation models.

periodic forcing—A periodic exogenous component of dynamics.

phase space—Multidimensional euclidean space in which each state variable is represented with its own axis.

phenomenological — Approaches or models that are not based on explicit ecological mechanisms.

population — For the purposes of this book (remember that I focus primarily on temporal aspects of population dynamics), I define population as a group of individuals of the same species that live together in a defined area of sufficient size to permit normal dispersal and migration behaviors. Thus, temporal changes in population abundance are primarily determined by birth and death processes, and emigration/immigration processes can be neglected without a serious loss of predictability.

primary data — Time-series data on population density of the focal species, providing the basis for characterizing the quantitative pattern of population dynamics. This quantified pattern of population fluctuations is what needs to be explained ("why do populations fluctuate as they do?").

probes — Quantitative measures of patterns in population dynamics.

process noise — Exogenous environmental stochasticity that affects the rate of population change. Not to be confused with measurement noise.

process order — The number of lagged densities affecting the realized per capita rate of population change.

pure resource-consumer and second-order systems — A pure resource-consumer system is one in which the resource per capita rate of change depends only on consumer density, while the consumer per capita rate of change depends only on resource density. An example is the Lotka-Volterra predation model. A pure resource-consumer system is a special case of a pure second-order model (a model lacking first-order feedbacks).

quasi-chaotic dynamics — Dynamics intermediate between chaotic and stable periodic oscillations; characterized by Λ_∞ near 0 (usually, in the range of ± 0.1).

quasiperiodicity — In continuous models, a torus-shaped attractor. In discrete models, an infinite set of points lying on a closed curve. The trajectory jumps around this curve without ever exactly repeating itself. Quasiperiodicity is a property of purely deterministic models.

quasi-stationarity—A dynamical system is quasi-stationary if the mechanisms generating dynamics do not appreciably change during the period that we analyze. Thus, we can employ analytical approaches that assume stationarity.

second-order oscillations—Oscillations arising in second-order dynamical systems, that is, systems in which population feedbacks operate with time lags of longer than one generation. Typical periods are in the range of 6–12 and more generation times. One diagnostic feature of second-order cycles is prolonged (two or more time steps) population declines.

semimechanistic—See mechanism in population ecology.

simple population dynamics—Dynamics characterized either by stochastic exponential growth/decline or by stability with noise.

stable limit cycle—In continuous models, a stable limit cycle is a closed curve in the phase space that attracts all trajectories within its domain of attraction. Once on the attractor, the trajectory is perfectly periodic (repeats itself exactly with a certain period). In discrete models, a stable limit cycle is an attractor consisting of a finite set of points, visited in turn by the trajectory. Stable limit cycles are a property of purely deterministic models.

stable periodic oscillations—Population dynamics resulting from combining an endogenous component that is either oscillatory damped stable point, stable limit cycle, or quasiperiodicity, and a stochastic exogenous component. Lyapunov exponent must be negative, implying trajectory stability, and periodicity must be statistically detectable.

stability with noise—Population dynamics resulting from combining an endogenous component characterized by monotonically damped stability, and a purely stochastic exogenous component.

state variables—The main dynamical quantities of interest to the model. The guts of the model describe how state variables interact. Typically, the number of equations in the model equals the number of state variables. In a differential equations model, temporal derivatives of all state variables appear on the left-hand sides of equations. An example of state variables for a predator-prey model is the population densities of prey and predators.

stationarity — A dynamical system is stationary when the mechanisms generating its fluctuations do not change with time.

stochastic exponential growth/decline — Population dynamics in which the per capita rate of population change is unaffected by any population feedbacks (density dependence).

structure of density dependence — Quantitative categorization of the pattern of population dynamics, employing such probes as process order, the shape of density dependence, trajectory stability, and signal/noise ratio.

trend — A long-term exogenously driven systematic change in the environment.

zero-order oscillations — See generation cycles.

References

Aarsen, L. W. 1997. On the progress of ecology. *Oikos* 80:177–178.

Abrams, P. A. 1977. Density-independent mortality and interspecific competition: A test of Pianka's niche overlap hypothesis. *American Naturalist* 111:539–552.

———. 1994. The fallacies of "ratio-dependent" predation. *Ecology* 75:1842–1850.

———. 1998. Apparent competition or apparent mutualism? Shared predation when populations cycle. *Ecology* 79:201–212.

———. 2000. The impact of habitat selection on the heterogeneity of resources in varying environments. *Ecology* 81:2902–2913.

Abrams, P. A., and L. R. Ginzburg. 2000. The nature of predation: Prey dependent, ratio dependent, or neither? *Trends in Ecology and Evolution* 15:337–341.

Abrams, P. A., and J. D. Roth. 1994. The effects of enrichment of three-species food chains with nonlinear functional responses. *Ecology* 75:1118–1130.

Aeschlimann, J. P. 1969. Contribution a l'étude de trois espèce d'Eulophides [Hym. Chalcidoidea] parasites de la tordeuse grise du mélèze, *Zeiraphera diniana* Guenee [Lep. Tortricidae] en Haut-Engadine. *Entomophaga* 14:261–320.

Ågren, G. I., and E. Bosatta. 1996. *Theoretical ecosystems ecology.* Cambridge University Press, Cambridge.

Akcakaya, H. R. 1992. Population cycles of mammals: Evidence for a ratio-dependent predation hypothesis. *Ecological Monographs* 62:119–142.

Akcakaya, R., R. Arditi, and L. R. Ginzburg. 1995. Ratio-dependent predation: An abstraction that works. *Ecology* 76:995–1004.

Anderson, R. M., and R. M. May. 1978. Regulation and stability of host-parasite population interactions. I. Regulatory processes. *Journal of Animal Ecology* 47:219–247.

———. 1980. Infectious diseases and population cycles of forest insects. *Science* 210:658–661.

————. 1991. *Infectious diseases of humans: Dynamics and control*. Oxford University Press, Oxford.

Andersson, M., and S. Erlinge. 1977. Influence of predation on rodent populations. *Oikos* 29:591–597.

Andrewartha, H. G. 1957. The use of conceptual models in population ecology. *Cold Spring Harbor Symposium on Quantitative Biology* 22:219–232.

Andrewartha, H. G., and L. C. Birch. 1954. *The distribution and abundance of animals*. University of Chicago Press, Chicago, Ill.

Arditi, R., and L. R. Ginzburg. 1989. Coupling in predator-prey dynamics: Ratio-dependence. *Journal of Theoretical Biology* 139:311–326.

Auer, C. 1977. Dynamik von Lärchenwicklerpopulationen längs des Alpenbogens. *Mitteilungen der Eidgenössischen Anstalt für forstliches Versuchswesen* 53:71–105.

Bailey, B., E. Ellner, and D. Nychka. 1997. Chaos with confidence: Asymptotics and applications of local Lyapunov exponents. In C. Cutler and D. T. Kaplan, editors, *Nonlinear dynamics and time series: Building a bridge between natural and statistical sciences*. American Mathematical Society, Providence, R.I.

Baltensweiler, W. 1958. Zur Kenntnuis der Parasiten des Grauen Larchenwichlers (*Zeiraphera griseana* Hubner) in Oberengadin. *Mitteilungen der Eidgenössen Anstalt für das forstliche Versuchswesen* 34:399–477.

————. 1977. Colour-polymorphism and dynamics of larch budmoth populations (*Zeiraphera diniana* Gn., Lep. Tortricidae). *Mitteilungen der Schweizerischen Entomologischen Gesellschaft* 50:15–23.

————. 1993a. A contribution to the explanation of the larch bud moth cycle, the polymorphic fitness hypothesis. *Oecologia* 93:251–255.

————. 1993b. Why the larch bud-moth cycle collapsed in the subalpine larch-cembran pine forests in the year 1990 for the first time since 1850. *Oecologia* 94:62–66.

Baltensweiler, W., and A. Fischlin. 1988. The larch budmoth in the Alps. Pages 331–351 in A. A. Berryman, editor, *Dynamics of forest insect populations: Patterns, causes, implications*. Plenum Press, New York.

Barras, S. J. 1970. Antagonism between *Dendroctonus frontalis* and the fungus *Ceratocystis minor*. *Annals of Entomological Society of America* 63:1187–1190.

Bartlett, M. S. 1966. *An introduction to stochastic processes*. Cambridge University Press, London.

Batzli, G. O. 1992. Dynamics of small mammal populations: A review. Pages 831–850 in D. R. McCullough and R. H. Barrett, editors, *Wildlife 2001: Populations*. Elsevier, London.

———. 1993. Food selection lemmings. Pages 831–850 in N. C. Stenseth and R. A. Ims, editors, *The biology of lemmings*. Academic Press, London.

———. 1996. Population cycles revisited. *Trends in Ecology and Evolution* 11:488–489.

Bazykin, A. D. 1974. Sistema Volterra i uravnenie Mihaelisa-Menten. Pages 103–143 in *Voprosy matematicheskoy genetiki*. Nauka, Novosibirsk, Russia.

Beal, J. A. 1927. Weather as a factor in southern pine beetle control. *Journal of Forestry* 25:741–742.

———. 1933. Temperature extremes as a factor in the ecology of the southern pine beetle. *Journal of Forestry* 31:328–336.

Beddington, J. R. 1975. Mutual interference between parasites or predators and its effect on searching efficiency. *Journal of Animal Ecology* 44:331–340.

Beddington, J. R., C. A. Free, and J. H. Lawton. 1976a. Dynamic complexity in predator-prey model framed in simple difference equations. *Nature* 225:58–60.

Beddington, J. R., M. P. Hassell, and J. H. Lawton. 1976b. The components of arthropod predation. II. The predator rate of increase. *Journal of Animal Ecology* 45:165–185.

Benz, G. 1974. Negative Ruckkopelung durch Raum- und Nahrungskonkurrenz sowie zyklische Veranderung der Nahrungsgrundlage als Regelsprinzip in der Popultionsdynamik des Grauen Larchenwicklers, *Zeiraphera diniana* (Guenee) (Lep. Tortricidae). *Zeitschrift für angewandte Entomologie* 76:196–228.

Berisford, C. W. 1980. Natural enemies and associated organisms. Pages 31–54 in R. C. Thatcher, J. L. Searcy, J. E. Coster, and G. D. Hertel, editors, *The southern pine beetle*. USDA–Forest Service Technical Bulletin 1631. US Department of Agriculture, Pineville, La.

Berryman, A. A. 1976. Theoretical explanation of mountain pine beetle dynamics in lodgepole pine forests. *Environmental Entomology* 5:1225–1233.

———. 1978. Population cycles of the douglas-fir tussock moth (Lepidoptera: Lymantriidae): The time-delay hypothesis. *Canadian Entomologist* 110:513–518.

———. 1991. Population theory: An essential ingredient in pest prediction, management, and policy making. *American Entomologist* 37:138–142.

———. 1999. *Principles of population dynamics and their application.* Stanley Thornes Publishers, Cheltenham, UK.

———. 2002. Population: A central concept for ecology? *Oikos* forthcoming.

Berryman, A. A., M. Lima, and B. A. Hawkins. subm. Population regulation, emergent properties, and a requiem for density dependence. *Oikos.*

Berryman, A. A., N. C. Stenseth, and A. S. Isaev. 1987. Natural regulation of herbivorous forest insect populations. *Oecologia* 71:174–184.

Berryman, A. A., N. C. Stenseth, and D. J. Wollkind. 1984. Metastability of forest ecosystems infested by bark beetles. *Researches in Population Ecology* 26:13–29.

Berryman, A. A., and P. Turchin. 2001. Identifying the density-dependent structure underlying ecological time series. *Oikos* 92:265–270.

Berryman, D., B. Bobée, D. Cluis, and J. Haemmerli. 1988. Nonparametric tests for trend detection in water quality time series. *Water Resources Bulletin* 24:545–556.

Billings, R. F. 1979. Detecting and aerially evaluating southern pine beetle outbreaks. *Southern Journal of Applied Forestry* 3:50–54.

———. 1990. Southern pine beetle: A decline in 1990? *Texas Forestry* 31:8–9.

Bjornstad, O. 2001. Chitty cycles—at last! *Trends in Ecology and Evolution* 16.

Boonstra, R. 1994. Population cycles in microtines: The senescence hypothesis. *Evolutionary Ecology* 8:196–219.

Boonstra, R., D. Hik, G. R. Singleton, and A. Tinnikov. 1998. The impact of predator-induced stress on the snowshoe hare cycle. *Ecological Monographs* 79:371–394.

Boutin, S. 1984. Effect of late winter food addition on numbers and movements of snowshoe hares. *Oecologia* 62:393–400.

———. 1992. Predation and moose population dynamics: A critique. *Journal of Wildlife Management* 56:116–127.

Bowers, R. G., M. Begon, and D. E. Hodgkinson. 1993. Host-pathogen population cycles in forest insects? Lessons from simple models reconsidered. *Oikos* 67:529–538.

Box, G.E.P., and D. R. Cox. 1964. An analysis of transformations. *Journal of Royal Statistical Society B* 26:211–252.

Box, G.E.P., and N. R. Draper. 1987. *Empirical model-building and response surfaces*. Wiley and Sons, New York.

Box, G.E.P., and G. M. Jenkins. 1976. *Time series analysis: Forecasting and control*. Holden Day, Oakland, Calif.

Boyce, M. S., and E. M. Anderson. 1999. Evaluating the role of carinvores in the greater Yellowstone ecosystem. Pages 265–283 in T. W. Clark, A. P. Curlee, S. C. Minta, and P. M. Kareiva, editors, *Carnivores in ecosystems*. Yale University Press, New Haven, Conn.

Brooks, R. T., H. R. Smith, and W. M. Healy. 1998. Small-mammal abundance at three elevations on a mountain in central Vermont, USA: A sixteen-year record. *Forest Ecology and Management* 110:181–193.

Brown, J. H. 1997. An ecological perspective on the challenge of complexity. EcoEssay Series Number 1. National Center for Ecological Analysis and Synthesis. Santa Barbara, Calif. http://www.nceas.ucsb.edu/fmt/doc?/nceas-web/resources/ecoessay/brown/.

Bulmer, M. G. 1974. A statistical analysis of the 10-year cycle in Canada. *Journal of Animal Ecology* 43:701–718.

Burnett, T. 1958. Effect of host distribution on the reproduction of *Encarsia formosa* Gahan (Hymenoptera: Chalcidoidea). *Canadian Entomologist* 90:179–191.

Burnham, K. P., and D. R. Anderson. 1998. *Model selection and inference: A practical information-theoretic approach*. Springer, New York.

Calder, W. A. 1983. An allometric approach to population cycles of mammals. *Journal of Theoretical Biology* 100:275–282.

Camus, P. A., and P. Lima. 2001. Populations, metapopulations, and the open-closed dilemma: The conflict between operational and natural population concepts. *Oikos*, forthcoming.

Cappuccino, N., and S. Harrison. 1996. Density-perturbation experiments for understanding population regulation. Pages 53–64 in R. B. Floyd, A. W. Sheppard, and P. J. De Barro, editors, *Frontiers of population ecology*. CSIRO.

Casdagli, M., S. Eubank, J. D. Farmer, and J. Gibson. 1991. State space reconstruction in the presence of noise. *Physica D* 51:52–98.

Caswell, H. 2000. *Matrix population models: Construction, analysis, and interpretation*. 2nd edition. Sinauer Associates, Sunderland, Mass.

Caswell, H., R. M. Nisbet, A. M. de Roos, and S. Tuljapurkar. 1997. Structured population models: Many methods, a few basic concepts. Pages 3–18 in S. Tuljapurkar and H. Caswell, editors, *Structured-population models in marine, terrestrial, and freshwater systems*. Chapman and Hall, New York.

Caughley, G., and J. Lawton. 1981. Plant-herbivore systems. Pages 132–166 in R. M. May, editor, *Theoretical ecology: principles and applications.* Sinauer Associates, Sunderland, Mass.

Chapman, R. N. 1928. Quantitative analysis of environmental factors. *Ecology* 9:111–122.

Charnov, E. L., and J. P. Finerty. 1980. Vole population cycles: A case for kin-selection? *Oecologia* 45:1–2.

Chatfield, C. 1989. *The analysis of time series: An introduction.* 4th edition. Chapman and Hall, London.

Cheng, B., and H. Tong. 1992. On consistent nonparametric order determination and chaos. *Journal of the Royal Statistical Society B* 54:427–449.

———. 1996. Orthogonal projection, embedding dimension, and sample size in chaotic time-series from a statistical perspective. Pages 1–30 in H. Tong, editor, *Chaos and forecasting: Proceedings of the Royal Society discussion meeting.* World Scientific, Singapore.

Chernyavsky, F. B., and I. V. Dorogoy. 1988. Relations between myophagous predators and lemmings in Arctic ecosystem (with Wrangel Island taken as an example). *Zhurnal Obschei Biologii* 49:813–824.

Chernyavsky, F. B., S. P. Kiryuschenko, and T. V. Kiryuschenko. 1981. Materials on the winter ecology of the brown (*Lemmus sibiricus*) and collared (*Dicrostonyx torquatus*) lemmings. Pages 99–122 in V. G. Krivosheev, editor, *Ecology of mammals and birds of Wrangel Island.* USSR Academy of Sciences, Vladivostok.

Chernyavsky, F. B., and A. V. Tkachev. 1982. *Population cycles of lemmings in the Arctic: Ecological and endocrine aspects* (in Russian). Nauka, Moscow.

Cherrett, J. M. 1988. Key concepts: The results of a survey of our members' opinions. Pages 1–16 in J. M. Cherrett, editor, *Ecological concepts.* Blackwell Scientific, Oxford.

Chesson, P. L. 1978. Predator-prey theory and variability. *Annual Review of Ecology and Systematics* 9:323–347.

———. 1981. Models for spatially distributed populations: The effect of within-patch variability. *Theoretical Population Biology* 19:288–325.

———. 1982. The stabilizing effect of a random environment. *Journal of Mathematical Biology* 15:1–36.

Chitty, D. 1967. The natural selection of self-regulatory behavior in animal populations. *Proceedings of the Ecological Society of Australia* 2:51–78.

———. 1996. *Do lemmings commit suicide? Beautiful hypothesis and ugly facts.* Oxford University Press, New York.

Cole, L. C. 1951. Population cycles and random oscillations. *Journal of Wildlife Management* 15:233–252.

Cooper, G. 2001. The balance of nature controversy in population ecology. *Biology and Philosophy*, in press.

Costantino, R. F., J. M. Cushing, B. Dennis, and R. A. Desharnais. 1995. Experimentally induced transitions in the dynamic behaviour of insect populations. *Nature* 375:227–230.

Costantino, R. F., R. A. Desharnais, J. M. Cushing, and B. Dennis. 1997. Chaotic dynamics in an insect population. *Science* 275:389–391.

Craighead, F. C. 1925. Bark beetle epidemics and rainfall deficiency. *Journal of Economic Entomology* 18:557–586.

Crawley, M. J. 1983. *Herbivory: The dynamics of animal-plant interactions.* University of California Press, Berkeley.

Crête, M. 1987. The impact of sport hunting on North American moose. *Swedish Wildlife Research Supplement* 1:553–563.

———. 1998. Ecological correlates of regional variation in life history of the moose *Alces alces*: Comment. *Ecology* 79:1836–1838.

Crête, M., and J. Bédard. 1975. Daily browse consumption by moose in the Gaspé Peninsula, Quebec. *Journal of Wildlife Management* 39:368–373.

Crowcroft, P. 1991. *Elton's ecologists.* University of Chicago Press, Chicago.

DeAngelis, D. L., R. L. Goldstein, and R. V. O'Neill. 1975. A model for trophic interaction. *Ecology* 56:881–892.

Debrot, S. 1981. Trophic relations between the stoat (*Mustela erminea*) and its prey, mainly the water vole (*Arvicola terrestris* Sherman). Pages 1259–1289 in J. A. Chapman and D. Pursley, editors, *Worldwide Furbearer Conference proceedings*, vol. II. Worldwide Furbearer Conference, Inc., Frostburg, Md.

Delucchi, V. 1982. Parasitoids and hyperparasitoids of *Zeiraphera diniana* (Lep., Tortricidae) and their role in population control in outbreak areas. *Entomophaga* 27:77–92.

Delucchi, V., and A. Renfer. 1977. The level of abundance of *Phytodietus griseanae* Kerrich (Hymenoptera: Ichneumonidae) determined by its host *Zeiraphera diniana* Guenee (Lep., Tortricidae) at high altitude. *Mitteilungen der Schweizerischen Entomologischen Gesellschaft* 50:233–248.

Den Boer, P. J., and J. Reddingius. 1996. *Regulation and stabilization paradigms in population ecology.* Chapman and Hall, London.

Dennis, B., R. A. Desharnais, J. M. Cushing, and R. F. Costantino. 1995. Nonlinear demographic dynamics: Mathematical models, statistical methods, and biological experiments. *Ecological Monographs* 65:261–281.

Dennis, B., and B. Taper. 1994. Density dependence in time series observations of natural populations: Estimation and testing. *Ecological Monographs* 64:205–224.

Dixon, A.F.G. 1990. Population dynamics and abundance of deciduous tree-dwelling aphids. Pages 11–23 in A. D. Watt, S. R. Leather, M. D. Hunter, and N. A. C. Kidd, editors, *Population dynamics of forest insects*. Intercept, Andover, Hampshire, UK.

Dobson, A. P., and P. J. Hudson. 1992. Regulation and stability of a free-living host-parasite system: *Trichostrongylus tenuis* in red grouse. II. Population models. *Journal of Animal Ecology* 61:487–498.

Dombrovsky, V. V. 1971. Principles of fluctuations in the abundance of *Microtus arvalis* Pall. in natural foci of tularemia in the Moscow province, associated with landscape features and human activity (in Russian). *Fauna and Ecology of Rodents* 10:199–216.

Dunn, J. P., and P. L. Lorio. 1993. Modified water regimes affect photosynthesis, xylem water potential, cambial growth, and resistance of juvenile *Pinus taeda* L. to *Dendroctonus frontalis* (Coleoptera: Scolytidae). *Environmental Entomology* 22:948.

Durant, S. M. 1998. Competition refuges and coexistence: an example from Serengeti carnivores. *Journal of Animal Ecology* 67:370–386.

Dwyer, G., J. Dushoff, J. S. Elkinton, and S. A. Levin. 2000. Pathogen-driven cycles in forest defoliators revisited: Building models from experimental data. *American Naturalist* 156:105–120.

Easterling, M. R., S. P. Ellner, and P. Dixon. 2000. Size-specific sensitivity: Applying a new structured population model. *Ecology* 81:694–708.

Eberhardt, L. L. 1997. Is wolf predation ratio-dependent? *Canadian Journal of Zoology* 75:1940–1944.

———. 1998. Applying difference equations to wolf predation. *Canadian Journal of Zoology* 76:380–386.

———. 2000. Reply: Predator-prey ratio dependence and regulation of moose populations. *Canadian Journal of Zoology* 78:511–513.

Ekerholm, P., L. Oksanen, and T. Oksanen. 2001. Long-term dynamics of voles and lemmings at the timberline and above the willow limit as a test of hypothesis on trophic interactions. *Ecography* 24:555–568.

Eckmann, J.-P., and D. Ruelle. 1985. Ergodic theory of chaos and strange attractors. *Review of Modern Physics* 57:617–656.

Edelstein-Keshet, L. 1984. *Mathematical theory for plant-herbivore systems.* Lefschetz Center for Dynamical Systems, Providence, R.I.

———. 1988. *Mathematical models in biology.* Random House, New York.

Edelstein-Keshet, L., and M. D. Rausher. 1989. The effects of inducible plant defenses on herbivore populations. I. Mobile herbivores in continuous time. *American Naturalist* 133:787–810.

Ellner, S. 1989. Inferring the causes of population fluctuations. Pages 286–307 in C. Castillo-Chavez, S. A. Levin, and C. A. Shoemaker, editors, *Mathematical approaches to problems in resource management and epidemiology,* Lecture Notes in Biomathematics, vol. 81. Springer-Verlag, Berlin.

Ellner, S., B. Bailey, G. Bobashev, A. R. Gallant, B. Grenfell, and D. W. Nychka. 1998. Noise and nonlinearity in measles epidemics: Combining mechanistic and statistical approaches to population modeling. *American Naturalist* 151:425–440.

Ellner, S., and P. Turchin. 1995. Chaos in a noisy world: New methods and evidence from time series analysis. *American Naturalist* 145:343–375.

Elton, C. 1949. Population interspersion: An essay on animal community patterns. *Journal of Ecology* 37:1–23.

Elton, C., and M. Nicholson. 1942. The ten-year cycle in numbers of the lynx in Canada. *Journal of Animal Ecology* 11:215–244.

Elton, C. S. 1924. Periodic fluctuations in the number of animals: Their causes and effects. *British Journal of Experimental Biology* 2:119–163.

Erlinge, S., G. Goransson, L. Hansson, G. Hogstedt, O. Liberg, I. N. Nilsson, T. von Schantz, and M. Sylven. 1983. Predation as a regulating factor on small rodent populations in southern Sweden. *Oikos* 40:36–52.

Errington, P. L. 1963. The phenomenon of predation. *American Scientist* 51:180–192.

Evsikov, V. I., and M. P. Mosjkin. 1994. Dynamics and homeostasis of natural animal populations. *Siberian Journal of Ecology* 4:323–337.

Fagerstrom, T. 1987. On theory, data, and mathematics in ecology. *Oikos* 50:258–261.

Falck, W., O. N. Bjornstad, and N. C. Stenseth. 1995. Bootstrap estimated uncertainty of the dominant Lyapunov exponent for Holarctic microtine rodents. *Proceedings of the Royal Society of London B* 261:159–165.

Finerty, J. P. 1979. Cycles in Canadian lynx. *American Naturalist* 114:453–455.

———. 1980. *The population ecology of cycles in small mammals.* Yale University Press, New Haven, Conn.

Fischlin, A. 1982. Analyse eines wald-insekten-systems: Der subalpine Lärchenarvenwald und der graue Lärchenwickler *Zeiraphera diniana* Gn. (Lep., Torticidae). Ph.D. dissertation no. 6977. ETH, Zurich.

Fischlin, A., and W. Baltensweiler. 1979. Systems analysis of the larch bud moth system. Part 1: the larch–larch bud moth relationship. *Mitteilungen der Schweizerischen Entomologischen Gesellschaft* 52:273–289.

Fisher, R. A. 1937. The wave of advance of advantageous genes. *Ann Eugen* 7:355–369.

Flamm, R. O., R. N. Coulson, and T. L. Payne. 1988. The southern pine beetle. Pages 531–553 in A. A. Berryman, editor, *Dynamics of forest insect populations*. Plenum Press, New York.

Forbes, G. J., and J. B. Theberge. 1996. Response by wolves to prey variation in central Ontario. *Canadian Journal of Zoology* 74:1511–1520.

Framstad, E., N. C. Stenseth, O. N. Bjornstad, and W. Falck. 1997. Limit cycles in Norwegian lemmings: Tensions between phase-dependence and density dependence. *Proceedings of the Royal Society of London B* 264:31–38.

Fryxell, J., J. Falls, and R. Brooks. 1998. Long-term dynamics of small-mammal population in Ontario. *Ecology* 79:213–228.

Fryxell, J. M. 1988. Population dynamics of Newfoundland moose using cohort analysis. *Journal of Wildlife Management* 52:14–21.

Fryxell, J. M., D.J.T. Hussell, A. B. Lambert, and P. C. Smith. 1991. Time lags and population fluctuations in white-tailed deer. *Journal of Wildlife Management* 55:377–385.

Fuller, T. K., and L. B. Keith. 1980. Wolf population dynamics and prey relationships in northeastern Alberta. *Journal of Wildlife Management* 44:583–602.

Garsd, A., and W. E. Howard. 1981. A 19-year study of microtine population fluctuations using time-series analysis. *Ecology* 62:930–937.

Gasaway, W., and J. W. Coady. 1974. Review of energy requirements and rumen fermentation in moose and other ruminants. *Naturaliste Canadien* 101:227–262.

Getz, L. L., J. E. Hoffman, L. Verner, F. R. Cole, and R. L. Lindroth. 1987. Fourteen years of population fluctuations of *Microtus ochrogaster* and *M. pennsylvanicus* in east central Illinois. *Canadian Journal of Zoology* 65:1317–1325.

Getz, W. 1991. A unified approach to multispecies modeling. *Natural Resource Modeling* 5:393–421.

Gilbert, B. S., and C. J. Krebs. 1991. Population dynamics of *Clethrionomys* and *Peromiscus* in southwestern Yukon, 1973–1989. *Holarctic Ecology* 14:250–259.

Gilpin, M. E. 1973. Do hares eat lynx? *American Naturalist* 107:727–730.

Ginzburg, L. R. 1970. On the dynamics and control of population age structure (in Russian). *Problemy Kibernetiki* 23:261–274.

———. 1986. The theory of population dynamics. I. Back to first principles. *Journal of Theoretical Biology* 122:385–399.

———. 1998. Assuming reproduction to be a function of consumption raises doubts about some popular predator-prey models. *Journal of Animal Ecololgy* 67:325–327.

Ginzburg, L. R., and P. Inchausti. 1997. Asymmetry of population cycles: Abundance-growth representation of hidden causes in ecological dynamics. *Oikos* 80:435–447.

Ginzburg, L. R., and D. E. Taneyhill. 1994. Population cycles of forest Lepidoptera: A maternal effect hypothesis. *Journal of Animal Ecology* 63:79–92.

Gleick, J. 1988. *Chaos: Making a new science*. Viking, New York.

Godfray, H.C.J. 1994. *Parasitoids: Behavioral and evolutionary ecology*. Princeton University Press, Princeton, N.J.

Gotelli, N. J. 1995. *A primer of ecology*. Sinauer Associates, Sunderland, Mass.

Goyer, R., and J. Hayes. 1991. Understanding the southern pine beetle. *Forests and People 1991* (4th quarter):10.

Green, R. G., and C. A. Evans. 1940. Studies on a population cycle of snowshoe hares in the Lake Alexander area. I. Gross annual censuses, 1932–1939. *Journal of Wildlife Management* 4:220–238.

Gurney, W.S.C., and R. M. Nisbet. 1998. *Ecological dynamics*. Oxford University Press, New York.

Hanski, I. 1990. Density-dependence, regulation, and variability in animal populations. *Philosophical Transactions of the Royal Society of London B* 330:141–150.

Hanski, I., L. Hansson, and H. Henttonen. 1991. Specialist predators, generalist predators, and the microtine rodent cycle. *Journal of Animal Ecology* 60:353–367.

Hanski, I., and H. Henttonen. 1996. Predation on competing rodent species: A simple explanation of complex patterns. *Journal of Animal Ecology* 65:220–232.

Hanski, I., H. Henttonen, and L. Hansson. 1994. Temporal variability and geographic patterns in the population density of microtine rodents: A reply to Xia and Boonstra. *American Naturalist* 144:329–342.

Hanski, I., H. Henttonen, E. Korpimäki, L. Oksanen, and P. Turchin. 2001. Small rodent dynamics and predation. *Ecology* 82:1505–1520.

Hanski, I., and E. Korpimäki. 1995. Microtine rodent dynamics in northern Europe: Parameterized models for the predator-prey interaction. *Ecology* 76:840–850.

Hanski, I., P. Turchin, E. Korpimäki, and H. Henttonen. 1993. Population oscillations of boreal rodents: Regulation by mustelid predators leads to chaos. *Nature* 364:232–235.

Hansson, L. 1987. An interpretation of rodent dynamics as due to trophic interactions. *Oikos* 50:308–318.

———. 1999. Intraspecific variation in dynamics: Small rodents between food and predation in changing landscapes. *Oikos* 86:159–169.

Hansson, L., and H. Henttonen. 1985. Gradients in density variations of small rodents: The importance of latitude and snow cover. *Oecologia* 67:394–402.

———. 1988. Rodent dynamics as a community process. *Trends in Ecology and Evolution* 3:195–200.

Harrison, G. W. 1995. Comparing predator-prey models to Luckinbill's experiment with *Didinium* and *Paramecium*. *Ecology* 76:357–374.

Hassell, M. P. 1978. *The dynamics of arthropod predator-prey systems.* Princeton University Press, Princeton, N.J.

Hassell, M. P., M. J. Crawley, H.C.J. Godfray, and J. H. Lawton. 1998. Top-down versus bottom-up and the Ruritanian bean bug. *Proceedings of the National Academy of Sciences of USA* 95:10661–10664.

Hassell, M. P., J. H. Lawton, and R. M. May. 1976. Patterns of dynamical behavior in single species populations. *Journal of Animal Ecology* 45:471–486.

Hassell, M. P., and G. C. Varley. 1969. New inductive population model for insect parasites and its bearing on biological control. *Nature* 223:1133–1136.

Hastings, A., and K. Higgins. 1994. Persistence of transients in spatially structured ecological models. *Science* 263:1133–1136.

Hastings, A., and T. Powell. 1991. Chaos in a three-species food chain. *Ecology* 72:896–903.

Haukioja, E., S. Neuvonen, S. Hanhimaki, and P. Niemelä. 1987. The autumnal moth in Fennoscandia. Pages 163–178 in A. A. Berryman, editor, *Dynamics of forest insect populations*. Plenum Press, New York.

Hendry, R., P. J. Bacon, R. Moss, S.C.F. Palmer, and J. McGlade. 1997. A two-dimensional individual-based model of territorial behaviour: Possible population consequences of kinship in red grouse. *Ecological Modelling* 105:23–39.

Henttonen, H., and I. Hanski. 2000. Population dynamics of small rodents in northern Fennoscandia. Pages 73–96 in J. N. Perry, R. H. Smith, I. P. Woiwod, and D. Morse, editors, *Chaos in real data*. Kluwer Academic Press, Dordrecht, The Netherlands.

Henttonen, H., T. Oksanen, A. Jortikka, and V. Haukislami. 1987. How much do weasels shape microtine cycles in the northern Fennoscandian taiga? *Oikos* 50:353–365.

Herren, H. R. 1976. Manipulation d'une population de la tordeuse grise du mélèze en vue d'augmenter l'efficacité de ses parasitoids. *Mitteilungen der Schweizerischen Entomologischen Gesellschaft* 49:307–308.

———. 1977. Le rôle des eulophides dans la gradation de la tordeuse grise de mélèze, *Zeiraphera diniana* Guenee (Lep., Tortricidae) en Haute-Engadine. Ph.D. dissertation no. 6037. ETH, Zurich.

Hik, D. 1995. Does risk of predation influence population dynamics? Evidence from the cyclic decline of snowshoe hares. *Wildlife Research* 22:115–129.

Hilborn, R., and M. Mangel. 1997. *The ecological detective*. Princeton University Press, Princeton, N.J.

Hodges, J. D., W. W. Elam, W. F. Watson, and N. E. Nebecker. 1979. Oleoresin characteristics and susceptibility of four southern pines to southern pine beetle (Coleoptera: Scolytidae) attacks. *Canadian Entomologist* 111:889–896.

Holling, C. S. 1959. The components of predation as revealed by a study of small mammal predation on the European pine sawfly. *Canadian Entomologist* 91:293–320.

———. 1965. The functional response of predators to prey density and its role in mimicry and population regulation. *Memoirs of the Entomological Society of Canada* 45:5–60.

Holsinger, K. E. 2000. Demography and extinction in small populations. Pages 55–74 in A. Young and G. Clarke, editors, *Genetics, demography, and variability of fragmented populations*. Cambridge University Press, Cambridge.

Holt, R. D. 1977. Predation, apparent competition, and the structure of prey communities. *Theoretical Population Biology* 12:197–229.

Holt, R. D., and J. H. Lawton. 1993. Apparent competition and enemy-free space in insect host-parasitoid communities. *American Naturalist* 142:623–645.

Hörnfeldt, B. 1994. Delayed density dependence as a determinant of vole cycles. *Ecology* 75:791–806.

Hudson, P., A. P. Dobson, and D. Newborn. 1999. Population cycles and parasitism: Response. *Science* 286:2425a.

Hudson, P. J. 1986. The effect of a parisitic nematode on the breeding production of the ref grouse. *Journal of Animal Ecology* 55:85–92.

———. 1992. *Grouse in space and time: The population biology of a managed gamebird.* Game Conservancy Trust, Fordingbridge, UK.

Hudson, P. J., and A. P. Dobson. 1996. Transmission dynamics and host-parasite interactions of *Trichostrongylus tenuis* in red grouse. *Journal of Parasitology* 83:194–202.

Hudson, P. J., A. P. Dobson, and D. Newborn. 1998. Prevention of population cycles by parasite removal. *Science* 282:2256–2258.

———. 2002. Parasitic nematodes and population cycles of red grouse. In A. A. Berryman, editor, *Population cycles: Evidence for trophic interactions.* Oxford University Press, Oxford.

Hudson, P. J., D. Newborn, and A. P. Dobson. 1992. Regulation and stability of a free-living host-prasite system: *Trichostrongylus tenius* in red grouse. I. Monitoring and parasite reduction experiments. *Journal of Animal Ecology* 61:477–486.

Hunter, A. F., and G. Dwyer. 1998. Outbreaks and interactive factors: Insect population explosions synthesized and dissected. *Integrative Biology* 1:166–177.

Hutchinson, G. E. 1948. Circular causal systems in ecology. *Annals of the New York Academy of Science* 50:221–246.

Inchausti, P., and L. R. Ginzburg. 1998. Small mammals cycles in northern Europe: Patterns and evidence for a maternal effect hypothesis. *Journal of Animal Ecology* 67:180–194.

Ivankina, E. V. 1987. Numerical dynamics and population structure of bank vole near Moscow. Ph.D. dissertation. Moscow University, Moscow.

Ivanter, E. V. 1981. Population dynamics. Pages 245–267 in N. V. Bashenina, editor, *The European red vole* (in Russian). Nauka, Moscow.

Ives, A. R., and D. L. Murray. 1997. Can sublethal parasitism destabilize predator-prey population dynamics? A model of showshoe hares, predators, and parasites. *Journal of Animal Ecology* 66:265–278.

Ivlev, V. S. 1961. *Experimental ecology of feeding fishes.* Yale University Press, New Haven, Conn.

Jassby, A. D., and T. M. Powell. 1990. Detecting changes in ecological time series. *Ecology* 71:2044–2052.

Jedrzejewski, W., and B. Jedrzejewska. 1996. Rodent cycles in relation to biomass and productivity of ground vegetation and predation in the Palearctic. *Acta Theriologica* 41:1–34.

Jones, J. W. 1914. *Fur farming in Canada.* Commission of Conservation, Ottawa.

Jorde, P. E., R. Boonstra, N. C. Stenseth, N. G. Yoccoz, and E. Tkadlec. 2002. Microtine rodent population cycles: Maternal effects and seasonality. Unpublished manuscript.

Kalkstein, L. S. 1981. An improved technique to evaluate climate–southern pine beetle relationships. *Forest Science* 27:579–589.

Keith, L. B. 1963. *Wildlife's ten-year cycle.* University of Wisconsin Press, Madison.

———. 1974. Some features of population dynamics in mammals. *Proceedings of the International Congress of Game Biologists* 11:17–58.

———. 1990. Dynamics of snowshoe hare populations. Pages 119–195 in H. H. Genoways, editor, *Current mammalogy.* Plenum Press, New York.

Keith, L. B., J. R. Cary, O. J. Rongstad, and M. C. Brittingham. 1984. Demography and ecology of a declining snowshoe hare population. *Wildlife Management* 90:1–43.

Keith, L. B., A. W. Todd, C. J. Brand, R. S. Adamcik, and D. H. Rusch. 1977. An analysis of predation during a cyclic fluctuation of snowshoe hares. *Proceedings of International Congress of Game Biologists* 13:151–175.

Keith, L. B., and L. A. Windberg. 1978. *A demographic analysis of the snowshoe hare cycle.* Wildlife Monographs 58.

Kendall, B. E., C. J. Briggs, W. W. Murdoch, P. Turchin, S. P. Ellner, E. McCauley, R. M. Nisbet, and S. N. Wood. 1999. Why do populations cycle? A synthesis of statistical and mechanistic modeling approaches. *Ecology* 80:1789–1805.

Kendall, B. E., J. Predergast, and O. N. Bjornstad. 1998. The macroecology of population dynamics: Taxonomic and biogeographic patterns in population cycles. *Ecology Letters* 1:160–164.

Kendall, B. E., W. M. Schaffer, and C. W. Tidd. 1993. Transient periodicity in chaos. *Physics Letters A* 177:13–20.

King, A. A., and W. M. Schaffer. 1999. The rainbow bridge: Hamiltonian limits and resonance in predator-prey dynamics. *Journal of Mathematical Biology* 39:439–469.

———. 2001. The geometry of a population cycle: A mechanistic model of snowshoe hare demography. *Ecology* 82:814–830.

King, E. W. 1972. Rainfall and epidemics of the southern pine beetle. *Environmental Entomology* 1:279–285.

Kingsland, S. E. 1995. *Modeling nature: Episodes in the history of population ecology*, Second edition. University of Chicago Press, Chicago, Ill.

Kinn, D. H., T. J. Perry, F. H. Guinn, B. L. Strom, and J. P. Woodring. 1994. Energy reserves of individual southern pine beetles (Coleoptera: Scolytidae) as determined by a modified phosphovanillin spectrophotometric method. *Journal of Entomological Science* 29:152–163.

Klemola, T., E. Koivula, E. Korpimäki, and K. Norrdahl. 2000. Experimental tests of predation and food hypotheses for population cycles of voles. *Proceedings of the Royal Society B* 267:351–356.

Klemola, T., K. Norrdahl, and E. Korpimäki. 2000. Do delayed effects of overgrazing explain population cycles in voles? *Oikos* 90:509–516.

Korpimäki, E. 1993. Regulation of multiannual vole cycles by density-dependant avian and mammalian predation. *Oikos* 66:359–363.

———. 1994. Rapid or delayed tracking of multi-annual vole cycles by avian predators? *Journal of Animal Ecology* 63:619–628.

Korpimäki, E., and K. Norrdahl. 1991a. Do breeding avian predators dampen population fluctuations of small mammals? *Oikos* 62:195–208.

———. 1991b. Numerical and functional responses of kestrels, short-eared owls, and long-eared owls to vole densities. *Ecology* 72:814–826.

———. 1998. Experimental reduction of predators reverses the crash phase of small-rodent cycles. *Ecology* 79:2448–2455.

Korpimäki, E., K. Norrdahl, and T. Rinta-Jaskari. 1991. Responses of stoats and least weasels to fluctuating food abundances: Is the low phase of the vole cycle due to mustelid predation? *Oecologia* 88:552–561.

Koshkina, T. V. 1966. On the periodic changes in the numbers of voles (as exemplified by the Kola Peninsula) (in Russian). *Bulletin of the Moscow Society of Naturalists, Biological Section* 71:14–26.

Kot, M., G. S. Sayler, and T. W. Schultz. 1992. Complex dynamics in a model microbial system. *Bulletin of Mathematical Biology* 54:619–648.

Krebs, C. J. 1978. A review of the Chitty hypothesis of population regulation. *Canadian Journal of Zoology* 56:2463–2480.

———. 1988. The experimental approach to rodent population dynamics. *Oikos* 52:143–149.

———. 1991. The experimental paradigm and long-term population studies. *Ibis* 133 suppl. 1:3–8.

———. 1995. Two paradigms of population regulation. *Wildlife Research* 22:1–10.

———. 1996. Population cycles revisited. *Journal of Mammalogy* 77:8–24.

Krebs, C. J., S. Boutin, and R. Boonstra. 2001. *Ecosystem dynamics in the boreal forest.* Oxford University Press, Oxford, UK.

Krebs, C. J., S. Boutin, R. Boonstra, A.R.E. Sinclair, J.N.M. Smith, M.R.T. Dale, K. Martin, and R. Turkington. 1995. Impact of food and predation on the snowshoe hare cycle. *Science* 269:1112–1115.

Krebs, C. J., B. S. Gilbert, S. Boutin, A.R.E. Sinclair, and J.N.M. Smith. 1986. Population biology of snowshoe hares. I. Demography of food-supplemented populations in southern Yukon, 1976–84. *Journal of Animal Ecology* 55:963–982.

Krebs, C. J., and J. H. Myers. 1974. Population cycles in small mammals. *Advances in Ecological Research* 8:267–399.

Kroll, J. C., and H. C. Reeves. 1978. A simple model for predicting annual numbers of southern pine beetle infestations in East Texas. *Southern Journal of Applied Forestry* 2:62–64.

Kunkel, K., and D. H. Pletscher. 1999. Species-specific population dynamics of cervids in a multipredator ecosystem. *Journal of Wildlife Management* 63:1082–1093.

Lack, D. 1954. *The natural regulation of animal numbers.* Oxford University Press, Oxford.

Lambin, X., C. J. Krebs, R. Moss, N. C. Stenseth, and N. G. Yoccoz. 1999. Population cycles and parasitism. *Science* 286:2425a.

Lambin, X., C. J. Krebs, R. Moss, and N. G. Yoccoz. 2002. Population cycles: Inferences from experimental modeling and time series approaches. In A. A. Berryman, editor, *Population cycles: Evidence for trophic interactions.* Oxford University Press, Oxford.

Lambin, X., S. J. Perry, and K. L. MacKinnon. 2000. Cyclic dynamics in field vole populations and generalist predation. *Journal of Animal Ecology* 69:106–118.

Lande, R. 1998. Demographic stochasticity and Allee effect on a scale with isotropic noise. *Oikos* 83:353–358.

Lawton, J. H. 1991. Ecology as she is done, and could be done. *Oikos* 61:289–290.

Lehtonen, A. 1998. Managing moose, *Alces alces*, population in Finland: Hunting virtual animals. *Annales Zoologici Fennici* 35:173–179.

Leigh, E. 1968. The ecological role of Volterra's equations. Pages 1–61 in M. Gerstenhaber, editor, *Some mathematical problems in biology*. American Mathematical Society, Providence, R.I.

Leslie, P. H. 1948. Some further notes on the use of matrices in population mathematics. *Biometrika* 325:213–245.

Lewis, M., and P. Kareiva. 1993. Allee dynamics and the spread of invading organisms. *Theoretical Population Biology* 43:141–158.

Lewontin, R. C. 1966. On the measurement of relative variability. *Systematic Zoology* 15:141–142.

Lidicker, W. Z. 1988. Solving the enigma of microtine cycles. *Journal of Mammalogy* 69:229–239.

———. 1991. In defense of a multifactorial perspective in population ecology. *Journal of Mammalogy* 72:631–635.

———. 2000. A food web/landscape interaction model for microtine rodent density cycles. *Oikos* 91:435–445.

Lindén, H. 1989. Characteristics of tetraonid cycles in Finland. *Finnish Game Research* 46:34–42.

Lindström, E. R., H. Andrén, P. Angelstam, G. Cederlund, B. Hörnfeldt, L. Jäderberg, P. A. Lemnell, B. Martinsson, K. Sköld, and J. E. Swenson. 1994. Disease reveals the predator: Sarcoptic mange, red fox predation, and prey populations. *Ecology* 75:1042–1049.

Lindström, J., E. Ranta, H. Kokko, P. Lundberg, and V. Kaitala. 2000. From arctic lemmings to adaptive dynamics: Charles Elton's legacy in population ecology. *Biological Reviews* 76:129–158.

Linit, M. J., and F. M. Stephen. 1983. Parasite and predator component of within-tree southern pine beetle (Coleoptera: Scolytidae) mortality. *Canadian Entomologist* 115:679–688.

Lopatin, V. N., and B. D. Abaturov. 2000. Mathematical modeling of trophically dependent cycle of reindeer (*Rangifer tarandus*) population (in Russian). *Zoologicheskiy Zhurnal* 79:461–470.

Lorenz, E. N. 1963. Deterministic nonperiodic flow. *Journal of Atmospheric Sciences* 20:130–141.

Lorio, P. L. 1986. Growth-differentiation balance: A basis for understanding southern pine beetle–tree interactions. *Forest Ecology and Management* 14:259–273.

Lorio, P. L., R. A. Sommers, C. A. Blanche, J. D. Hodges, and T. E. Nebeker. 1990. Modeling pine resistance to bark beetles based on growth and differentiation balance principles. Pages 402–409 in R. K. Dixon, R. S. Meldahl, G. A. Ruark, and W. G. Warren, editors, *Process modeling of forest growth responses to environmental stress.* Timber Press, Portland, Oreg.

Lotka, A. 1925. *Elements of physical biology.* Williams and Wilkins, Baltimore.

Ludwig, D., D. D. Jones, and C. S. Holling. 1978. Qualitative analysis of insect outbreak systems: The spruce budworm and forest. *Journal of Animal Ecology* 47:315–332.

MacArthur, R. H. 1972. *Geographical ecology.* Harper and Row, New York.

MacKenzie, J.M.D. 1952. Fluctuations in the numbers of British tetraonids. *Journal of Animal Ecology* 21:128–153.

MacLulich, D. A. 1957. The place of chance in population processes. *Journal of Wildlife Management* 21:293–299.

Malthus, T. R. 1798. *An essay on the principle of population.* J. Johnson, London.

Marcström, V., N. Höglund, and C. J. Krebs. 1990. Periodic fluctuations in small mammals at Boda, Sweden, from 1961 to 1988. *Journal of Animal Ecology* 59:753–761.

Martinat, P. J. 1987. The role of climatic variation and weather in forest insect outbreaks. Pages 241–262 in P. Barbosa and J. C. Schultz, editors, *Insect outbreaks.* Academic Press, San Diego, Calif.

Matthiopoulos, J., R. Moss, and X. Lambin. 1998. Models of red grouse cycles: A family affair? *Oikos* 82:574–590.

———. 2002. The kin facilitation hypothesis in red grouse population cycles: Territorial dynamics of the family cluster. *Ecological Modelling* 147:291–307.

———. 2000. The kin-facilitation hypothesis for red grouse population cycles: Territory sharing between relatives. *Ecological Modelling* 127:53–63.

May, R. M. 1973a. On relationships among various types of population models. *American Naturalist* 107:46–57.

———. 1973b. Stability in randomly fluctuating versus deterministic environments. *American Naturalist* 107:621–650.

———. 1974a. Biological populations with nonoverlapping populations: Stable points, stable cycles, and chaos. *Science* 186:645–647.

———. 1974b. *Stability and complexity in model ecosystems.* 2nd edition. Princeton University Press, Princeton, N.J.

———. 1976. Simple mathematical models with very complicated dynamics. *Nature* 261:459–467.

———. 1981. Models for single populations. Pages 5–29 in R. M. May, editor, *Theoretical ecology: Principles and applications.* 2nd edition. Sinauer Associates, Sunderland, Mass.

———. 1999. Crash tests for real. *Nature* 398:371–372.

May, R. M., and R. M. Anderson. 1978. Regulation and stability of host-parasite population interactions. II. Destabilizing processes. *Journal of Animal Ecology* 47:249–267.

Maynard Smith, J. 1974. *Models in ecology.* Cambridge University Press, Cambridge.

McCaffrey, D., S. Ellner, D. W. Nychka, and A. R. Gallant. 1992. Estimating the Lyapunov exponent of a chaotic system with nonlinear regression. *Journal of American Statistical Association* 87:682–695.

McCallum, H., N. Barlow, and J. Hone. 2001. How should pathogen transmission be modeled? *Trends in Ecology and Evolution* 16:295–300.

McClelland, W. T., and F. P. Hain. 1979. Survival of declining *Dendroctonus frontalis* populations during a severe and nonsevere winter. *Environmental Entomology* 8:231–235.

McIntosh, R. P. 1985. *The background of ecology: Concept and theory.* Cambridge University Press, Cambridge.

McNair, J. N. 1986. The effect of refuges on predator-prey interactions: A reconsideration. *Theoretical Population Biology* 29:38–63.

McRoberts, R. E., D. L. Mech, and R. O. Peterson. 1995. The cumulative effect of consecutive winters' snow depth on moose and deer populations: A defence. *Journal of Animal Ecology* 64:131–135.

Mech, L. D. 1966. *The wolves of Isle Royale.* US Department of the Interior, National Park Service, Washington, D.C.

———. 1977. Population trend and winter deer consumption in a Minnesota wolf pack. Pages 55–83 in R. Phillips and C. Jonkel, editors, *Proceedings of 1975 Predator Symposium.* Bulletin of Montana Forest Conservation Experiment Station, Missoula.

Mech, L. D., R. E. McRoberts, R. O. Peterson, and R. E. Page. 1987. Relationship of deer and moose populations to previous winters' snow. *Journal of Animal Ecology* 56:615–627.

Messier, F. 1991. The significance of limiting and regulating factors on the demography of moose and white-tailed deer. *Journal of Animal Ecology* 56:615–627.

———. 1994. Ungulate population models with predation: A case study with the North American moose. *Ecology* 75:478–488.

———. 1995. Is there evidence for a cumulative effect of snow on moose and deer populations? *Journal of Animal Ecology* 64:136–140.

Messier, F., and M. Crête. 1985. Moose-wolf dynamics and the natural regulation of moose populations. *Oecologia* 65:503–512.

Metz, J., and O. Diekmann. 1986. *The dynamics of physiologically structured populations.* Springer-Verlag, New York.

Michaels, P. J. 1984. Climate and southern pine beetle in Atlantic coastal and piedmont regions. *Forest Science* 30:143–156.

Michaels, P. J., P. J. Stenger, and D. E. Sappington. 1986. Modelling changes in the epidemic range of southern pine beetle using temperature and objective moisture indicators. *Theoretical and Applied Climatology* 37:39–50.

Middleton, A. D. 1934. Periodic fluctuations in British game populations. *Journal of Animal Ecology* 3:231–249.

Moran, P.A.P. 1953. The statistical analysis of the Canada lynx cycle. I. Structure and prediction. *Australian Journal of Zoology* 1:163–173.

Moss, R., and A. Watson. 1985. Adaptive value of spacing behaviour in population cycles of red grouse and other animals. Pages 275–294 in R. M. Sibley and R. H. Smith, editors, *Behavioural ecology.* Blackwell Scientific, Oxford.

———. 2001. Population cycles in birds of the grouse family (Tetraonidae). *Advances in Ecological Research* 32:53–111.

Moss, R., A. Watson, and R. Parr. 1996. Experimental prevention of a population cycle in red grouse. *Ecology* 77:1512–1530.

Moss, R., A. Watson, I. B. Trenholm, and R. Parr. 1993. Caecal threadworms *Trichostrongylus* tenuis in red grouse *Lagopus lagopus scoticus*: Effects of weather and host density upon estimated worm burdens. *Parasitology* 107:199–209.

Mountford, M. D., A. Watson, R. Moss, R. Parr, and P. Rothery. 1990. Land inheritance and population cycles of red grouse. Pages 78–83 in A. N. Lance and J. H. Lawton, editors, *Red grouse population processes.* Royal Society for Protection of Birds, Sandy, Bedsfordshire, UK.

Mueller, A. D., and A. Joshi. 2000. *Stability in model populations*. Princeton University Press, Princeton, N.J.

Murdoch, W. W. 1969. Switching in general predators: Experiments on predator specificity and stability of prey populations. *Ecological Monographs* 39:335–354.

———. 1994. Population regulation in theory and practice. *Ecology* 75:271–287.

Murray, B. G. 1992. Research methods in physics and biology. *Oikos* 64:594–596.

———. 2000. Universal laws and predictive theory in ecology and evolution. *Oikos* 89:403–408.

Murray, D. L., J. R. Cary, and L. B. Keith. 1997. Interactive effects of sublethal nematodes and nutritional status on snowshoe hare vulnerability to predation. *Journal of Animal Ecology* 66:250–264.

Myasnikov, Y. A. 1976. Distribution and numerical oscillations of rodents, lagomorpha, and insectivor of Tula district (in Russian). *Fauna and Ecology of Rodents* 13:164–236.

Myllimäki, A., A. Paasikallio, and V. Kanervo. 1971. Removal experiments on small quadrats as a means of rapid assessment of the abundance of small mammals. *Annales Zoologici Fennici* 8:177–185.

Nazarov, A. A. 1988. An investigation into chronological characteristics of wolf population in Russian Federation (in Russian). Pages 80–90 in A. A. Nazarov, editor, *Hronologicheskie Izmeneninya Chislennosti Ohotnich'ih zhivotnyh v RSFSR* (Numerical dynamics of game animals in Russian Federation). Central Research Laboratory of Game Management and wildlife Refuges, Moscow.

Nicholson, A. J. 1933. The balance of animal populations. *Journal of Animal Ecology* 2:132–178.

———. 1954. An outline of the dynamics of animal populations. *Australian Journal of Zoology* 2:9–65.

———. 1957. The self-adjustment of populations to change. *Cold Springs Harbor Symposium on Quantitative Biology* 22:153–173.

Nisbet, R., and W. Gurney. 1982. *Modelling fluctuating populations*. Wiley, New York.

Norrdahl, K., and E. Korpimäki. 1995. Mortality factors in a cyclic vole population. *Proceedings of the Royal Society of London* 261:49–53.

Nychka, D., B. Bailey, S. Ellner, P. Haaland, and M. O'Connell. 1996. *FUNFITS: Data analysis and statistical tools for estimating functions*. Report no. 2289, Institute of Statistics, North Carolina State University, Raleigh.

Nychka, D., S. Ellner, D. McCaffrey, and A. R. Gallant. 1992. Finding chaos in noisy systems. *Journal of the Royal Statistical Society B* 54:399–426.

Oksanen, L. 2001. Logic of experiments in ecology: Is pseudoreplication a pseudoissue? *Oikos* 94:27–38.

Oksanen, L., S. D. Fretwell, J. Arruda, and P. Niemela. 1981. Exploitation ecosystems in gradients of primary productivity. *American Naturalist* 118:240–261.

Oksanen, L., and T. Oksanen. 1992. Long-term microtine dynamics in north Fennoscandian tundra—the vole cycle and the lemming chaos. *Ecography* 15:226–236.

Omlin, F. X. 1977. Zur populationsdynamischen Wirkung der durch Raupenfrass und Dungung veranderten Nahrungsbasis auf den Grauen Larchenwickler *Zeiraphera diniana* Gn. (Lep: Torticidae). Ph.D. dissertation no. 6064. ETH, Zurich.

Ostfeld, R. S., C. D. Canham, and S. R. Pugh. 1993. Intrinsic density-dependent regulation of vole populations. *Nature* 366:259–261.

Ostfeld, R. S., and F. Keesing. 2000. Pulsed resources and community dynamics of consumers in terrestrial ecosystems. *Trends in Ecology and Evolution* 15:232–237.

Packard, N. H., J. P. Crutchfield, J. D. Farmer, and R. S. Shaw. 1980. Geometry from a time series. *Physical Reviews Letters* 45:712.

Park, T. 1948. Experimental studies of interspecies competition. I. Competition between population of flour beetles *Tribolium confusum* Duval and *Tribolium castaneum* Herbst. *Ecological Monographs* 18:265–308.

Parton, W. J., J.M.O. Scurlock, D. S. Ojima, T. G. Gilmanov, R. J. Scholes, D. S. Schimel, T. Kirchner, J. C. Menaut, T. Seastedt, E. Garcia Moya, A. Kamnalrut, and J. I. Kinyamario. 1993. Observations and modeling of biomass and soil organic matter dynamics for the grassland biome worldwide. *Global Biogeochemical Cycles* 7:785–809.

Payne, T. L. 1980. Life history and habits. Pages 7–30 in R. C. Thatcher, J. L. Searcy, J. E. Coster, and G. D. Hertel, editors, *The Southern Pine Beetle*. USDA–Forest Service Technical Bulletin 1631. US Department of Agriculture, Pineville, La.

Pearl, R., and L. J. Reed. 1920. On the rate of growth of the population of the United States since 1790 and its mathematical representation. *Proceedings of the National Academy of Sciences of the USA* 6:275–288.

Pease, J. L., R. H. Vowles, and L. B. Keith. 1979. Interaction of snow-shoe hares and woody vegetation. *Journal of Wildlife Management* 43:43–60.

Perry, J. N., I. P. Woiwod, and I. Hanski. 1993. Using response-surface methodology to detect chaos in ecological time series. *Oikos* 68:329–339.

Peterson, R. O. 1999. Wolf-moose interaction on Isle Royale: The end of natural regulation? *Ecological Applications* 9:10–16.

Peterson, R. O., and R. E. Page. 1988. The rise and fall of Isle Royale wolves, 1975–1986. *Journal of Mammalogy* 69:89–99.

Pickett, S.T.A., J. Kolasa, and C. G. Jones. 1994. *Ecological understanding.* Academic Press, San Diego, Calif.

Pimm, S. L. 1991. *The balance of nature?* University of Chicago Press, Chicago, Ill.

Pinter, A. J. 1988. Multiannual fluctuations in precipitation and population dynamics of the montane vole, *Microtus montanus. Canadian Journal of Zoology* 66:2128–2132.

Pitelka, F. A. 1957. Some characteristics of microtine cycles in the Arctic. Pages 73–88 in H. P. Hansen, editor, *Arctic biology.* Oregon State University Press, Corvalis.

————. 1976. Cyclic pattern in lemming populations near Barrow, Alaska. Pages 199–215 in M. E. Britton, editor, *Alaskan arctic tundra.* Arctic Institute of North America, Arlington, Va.

Post, E., R. O. Peterson, N. C. Stenseth, and B. E. McLaren. 1999. Ecosystem consequences of wolf behavioural response to climate. *Nature* 401:905–907.

Post, E., and N. C. Stenseth. 1998. Large-scale climatic fluctuation and population dynamics of moose and white-tailed deer. *Journal of Animal Ecology* 67:537–543.

————. 1999. Climatic variability, plant phenology, and northern ungulates. *Ecology* 80:1322–1339.

Potapov, E. R. 1997. What determines the population density and reproductive success of rough-legged buzzards, *Buteo lagopus*, in the Siberian tundra? *Oikos* 78:362–376.

Potts, G. R., S. C. Tapper, and P. J. Hudson. 1984. Population fluctuations in red grouse: Analysis of bag records and a simulation model. *Journal of Animal Ecology* 54:21–36.

Price, T. S., C. Dogget, J. M. Pye, and T. P. Holmes. 1992. A history of southern pine beetle outbreaks in the southeastern United States. *Report of the Southern Forest Insect Working Group.* Georgia Forestry Commission, Macon.

Pucek, Z., W. Jedrzejewski, B. Jedrzejewska, and M. Pucek. 1993. Rodent population dynamics in a primeval deciduous forest (Bialowieza National Park) in relation to weather, seed crop, and predation. *Acta Theriologica* 38:199–232.

Quenette, P. Y., and J. F. Gerard. 1993. Why biologists do not think like Newtonian physicists. *Oikos* 68:361–363.

Quinn, T. J., and R. B. Deriso. 1999. *Quantitative fish dynamics.* Oxford University Press, New York.

Rand, D. A., and H. B. Wilson. 1991. Chaotic stochasticity: A ubiquitous source of unpredictability in epidemics. *Proceedings of the Royal Society of London B* 246:179–184.

Reeve, J. D. 1997. Predation and bark beetle dynamics. *Oecologia* 112:48–54.

Reeve, J. D., M. P. Ayres, and P. L. Lorio. 1995. Host suitability, predation, and bark beetle population dynamics. Pages 339–357 in N. Cappuccino and P. Price, editors, *Population dynamics: New approaches and synthesis.* Academic Press, New York.

Reeve, J. D., D. J. Rhodes, and P. Turchin. 1998. Scramble competition in the southern pine beetle, *Dendroctonus frontalis. Ecological Entomology* 23:433–443.

Reeve, J. D., J. A. Simpson, and J. S. Fryar. 1996. Extended development in *Thanasimus dubius* (F.) (Coleoptera: Cleridae), a predator of southern pine beetle. *Journal of Entomological Science* 31:123–131.

Reeve, J. D., P. Turchin, and A. D. Berryman. 2002. Evidence for predator-prey cycles in a bark beetle. In A. A. Berryman, editor, *Population cycles: Evidence for trophic interactions.* Oxford University Press, Oxford.

Reid, D. G., C. J. Krebs, and A. Kenney. 1995. Limitation of collared lemming population growth at low densities by predation mortality. *Oikos* 73:387–398.

Renfer, A. 1974. Caracteristiques biologiques et efficacite de *Phytodietus griseanae* Kerrich (Hym Ichneumonidae) parasitoide de *Zeiraphera diniana* Guenee (Lep Tortricidae) en haute montagne. Ph.D. dissertation no. 5278. ETH, Zurich.

————. 1975. Characteristiques biologiques de *Phytodietus griseanae* (Hym., Ichneumonidae) parasitoide de la tordeuse grise du mélèze *Zeiraphera diniana* (Lep., Tortricidae) en haute montagne. *Annales de la Société Entomologique de France* 11:425–455.

Rodriguez, D. J. 1989. A model of population dynamics for the fruit fly *Drosophila melanogaster* with density dependence in more than one life stage and delayed density dependent effects. *Journal of Animal Ecology* 58:349–365.

Romankow-Zmudovska, A., and B. Grala. 1994. Occurrence and distribution of the common vole, *Microtus arvalis* (Pallas), in legumes and seed grasses in Poland between 1977 and 1992. *Polish Ecological Studies* 20:503–508.

Rosenzweig, M. L. 1969. Why the prey curve has a hump. *American Naturalist* 103:81–87.

Rosenzweig, M. L., and R. H. MacArthur. 1963. Graphical representation and stability conditions of predator-prey interaction. *American Naturalist* 97:209–223.

Roughgarden, J. 1998. *Primer of ecological theory*. Prentice Hall, Upper Saddle River, N.J.

Royama, T. 1977. Population persistence and density dependence. *Ecological Monographs* 47:1–35.

————. 1981. Fundamental concepts and methodology for the analysis of animal population dynamics, with particular reference to univoltine species. *Ecological Monographs* 51:473–493.

————. 1992. *Analytical population dynamics*. Chapman and Hall, London.

Saether, B. E. 1997. Environmental stochasticity and population dynamics: A search for mechanisms. *Trends in Ecology and Evolution* 12:143–149.

Saether, B. E., R. Andersen, O. Hjelford, and M. Heim. 1996. Ecological correlates of regional variation in life history of the moose *Alces alces*. *Ecology* 77:1493–1500.

————. 1998. Ecological correlates of regional variation in life history of the moose *Alces alces: Reply. Ecology* 79:1838–1839.

Saitoh, T. 1987. A time series and geographical analysis of population dynamics of the red-backed vole in Hokkaido, Japan. *Oecologia* 73:382–388.

Saucy, F. 1988. Description des cycles pluriannuels d'*Arvicola terrestris* Schermann en Suisse occidentale par la méthode de l'analyse des séries temporelles. *EPPO Bulletin* 18:401–413.

Schaffer, W. M. 1985. Order and chaos in ecological systems. *Ecology* 66:93–106.

Schaffer, W. M., S. Ellner, and M. Kot. 1986. Effects of noise on some dynamical models in ecology. *Journal of Mathematical Biology* 24:479–523.

Schaffer, W. M., and M. Kot. 1985. Nearly one-dimensional dynamics in an epidemic. *Journal of Theoretical Biology* 112:403–427.

Scheffer, V. B. 1951. The rise and fall of a reindeer herd. *Scientific Monthly* 73:356–362.

Schmitz, O. J., and A.R.E. Sinclair. 1997. Rethinking the role of deer in forest ecosystems dynamics. Pages 201–223 in W. J. McShea and H. B. Underwood, editors, *The science of overabundance: Deer ecology and population management.* Smithsonian Institution Press, Washington, D.C.

Schoener, T. W. 1976. Alternatives to Lotka Volterra competition: Models of intermediate complexity. *Theoretical Population Biology* 10:309–333.

Schultz, A. M. 1969. A study of an ecosystem: the arctic tundra. Pages 77–93 in G. M. Van Dyne, editor, *The ecosystem concept in natural resource management.* Academic Press, New York.

Sharov, A., and J. J. Colbert. 1996. A model for testing hypotheses of gypsy moth, *Lymantria dispar* L., population dynamics. *Ecological Modelling* 84:31–51.

Shehan, K. A. 1988. Status of the gypsy moth life system model. Pages 166–171 in A. R. Miller, editor, *Proceedings, National Gypsy Moth Review.* West Virginia Agricultural Department, Charleston.

Sinclair, A.R.E. 1991. Science and practice of wildlife management. *Journal of Wildlife Management* 55:767–773.

Sinclair, A.R.E., J. M. Gosline, G. Holdsworth, C. J. Krebs, S. Boutin, J.N.M. Smith, R. Boonstra, and M. Dale. 1993. Can the solar cycle and climate synchronize the snowshoe hare cycle in Canada? Evidence from tree rings and ice cores. *American Naturalist* 141:173–198.

Sinervo, B., E. Svensson, and T. Comendant. 2000. Density cycles and an offspring quantity and quality game driven by natural selection. *Nature* 406:985–988.

Skellam, J. G. 1951. Random dispersal in theoretical populations. *Biometrika* 38:196–218.

Southern, H. N. 1979. The stability and instability of small mammal populations. Pages 103–134 in D. M. Stoddard, editor, *Ecology of small mammals.* Chapman and Hall, London.

Spalinger, D. E., and N. T. Hobbs. 1992. Mechanisms of foraging in mammalian herbivores: New models of functional response. *American Naturalist* 140:325–348.

St. George, R. A. 1930. Drought affected and injured trees attractive to bark beetles. *Journal of Economic Entomology* 23:825–828.

Steen, H. 1995. Untangling the causes of disappearance from a local population of root voles, *Microtus oeconomus*: A test of the regional synchrony hypothesis. *Oikos* 73:65–72.

Stenseth, N., K. S. Chan, H. Tong, R. Boonstra, S. Boutin, C. J. Krebs, E. Post, M. O'Donoghue, N. G. Yoccoz, M. C. Forchhammer, and J. W. Hurrell. 1999. Common dynamic structures of Canada lynx populations within three climatic regions. *Science* 285:1071–1073.

Stenseth, N. C., W. Falck, O. N. Bjornstad, and C. J. Krebs. 1997. Population regulation in snowshoe hare and Canadian lynx: Asymmetric food web configurations between hare and lynx. *Proceedings of the National Academy of Sciences of the USA* 94:5147–5152.

Stenseth, N. C., and E. Framstad. 1980. Reproductive effort and optimal reproductive rates in small rodents. *Oikos* 34:23–34.

Stenseth, N. C., and R. A. Ims. 1993a. The history of lemming research: From the Nordic Sagas to *The Biology of Lemmings*. Pages 3–34 in N. C. Stenseth and R. A. Ims, editors, *The Biology of Lemmings*. Linnean Society, London.

————. 1993b. Population dynamics of lemmings: Temporal and spatial variation—an introduction. Pages 61–96 in N. C. Stenseth and R. A. Ims, editors, *The Biology of Lemmings*. Linnean Society, London.

Stenseth, N. C., and A. Lomnicki. 1990. On the Charnov-Finerty hypothesis: The unproblematic transition from docile to aggressive and the problematic transition from aggressive to docile. *Oikos* 58:234–238.

Stephen, F. M., M. P. Lih, and G. W. Wallis. 1989. Impact of arthropod natural enemies on *Dendroctonus frontalis* (Coleoptera: Scolytidae) mortality and their potential role in infestation growth. Pages 169–185 in D. L. Kulhavy and M. C. Miller, editors, *Potential for biological control of* Dendroctonus *and* Ips *bark beetles*. Stephen F. Austin State University, Nacogdoches, Tex.

Stiling, P. 1999. *Ecology: Theories and applications*. 3rd edition. Prentice Hall, Upper Saddle River, N.J.

Summers, R. W., L. G. Underhill, and E. E. Syroechkovski. 1998. The breeding productivity of dark-bellied geese amd curlew sandpipers in relation to changes in the numbers of arctic foxes and lemmings on the Taimyr Peninsula, Siberia. *Ecography* 21:573–580.

Takens, F. 1981. Detecting strange attractors in turbulence. Pages 366–381 in D. A. Rand and L. S. Young, editors, *Dynamical systems and turbulence*. Springer-Verlag, New York.

Tanner, J. T. 1975. The stability and the intrinsic growth rates of prey and predator populations. *Ecology* 56:855–867.

Tast, J. 1974. The food and feeding habits of the root vole, *Microtus oecnomus*, in Finnish Lapland. *Aquilo Seriae Zooligica* 15:25–32.

Thatcher, R. C., and L. S. Pickard. 1964. Seasonal variations in activity of the southern pine beetle in East Texas. *Journal of Economic Entomology* 57:840–842.

Thatcher, R. C., J. L. Searcy, J. E. Coster, and G. D. Hertel, editors. 1980. *The southern pine beetle*. USDA–Forest Service Technical Bulletin 1631. US Department of Agriculture, Pineville, La.

Thirgood, S. J., S. M. Redpath, D. T. Haydon, P. Rothery, I. Newton, and P. J. Hudson. 2000a. Habitat loss and raptor predation: Disentangling long- and short-term causes of red grouse declines. *Proceedings of the Royal Society B* 267:651–656.

Thirgood, S. J., S. M. Redpath, P. Rothery, and N. Aebischer. 2000b. Raptor predation and population limitation in red grouse. *Journal of Animal Ecology*, forthcoming.

Tidd, C. W., L. F. Olsen, and W. M. Schaffer. 1993. The case for chaos in childhood epidemics. II. Predicting historical epidemics from mathematical model. *Proceedings of the Royal Society of London B* 254:257–273.

Tong, H. 1977. Some comments on the Canadian lynx data. *Journal of the Royal Statistical Society A* 140:432–436.

———. 1990. *Nonlinear time-series: A dynamical system approach*. Oxford University Press, Oxford.

———. 1995. A personal overview of nonlinear time-series analysis from a chaos perspective. *Scandinavian Journal of Statistics* 22:399–445.

Trostel, K., A.R.E. Sinclair, C. J. Walters, and C. J. Krebs. 1987. Can predation cause the 10-year hare cycle? *Oecologia* 74:185–192.

Tuljapurkar, S., and H. Caswell. 1997. *Structured population models in marine, terrestrial, and freshwater systems*. Chapman and Hall, New York.

Turchin, P. 1990. Rarity of density dependence or population regulation with lags? *Nature* 344:660–663.

——. 1991. Reconstructing endogenous dynamics of a laboratory *Drosophila* population. *Journal of Animal Ecology* 60:1091–1098.

——. 1993. Chaos and stability in rodent population dynamics: Evidence from nonlinear time-series analysis. *Oikos* 68:167–172.

——. 1995a. Chaos in microtine populations. *Proceedings of the Royal Society of London B* 262:357–361.

——. 1995b. Population regulation: Old arguments and a new synthesis. Pages 19–40 In N. Cappuccino and P. Price, editors, *Population dynamics*. Academic Press, New York.

——. 1996. Nonlinear time-series modeling of vole population fluctuations. *Researches in Population Ecology* 38:121–132.

——. 1998. *Quantitative analysis of movement: Measuring and modeling population redistribution in animals and plants.* Sinauer Associates, Sunderland, Mass.

——. 1999. Population regulation: A synthetic view. *Oikos* 84:153–159.

Turchin, P., and G. Batzli. 2001. Availability of food and the population dynamics of arvicoline rodents. *Ecology* 82:1521–1534.

Turchin, P., and S. P. Ellner. 2000a. Living on the edge of chaos: Population dynamics of Fennoscandian voles. *Ecology* 81:3099–3116.

——. 2000b. Modelling time-series data. Pages 33–48 in J. N. Perry, R. H. Smith, I. P. Woiwod, and D. Morse, editors, *Chaos in real data*. Kluwer Academic Press, Dordrecht, The Netherlands.

Turchin, P., and I. Hanski. 1997. An empirically-based model for the latitudinal gradient in vole population dynamics. *American Naturalist* 149:842–874.

——. 2001. Contrasting alternative hypotheses about rodent cycles by translating them into parameterized models. *Ecology Letters* 4:267–276.

Turchin, P., P. L. Lorio, A. D. Taylor, and R. F. Billings. 1991. Why do populations of southern pine beetles (Coleoptera: Scolytidae) fluctuate? *Environmental Entomology* 20:401–409.

Turchin, P., L. Oksanen, P. Ekerholm, T. Oksanen, and H. Henttonen. 2000. Are lemmings prey or predators? *Nature* 405:562–565.

Turchin, P., and R. S. Ostfeld. 1997. Effects of density and season on the population rate of change in the meadow vole. *Oikos* 78:355–361.

Turchin, P., and A. D. Taylor. 1992. Complex dynamics in ecological time series. *Ecology* 73:289–305.

Turchin, P., A. D. Taylor, and J. D. Reeve. 1999. Dynamical role of predators in population cycles of a forest insect: An experimental test. *Science* 285:1068–1071.

Turchin, P., and W. T. Thoeny. 1993. Quantifying dispersal of southern pine beetles with mark-recapture experiments and a diffusion model. *Ecological Applications* 3:187–198.

Underhill, L. G., R. P. Prys-Jones, E. E. Syroechkovski, N. M. Groen, V. Karpov, H. G. Lappo, M.W.J. Van Roomen, A. Rybkin, H. Schekkerman, H. Spiekman, and R. W. Summers. 1993. Breeding of waders (Charadrii) and brent geese *Branta bernicla bernicla* at Pronchishcheva Lake, northeastern Taymyr, Russia, in a peak and a decreasing lemming year. *Ibis* 135:277–292.

Van Ballenberghe, V. 1983. Rate of increase in moose populations. *Alces* 19:98–117.

Van Ballenberghe, V., and W. B. Ballard. 1994. Limitation and regulation of moose populations: The role of predation. *Canadian Journal of Zoology* 72:2071–2077.

Varley, G. C., G. R. Gradwell, and M. P. Hassell. 1973. *Insect population ecology: An analytical approach.* University of California Press, Berkeley.

Vivas, H. J., and B. E. Saether. 1987. Interactions between a generalist herbivore, the moose *Alces alces*, and its food resources: An experimental study of winter foraging behaviour in relation to browse availability. *Journal of Animal Ecology* 56:509–520.

Volterra, V. 1926. Fluctuations in the abundance of a species considered mathematically. *Nature* 118:558–600.

———. 1931. *Leçons sur la théorie mathématique de la lutte pour la vie.* Gauthiers-Vilars, Paris.

Wahba, G. 1990. *Spline models for observational data.* Society for Applied Mathematics, Philadelphia, Pa.

Ward, R.M.P., and C. J. Krebs. 1985. Behavioural response of lynx to declining snowshoe hare abundance. *Canadian Journal of Zoology* 63:2817–2824.

Warkowska-Dratnal, H., and N. C. Stenseth. 1985. Dispersal and the microtine cycle: Comparison of two hypotheses. *Oecologia* 65:468–477.

Watson, A., R. Moss, and S. Rae. 1998. Population dynamics of Scottish rock ptarmigan cycles. *Ecology* 79:1174–1192.

Watson, A., R. Moss, P. Rothery, and R. Parr. 1984. Demographic causes and predictive models of population fluctuations in red grouse. *Journal of Animal Ecology* 53:639–662.

Weiner, J. 1995. On the practice of ecology. *Journal of Ecology* 83:153–158.

Wielgolaski, F. E. 1975. Productivity of tundra ecosystems. Pages 1–12 in D. E. Reichle, J. F. Franklin, and D. W. Goodall, editors, *Productivity of world ecosystems*. National Academy of Sciences, Washington, D.C.

Wigner, E. P. 1970. *Symmetries and reflections: Scientific essays of Eugene P. Wigner*. MIT Press, Cambridge, Mass.

Wilkens, R. T., M. P. Ayres, P. L. Lorio, and H. D. Hodges. 1998. Environmental effects on pine tree carbon budgets and resistance to bark beetles. Pages 591–616 in R. A. Mickler and S. Fox, editors, *The productivity and sustainability of southern forest ecosystems in a changing environment*. Springer-Verlag, New York.

Williams, G. R. 1954. Population fluctuations in some Northern Hemisphere game birds (Tetraonidae). *Journal of Animal Ecology* 23:1–34.

Williams, J. 1985. Statistical analysis of fluctuations in red grouse bag data. *Oecologia* 65:269–272.

Wolda, H. 1989. The equilibrium concept and density dependence tests. What does it all mean? *Oecologia* 81:430–432.

Wolda, H., and B. Dennis. 1993. Density dependence tests, are they? *Oecologia* 95:581–591.

Wood, S. N. 2001. Partially specified ecological models. *Ecological Monographs* 71:1–25.

Wyman, L. 1924. Bark beetle epidemics and rainfall deficiency. *US Forest Service Bulletin* 8:2–3.

Yodzis, P. 1989. *Introduction to theoretical ecology*. Harper and Row, New York.

Zimmer, C. 1999. Life after chaos. *Science* 284:83–86.

Index